Metallic Resources 1

SCIENCES

Geoscience, Field Director – Yves Lagabrielle

Natural Resources: Applied Basic Research,
Subject Head – Philippe Boulvais

Metallic Resources 1

Geodynamic Framework and Remarkable Examples in Europe

Coordinated by
Sophie Decrée

iSTE

WILEY

First published 2023 in Great Britain and the United States by ISTE Ltd and John Wiley & Sons, Inc.

ISTE Ltd
27-37 St George's Road
London SW19 4EU
UK

www.iste.co.uk

John Wiley & Sons, Inc.
111 River Street
Hoboken, NJ 07030
USA

www.wiley.com

Library of Congress Control Number: 2023935886

British Library Cataloguing-in-Publication Data
A CIP record for this book is available from the British Library
ISBN 978-1-78945-135-1

ERC code:
PE10 Earth System Science
PE10_10 Mineralogy, petrology, igneous petrology, metamorphic petrology

Contents

1

The Rare Earth Resources of Europe and Greenland: Mining Potential and Challenges

Nicolas CHARLES[1], Johann TUDURI[1,2], Gaétan LEFEBVRE[1], Olivier POURRET[3], Fabrice GAILLARD[2] and Kathryn GOODENOUGH[4]

[1] *BRGM, Orléans, France*

[2] *ISTO, CNRS, University of Orléans, France*

[3] *AGHYLE, UniLaSalle, Beauvais, France*

[4] *British Geological Survey, Edinburgh, United Kingdom*

1.1. Introduction

In the past few years, 17 metal elements have absorbed the principal attention of industrialized countries, which are now aiming for a digital transformation and energy transition. Rare earths, which were little used in industry at the outset, are strategic metals today, a kind of vitamin for digital and green technologies (Bru et al. 2015; EC 2019a, 2019b; Blengini et al. 2020; Dushyantha et al. 2020). Used anecdotally in the 1950s and 1960s within heavy mineral placers in Australia, India, South Africa or Brazil, their attractiveness grew, and the United States became their main producer from the 1970s to 1980s with its Mountain Pass deposit.

For a color version of all figures in this chapter, see www.iste.co.uk/decree/metallic1.zip.

Metallic Resources 1,
coordinated by Sophie DECRÉE.

At the end of the 1980s, China then became the main global producer of rare earths. In 1992, the first secretary of the Chinese Communist Party, Deng Xiaoping, was pleased to say that “the Middle East has oil; China has rare earths”. This is still the case for China today with its hegemonic position, which it owes to the use of gigantic deposits such as Bayan Obo, but above all to its expertise in rare earth separation.

At the outset, everyone seemed to accommodate this situation, with importing countries benefiting from the low price, while also avoiding the environmental and social impacts of mining these metals on their own territory.

It took a diplomatic incident involving a small island in the East China Sea for the whole situation to change in September 2010. China declared an embargo on the export of rare earths to Japan, a major importer of rare earths and a country at the forefront of digital technologies. Then, the lowering of Chinese export quotas lit the keg in 2011, with the rapid increase in the price of rare earths (as much as 10,000% for dysprosium).

Many countries were already aware of the vulnerability of their supply chain, principally in Europe (EC 2011, 2017; Sebastiaan et al. 2017; Lauri et al. 2018; Blengini et al. 2020). Rare earths were already strategic raw materials, and their criticality for European industry was classed as a priority. Therefore, for some years mining for rare earths benefited from investments intended for finding alternatives to the Chinese deposits.

In addition to the study of primary deposits (Goodenough et al. 2016; EURARE 2017; Tuduri et al. 2020), the recycling path was also considered (Ahonen et al. 2015; Guyonnet et al. 2015). It transpired that Europe and Greenland (a constituent territory of Denmark) had real potential for rare earth mining that could support the needs of the Old Continent for several decades, or even beyond. Europe was the continent where rare earths were discovered in 1787, but to date, there is no mine for these metals, and 95% of their imports depend on China (Gislev and Grohol 2018).

Geological potential is not all, however, and an entire value chain must be developed, from the extraction of the minerals, or rather of rare earth minerals, which require processing techniques adapted for each of them, to their transformation into high added value products (Ahonen et al. 2015).

It is here that the challenges begin, because the investments to be considered are colossal. In addition, what about the environmental, social and societal acceptability of mining activity in Europe? Although Europe has remarkable rare earth potential, can we envisage exploiting it? This original summary provides a panorama of rare earths in Europe through their physicochemical, mineralogical, socioeconomical, geodynamic and metallogenetic aspects.

1.2. The extreme diversity of rare earths

1.2.1. *Rare earth elements*

1.2.1.1. *Definitions*

"Rare earths" is a collective name given, for historical and practical reasons, to a set of metal elements in the periodic table (Figure 1.1). Rare earth elements (REEs) are elements called "trace elements" in most natural environments. REEs are a group of 17 metal elements that are coherent in terms of ionic rays, charge and coordination. According to the definitions (Henderson 1984; Lipin and Mckay 1989; Jones et al. 1996; Atwood 2012), the REEs include lanthanides (from ^{57}La to ^{71}Lu), yttrium (^{39}Y) and sometimes scandium (^{21}Sc).

Scandium's membership in this group has given rise to debate since its behavior in geological environments is generally different from that of the other REEs. Similarly, promethium (^{61}Pm), with its extremely short half-life, is almost non-existent in nature.

Depending on the specific electron configurations within each atom, the REEs can be divided into two subgroups:

– the light rare earth elements (LREE) with La-Ce-Pr-Nd-Sm-Eu-Gd;

– the heavy rare earth elements (HREE) with Tb-Dy-Ho-Er-Tm-Yb-Lu-Y.

It should be noted, however, that the membership of some REEs to one of the two subgroups is still debated (Gupta and Krishnamurthy 2005; Chakhmouradian and Wall 2012; McLennan and Taylor 2012; Weng et al. 2013; Zepf 2013). There is therefore a classification specific to the mining industry that differs from that recommended by the International Union of Pure and Applied Chemistry (IUPAC). For economic reasons linked to geological abundance, to the costs of processing the ore, to separation techniques, etc., many actors in the mineral industry only consider these as LREE: La-Ce-Pr-Nd.

These same industrial actors separate and sell their concentrations of rare earths in the form of oxides, and it is usual to consider these elements in terms of rare earth oxides (REO). Thus, the terms light rare earth oxides (LREOs) and heavy rare earth oxides (HREOs) are generally considered over the course of any value chain, from mining exploration to their separation, by mining extraction.

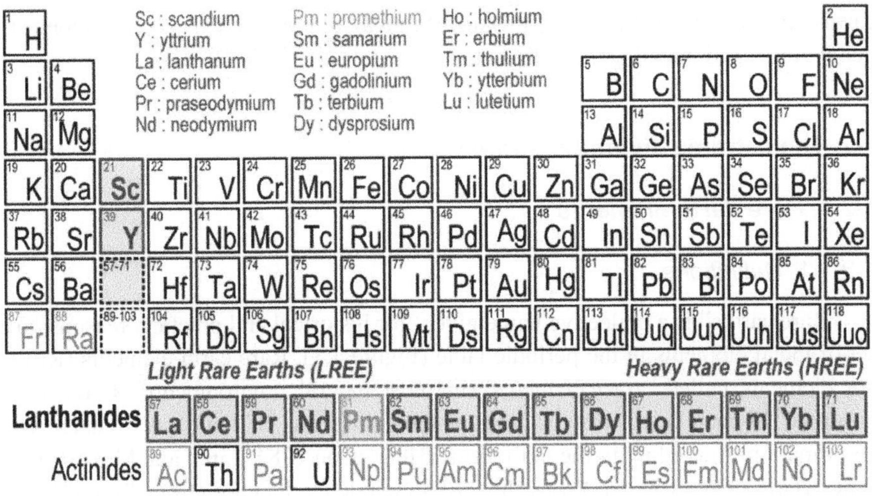

Figure 1.1. *Rare earth elements in the periodic table of elements*

1.2.1.2. *The physical and chemical properties of the rare earths*

The REEs have physical-chemical properties much sought after by manufacturers, thanks in particular to their internal electron structure linked to their 4f electron subshell. The REEs in neutral metallic form have fairly similar atomic radii, except for europium and ytterbium, whose larger atomic radii explain their lower densities. Their hardnesses are variable, ranging from soft (La, Nd, Yb) to hard (Ho, Er, Lu).

The REEs are moderately fusible, and their fusion temperatures range between 799°C (Ce) and 1,663°C (Lu). The spectral properties of REEs are most remarkable for absorption (coloration), as well as for transmission (luminescence), which explains their current use in phosphors and lasers. It is the electrons' level of mobility in the atoms' energy levels that forms the basis of this property. Subject to powerful rays (e.g. UV rays), an REE atom surrounded by ligands (oxides or molecules) will be excited electronically, and de-excitation of the REE atom will result in the emission of light with wavelength emission peaks specific to the REE element (e.g. Y and Eu: red and blue; Tb and Tm: green; Ce: yellow).

The other remarkable physical property of some REEs is their magnetism, which is the origin of their use for making high-performance permanent magnets (high magnetization and remanence). REEs paired with transition elements (Fe, Co) allow the creation of the most coercive magnets manufactured at the industrial scale (Sm-Co, Nd-Fe-B).

From a chemical point of view, REEs are reducing metals (deoxidation and desulfurization properties). Most take between a few days and a few months to oxidize rapidly in air and at ambient temperature (Eu, La, Ce, Pr, Nd). The HREEs resist oxidation for several years (Y, Dy, Ho, Er). Finely ground, Ce allied with Fe (ferrocerium) burns in air. This was one of the first applications of REEs with flint-spark-lighters formed of ferrocerium. In minerals, REEs are in trivalent cation form. Although the ionic radius varies from one rare earth to another, their chemical properties remain remarkably homogeneous, especially in solution. Although many substitutions between REEs are produced within the crystalline network of minerals, some mineral species are better adapted and more favorable to receiving a range of ionic radii. This is the case with "ceric earth" minerals, which are more inclined to incorporate Ce, La, Nd, Pr, (Sm), (Eu) and (Gd), such as monazite and bastnäsite. Other minerals called "yttric earths" are more favorable to Y, Tb, and Dy, such as xenotime and gadolinite.

1.2.1.3. *A brief history of rare earths*

The discovery of rare earths began in 1787 in Sweden. Remember in passing that the word "earths" was the name given by chemists to the metal oxides supposed to be simple bodies. The epithet "rare" recalls the difficulties encountered by chemists in isolating them and their supposedly low concentration in ores. Carl Axel Arrhenius, a Swedish chemist, described for the first time a dense, black mineral in the pegmatites of Ytterby. They were sent for analysis to the Finnish chemist Johan Gadolin. In 1794, he discovered a new "earth" contained in this mineral, *yttria*, which would give its name to ytterbite, since it was known by the name gadolinite. However, a reddish mineral was discovered earlier, in 1751, by the Swedish mineralogist Axel Fredrik Cronstedt in the copper mine at Bastnäs. It was only in 1803 that the Swedish chemists Jöns Jakob Berzelius and Wilhelm Hisinger isolated a new "earth" with properties similar to yttria. This new earth was called *ceria*, a reference to the asteroid Ceres that had just been discovered, and would give its name to the mineral bastnäsite.

Throughout the 19th century, different chemists discovered that some previously isolated earths actually contained several other chemical elements with similar properties. For example, in 1839, Carl Gustaf Mosander discovered that cerium oxide is also formed of an oxide of another element that he would call lanthanum (from the Greek *lanthano*: to be hidden). In 1843, he discovered that ytterbium was in fact a mixture containing two other elements, terbium and erbium. The final REE element, lutetium, was discovered in 1907. Finally, it should be noted that promethium was discovered in 1947 in uranium fission products from the Oak Ridge reactor (United States).

REEs were thus discovered by gradual separation, some from cerium (ceric earths) and others from yttrium (yttric earths). It should be noted that gadolinium would be identified in both types of earth. In order of discovery (Figure 1.2): Y (1794), Ce (1803), La (1839), Er (1843), Tb (1843), Sc (1876), Yb (1878), Ho (1879), Sm (1879), Tm (1879), Gd (1880), Nd (1885), Pr (1885), Dy (1886), Eu (1901), Lu (1907) and Pm (1947).

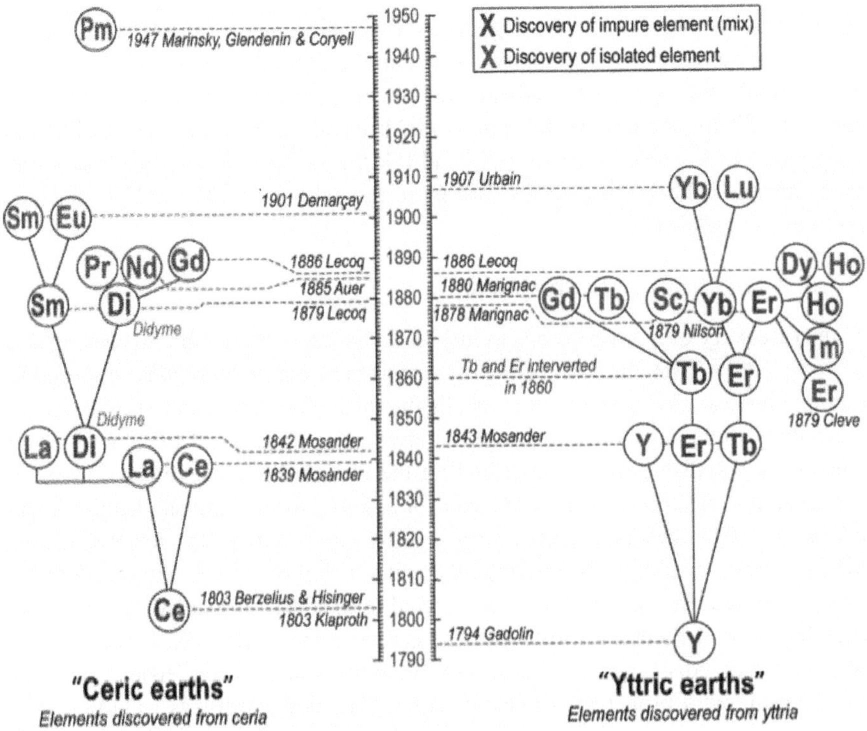

Figure 1.2. *Filiation of the successive discoveries of REEs*[1,2]

1.2.2. *Rare earth minerals*

From a geological point of view, REEs are not rare. Their natural abundance in the Earth's crust is at least equivalent to that of other base metals (Zn, Cu, Pb, Ni, Co) and much less rare than previous metals (Ag, Au). On the other hand, although LREEs are the most abundant, they are found at some of the highest concentrations

1 See: asianmetals.com.

2 See: vanderkrogt.net.

in numerous primary deposits. Conversely, the HREEs are clearly less abundant, much like the rare metals (Sn, W, Ta) and are therefore much rarer, including in deposits of these elements (except Y). For this reason, HREEs are considered to be more strategically important than LREEs, hence their much higher market value. Thus, the HREO/LREO ratio is a critical parameter for evaluating a deposit.

In fact, it is the low quantity of natural minerals containing REEs that explains the term "rare". Although more than 200 REE mineral species are known today (this number remains low), only some of them (Table 1.1) are of commercial interest (Gupta and Krishnamurthy 2005). In fact, although REEs are often incorporated into the crystal lattice of carbonates, oxides, silicates or even phosphates as a substitution for other more common chemical elements, the mineral industry values only a very small number of REE ores: bastnäsite (fluorocarbonate), monazite and xenotime (phosphates), and loparite (oxide). REEs have also been extracted from apatite in Russia and South Africa, but this production was extremely limited.

Except for xenotime, which is enriched with HREEs, all of these minerals are essentially characterized by their richness in LREEs. Since lanthanides share similarities with elements from the actinide group, the REE minerals also and almost systematically include radioactive elements (e.g. Th and/or U). Uraninite, as well as thorite may also accompany REE minerals. The presence of these radioactive elements is a restriction on mining project development. In fact, radioactivity accompanies every stage of the ore enrichment process, right up to metallurgy. These minerals and radioactive elements are a form of waste that must be managed rigorously. For these reasons, it is necessary to focus on the U and Th contents in deposits, particularly on the ThO_2/REO and U_3O_8/REO ratios of the minerals that form the ore.

Extracted from the main Chinese mines of Bayan Obo, Maoniuping Weishan and Dalucao, as well as from the Mountain Pass mine in the United States, bastnäsite is the main REE ore (Verplanck et al. 2016; Jia and Liu 2020). Since the 1970s, so-called ionic clays, or lateritic ion-adsorption clays, have formed an important source of HREEs (Bao and Zhao 2008). This production, located uniquely in southern China in Longnan (Jiangxi), but also in the provinces of Guangdong and Guangxi, is favored because of the low cost of labor and relatively simple extraction procedures using in situ lixiviation with neutral or acidic solutions.

However, with extremely low ore grades or 0.03–0.35% REO (Chi and Tian 2008; Zhou et al. 2017), such procedures question the balance between economic feasibility and protection of the environment. Nevertheless, the high proportion of HREEs coupled with the very low concentration of radioactive elements (U, Th) associated with this type of deposit make it an attractive challenge in mineral exploration.

Mineral	Chemical formula	Concentration (%)		
		RE_2O_3	ThO_2	UO_2
Allanite	$([REE],Ca)_2(Al,Fe)_3Si_3O_{12}(OH)$	2.5–17	<3	–
Ancylite	$Sr[LREE](CO_3)_2(OH).H_2O$	46–53	<0.4	<0.1
Apatite	$Ca_5(PO_4)_3(F,Cl,OH)$	<<2	–	<0.05
Bastnäsite	$[LREE]CO_3(F,OH)$	58–75	<2.8	<0.1
Clays	REE adsorbed on kaolinite/halloysite $Al_2Si_2O_5(OH)_4$	<<4	<0.01	<0.001
Eudialyte	$(Na,[REE])_{15}(Ca,[REE])_6(Fe,Mn)_3(Si,Nb)_2(Zr,Ti)_3Si_{24}O_{72}(OH,F,Cl,H_2O)_6$	1–10	–	<0.1
Euxenite	$([REE],U,Th)(Nb,Ta,Ti)_2O_6$	16–30	<4.3	3–9
Fergusonite	$[REE]NbO_4$	43–52	<8	<13.6
Gadolinite	$[REE]_2FeBe_2Si_2O_{10}$	45–54	<0.4	–
Loparite	$(Na,[LREE],Ca,Sr,Th)(Ti,Nb,Ta)O_3$	28–37	1.6	0.03
Monazite	$([LREE],Th,Ca)(P,Si)O_4$	35–71	<20	<16
Parisite	$Ca[LREE]_2(CO_3)_3F_2$	50–59	<4	<0.3
Pyrochlore	$(Ca,Na,U,[REE])_2(Nb,Ta)_2O_6(OH,F)$	<22	<4	<27
Steenstrupine	$Na_{14}[LREE]_6Mn_2Fe_2(Zr,Th)(PO_4)_7Si_{12}O_{36}.3(H_2O)$	<31	<6	<1
Thorite	$(Th,U,[REE])SiO_4$	<3	65–81	10–16
Uraninite	UO_2	<1.5	<12.2	50–98
Xenotime	$([HREE],Zr,U)(P,Si)O_4$	54–74	<8.4	<5.8
Zircon	$(Zr,[HREE],Th,U)SiO_4$	<19	0.01–0.8	0.01–4

Table 1.1. *Minerals valued by the extraction industry (Tuduri et al. 2015)*

The last decade has seen the emergence of a substantial market for the exploration of REEs, intended to find resources outside of China. Although different projects are in the process of valuation, many of them are innovative projects since they are not conventional and thus present economic risks. Some projects correspond to new deposit models, and others seek to develop new minerals: pyrochlore, fergusonite, oxides and numerous silicates (e.g. allanite, eudialyte, steenstrupine). These mineral stages, especially silicates, have the advantage of being enriched with HREOs.

1.3. The economy of rare earths in the world and their place in Europe

1.3.1. *The application domains for rare earths*

The first use of REEs was attested in 1885, when cerium was used in gas sleeves in the city of Vienna. The usage of REEs was very limited until the 1960s (ferrocerium in lighter flints), then more diversified with applications in technology from the 1970s (color cathode screens, lasers, phosphors, etc.) and 1980s (permanent magnets).

REEs have exceptional magnetic, electronic, optical and catalytic properties, and are particularly useful today for ever more diversified, well-developed technologies, especially in the defense industry, electronics and renewable energies (Figures 1.3 and 1.4) (Balaram 2019; Dushyantha et al. 2020). REEs are used in different chemical forms: metals, alloys, oxides and chlorides.

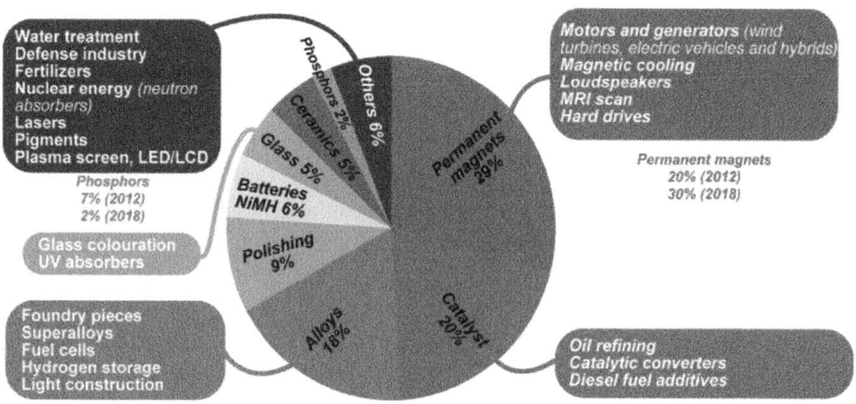

Figure 1.3. *Global distribution of REE use by sector in 2018*

Element	Applications
Y	Lighting, ceramics, metal alloys, lasers, catalysts
La	Catalyst for oil cracking, NiMH batteries, metal alloys, optics, lighting, mischmetal
Ce	Glass polishing, metal alloys, auto-catalysts, optics, lighting, mischmetal
Pr	Permanent magnets, lighting, ceramics, mischmetal, metal alloys, oxydation catalysts
Nd	Permanent magnets, ceramics, metal alloys, mischmetal, lighting, lasers
Pm	Nuclear batteries, defence, X-rays source in measurement instruments
Sm	Permanent magnets, optics, medicine, nuclear industry
Eu	Lighting, optics, nuclear industry
Gd	Permanent magnets, metal alloys, lighting, medical imaging, lasers
Tb	Lighting, permanent magnets, metal alloys, electronic components doping
Dy	Permanent magnets, metal alloys, optics, nuclear industry
Ho	Glass coloration, laser crystal doping
Er	Glass coloration, lighting, telecommunication optics, medical lasers
Tm	Laser fibres and lighting doping
Yb	Laser crystals component, seismic measurements, metals
Lu	Medical tomography, ceramics, metal alloys, lasers, catalyst

Figure 1.4. *The main applications for each REE (Bru et al. (2015) and references therein; Sebastiaan et al. (2017))*

REEs made it possible to increase the performance of technological products, while still ensuring miniaturization, such as with magnets, and they deserve the name "the vitamins of modern industry" (Balaram 2019). For some years, REEs have excited additional interest in their use in manufacturing green technologies (wind turbines, solar panels), which are indispensable for ensuring energy transition (Judge et al. 2017). Thus, the use of Pr, Gd, Eu and Er in nanoparticle form in solar panels makes it possible to improve the capacity to convert photons into energy. Pr, Nd, Tb and Dy are used in permanent magnets in wind turbines and allow them to be lightened drastically (a 100 g NdFeB magnet is equivalent to a 1 kg traditional magnet), also allowing the miniaturization of the engine, while still ensuring increased performance, even at low wind speeds, as well as reduced maintenance.

Nevertheless, the market volatility of REE has led substitution elements to be found (Bru et al. 2015; Pavel et al. 2017). Other less well-known REE domains are zootechnics and agriculture since REEs play an essential role in the working and structure of the molecules of biological systems, especially La and Ce (Abdelnour et al. 2019). REEs are thus incorporated into animal food and fertilizer. REEs can also be used for military purposes, in the form of effective permanent magnets

(SmCo, NdFeB), useful in guiding systems for missiles, fighter planes and drones, or for on-board electrical engines. Tb, Er and Gd are useful for night vision optical lenses. Y is used in the manufacture of highly resistant ceramics (tanks, bullet-proof vests). Eu and Lu allow radar signals to be amplified or are useful in sonar detection.

The sectors where REEs are used have evolved extensively since 2010, the year of the global crisis triggered by the restriction of Chinese exports and a momentary drop in global consumption (Bru et al. 2015). Since 2012, the growth of this consumption has resumed and exceeded 100,000 t REO (Figure 1.6), particularly due to the permanent magnet sector. The main technology used is NdFeB magnet technology, using Nd and Pr in particular and, to a lesser extent, Dy and Tb for high-performance applications. The application that consumes the most REEs globally is NdFeB magnets, with approximately 30% of REE use in tonnage in 2018 and nearly 53% of the total value of the REE market. This demand is increasing by 10% per year. There are multiple sectors for the use of these magnets, such as the domain of high-efficiency electrical engines, where they allow miniaturization (electronics, robotics), and lightning of equipment (offshore wind farms, electrical vehicle engines, etc.)

Other sectors using REEs are becoming a minority in proportion (Bru et al. 2015), either due to restricted or specific uses (the defense industry, medical lasers, etc.), or on the contrary, usage with low added value for applications where performance is of less importance (polishing powders catalysts in cars, metallurgical alloys).

1.3.2. *The evolution of prices*

There is a great disparity between the price of LREEs, which are very abundant, and that of HREEs, which are reserved for niche applications because of their rarity. These prices thus range from single to triple digits or more depending on the elements (Figure 1.5).

These prices have also varied greatly over time, with prices multiplying by several dozen between 2002 and 2003, reaching a peak in July 2011, before a great drop, then a new and relative equilibrium after 2015. The very sharp rise in the prices of all REEs began at the start of 2010, was amplified after February 2011, and reached its height in mid-July 2011. On July 14, 2011, the price of dysprosium-metal reached 3,410 US$/kg, a multiplication by a factor of 106 compared to the average price of this metal from 2002–2003 (then, it was 32.1 US$/kg), which is an increase of 10,500%, which is probably an absolute record for all raw materials combined.

The sky-rocketing of prices in 2010–2011 was triggered by the conjunction of two main factors:

– the decision to reduce Chinese export quotas considerably (these exceeded approximately 60 kt from 2006–2007, 57 kt in 2008, 50 kt in 2009 and then 30 kt/year from 2010);

– the increase in demand (after a marked drop in 2009 in the context of the slump following the economic crisis) and in anticipation of furthering this increase, particularly for permanent magnets (the push for renewable energies, including wind power; anticipating, at that time, the striking take-off of the electric vehicle market, etc.), and phosphors (the increasingly widespread replacement of incandescent bulbs by compact fluorescent bulbs).

The incident at the Senkaku/Diaoyu Islands, in September 2010 between China and Japan, led China to declare an embargo on the export of REEs to Japan in October 2010. Cited by many as the trigger for the spike in the price of REEs, this event, in reality, only accentuated the rise in prices that had already begun several months before and that continued for several months afterwards. Threats to world supplies outside China following this event led manufacturer-users to buy to create preventive stocks, thus contributing to a price increase, the whole accentuated by a speculative factor. Following a classical "bubble" mechanism, soaring prices led to a drop in demand (Figure 1.6). This drop in consumption led to the "bursting" of this bubble and a near-continuous fall in prices for two years (between mid-2011 and mid-2013).

After occasional rises in 2014 and 2017 linked to imbalances in supply and demand and the effect of advertising (Lefebvre 2017), the prices of different REEs remained at almost unchanged levels until 2020. Expressed in metal form at 99% FOB China, La and Ce remained at almost unchanged levels in 2018 and 2019, at 6 $/kg on average. The price of Tb, then more expensive than REEs, became established at approximately 600 $/kg, whereas Dy is exchanged at 250 $/kg on average. Finally, the prices of Nd and Pr have stabilized today at 55 $/kg and 100 $/kg, respectively (compared to 85 $/kg on average from 2016 and 2017 for Pr, illustrating its growing role in permanent magnets).

The health crisis linked to the global Covid-19 pandemic had a limited impact on the rare earths market in China. Although 70–80% of rare earth transformation capacities underwent temporary interruptions in January and February 2020, linked to lockdown (in particular, capacities for processing HREEs in the southern regions, of the provinces of Jiangxi and Guangdong, close to the city of Wuhan), most of them were already functioning at less than full capacity or involved illegal or obsolete sites. There was therefore excess capacity in the country for the production of rare earths, which limited the impact of the health crisis on the sector. Only dysprosium and terbium, the HREE elements involved in high-performance, permanent NdFeB magnets (in temperature conditions higher than 200°C), saw high prices between January and March 2020: 15.5% for metal terbium (762 $/kg) and 20% for metal dysprosium (372 $/kg), respectively, reflecting the low stocks available rather than an underlying trend.

The increase in demand linked to the take-off in sales of electrical vehicles and renewable energies, faced with limitations mine in Chinese production capacities (Figure 1.6), could lead to a new rise in prices between now and 2025. On average, demand for REEs is progressing from 8 to 10% per year and should continue at this rate, driven by the rise in electrical vehicles, as well as by the rise in electronics and robotics. At such a rate, global REE consumption could double in less than 10 years.

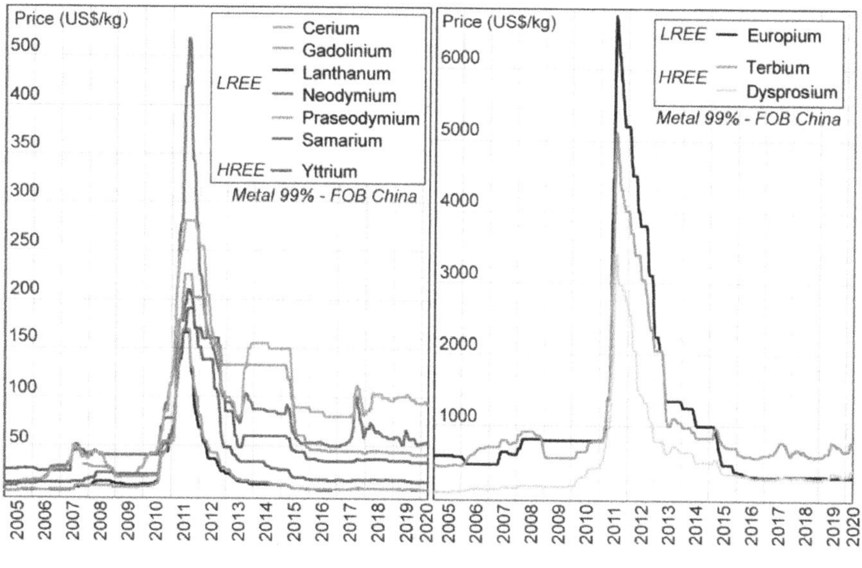

Figure 1.5. *Evolution of prices for the main REEs (source: BRGM)*

The case of electrical vehicles is eloquent. Although sales were in the order of two million units in 2018, many scenarios predict an exponential increase in 2030 between 20 and 38 million/year, as estimated by the International Energy Agency (New Policies Scenarios and EV30). Indeed, according to data from the consultant Roskill (2018), 90% of new electric vehicles are equipped with engines using permanent REE magnets (dominated by Chinese models, but also Tesla for its Model 3 Long Range) compared to induction motor technology, chosen especially by BMW to eliminate the need for REEs. Consequently, with an average value of 2 kg of permanent magnets mounted on each vehicle, which is 750 g of Nd-Pr alloy, this new market would reach between 15,000 and 28,500 tons of Nd-Pr allowance, compared to 3,000 tons in 2018.

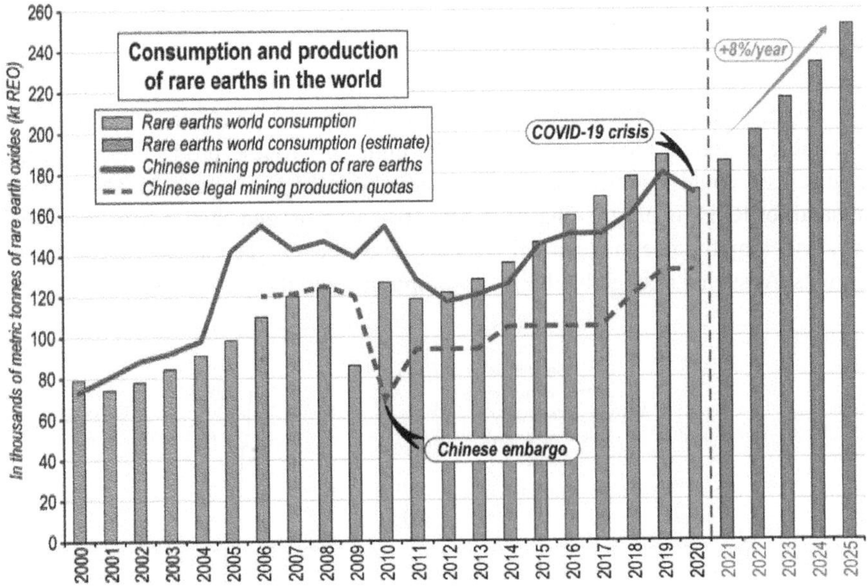

Figure 1.6. *Global consumption and production of REEs between 2000 and 2020 and estimates between 2021 and 2025 (source: BRGM)*

1.3.3. *Europe in the rare earth economy*

Europe consumes 10% of the REEs produced in the world. Moreover, this is one of the most promising markets for the development of electrical mobility and for the installation of offshore wind farms. Legitimate questions arise about potential sources of REEs to ensure their needs. In 2018, global production was estimated at 170,000 t REO. This figure represents a low range, since only official Chinese government quotas are considered (120,000 t in 2018, an increase of 14% compared to 2017), and not the country's illegal production (estimated between 40,000 and 50,000 tons). China's contribution to global production should comprise between 70 and 90% according to estimates.

Vertical integration of the whole REE production chain up to permanent magnets is one of China's major competitive advantages and raises the important question of the vulnerability of supplies for the European industry (Figure 1.7), not only of the raw material, but also of the products that are derived from it (Figure 1.8). To attenuate the concomitant industrial risks, the EU is encouraging the United States to develop and diversify their supply sources, whether primary (mining) or secondary, thanks to reuse, recycling and waste reduction (Guyonnet et al. 2015; Rollat et al. 2016).

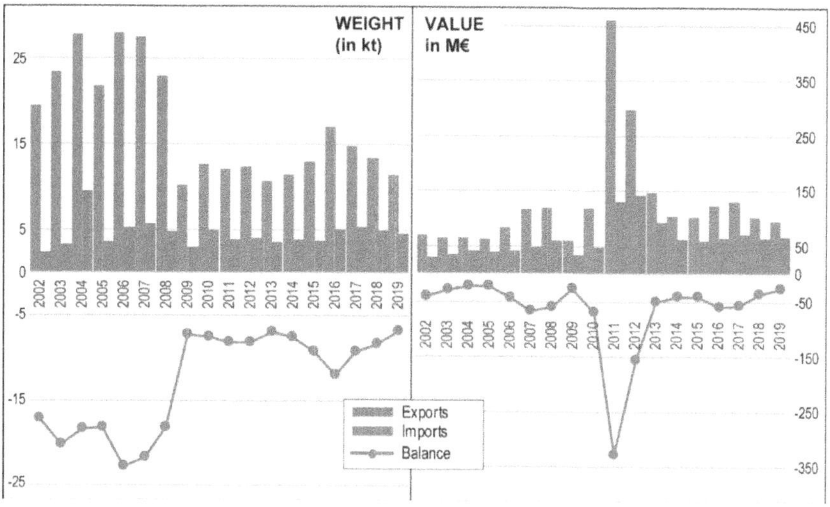

Figure 1.7. *Imports, exports and trade balance of the European Union for REEs (metals, alloys and compounds) expressed in weight (kt-thousands of tons) and in value (M€-million euros) between 2002 and 2019 (source: Eurostat). Codes CN8 no. 2846 and 280530*

One of the main challenges for the European continent, beyond commissioning these deposits, is to develop expertise in transforming REEs at the separation and refining stages, inspired by recent dynamics from Australia and the United States.

Thus, the Australians declared some projects as being of "national interest" and confirmed in March 2020 the desire to form part of a financial consortium for the development of the Dubbo project in New South Wales (Alkane 2020). In July 2020, via the Pentagon, Americans announced 40 million US$ worth of financial assistance to build two heavy rare earth separation factories on American soil (Scheyder 2019, 2020).

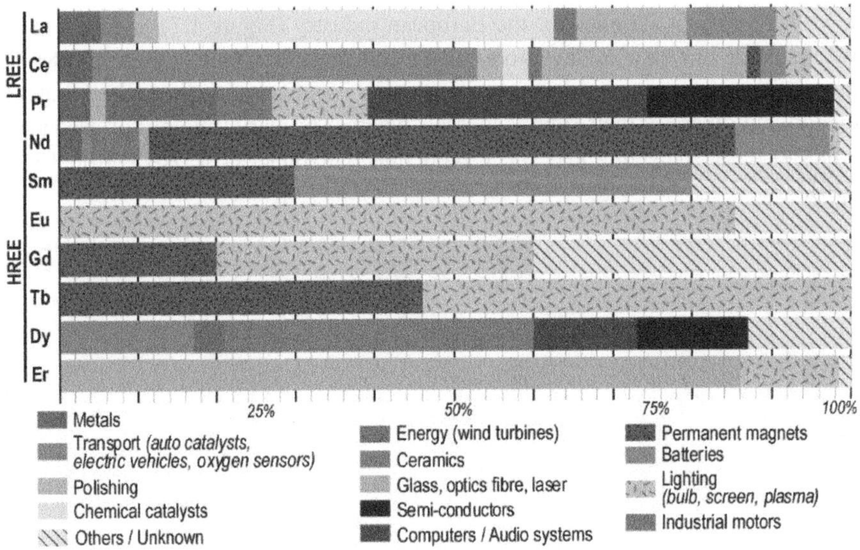

Figure 1.8. *Use of REEs in the European Union in 2016 (Sebastiaan et al. (2017) and references therein)*

1.4. Classification of rare earth deposits

REE deposits can be divided into two major categories (Figure 1.9):

– primary or endogenous deposits associated with magmatic and hydrothermal processes;

– secondary exogenous deposits linked to sedimentation and/or climatic processes (Walters et al. 2011; Chakhmouradian 2012a, 2012b; Charles et al. 2013; Tuduri et al. 2015; Goodenough et al. 2016; Dushyantha et al. 2020). Despite the great variety of deposits, only three types are in use:

- carbonatites (48% of production globally),
- alkaline magmatism deposits (2%),
- ionic clay deposits (36%),
- lateritic deposits (12%),
- placers (2%).

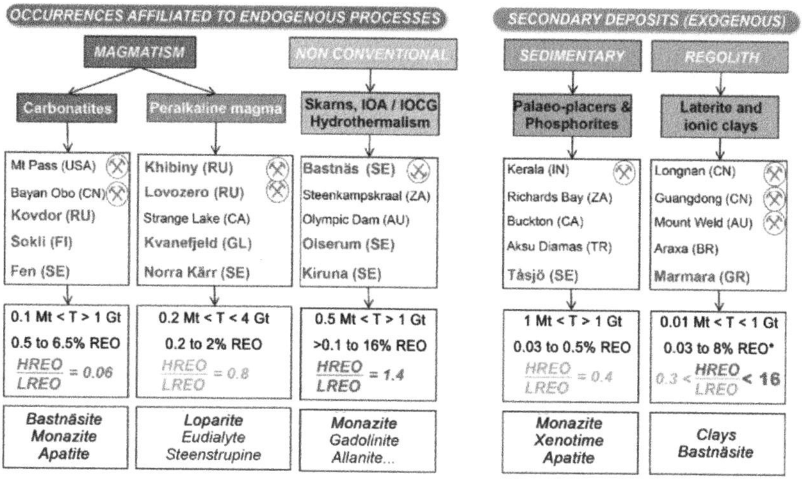

Figure 1.9. *Characteristics of the main deposit models*

COMMENT ON FIGURE 1.9.– *In blue, the major deposits located in Europe or Greenland. T: tonnage; *the 8% value corresponds to the Mount Weld deposit, and the corresponding HREO/LREO ratio is 0.3. Active mines are marked with a mining symbol. Bastnäs is a part of mining heritage. Abbreviations: AU (Australia), BR (Brazil), CA (Canada), CN (China), FI (Finland), GL (Greenland), GR (Greece), IN (India), RU (Russia), SE (Sweden), TR (Turkey), US (United States), ZA (South Africa).*

1.4.1. *Primary endogenous deposits*

REE carbonatites are characterized by intrusive bodies formed of carbonate rocks that are low in silica (< 10% weight SiO_2), but often enriched in REEs and accompanied by additional Nb, Ba, Sr, F, U, Th, Ti, Zr, and P minerals, as well as base metals (Cu, Pb, Zn). The carbonatites generally have REO levels above 1% and may exceed 5%, as in Bayan Obo (China) or the Mountain Pass mine (United States). Enrichment with REEs is marked by LREE levels in bastnäsite, monazite, apatite and allanite. The HREO/LREO ratio is low and generally below 0.1 (Figure 1.9).

REE deposits may also be associated with the alkaline complexes formed by magmatic alkaline rocks, generally undersaturated and moderately enriched with silica (35–60% weight SiO_2). These types of rocks (e.g. nepheline syenite) are characterized by an abundant and rich mineralogy that, in addition to REEs and Nb, are enriched with Zr, Ta, Be, Ti, Li, U and Th.

In contrast, REE levels associated with alkaline complexes are lower than those in carbonatites and generally < 2% (e.g. Norra Kärr in Sweden and Kringlerne in Greenland at 0.6% REO). These types of deposits are frequently enriched in HREEs (in comparison to carbonatites). Therefore, the HREO/LREO ratio, although it is generally very variable, is generally > 0.15 and sometimes > 1. The main, useful minerals are loparite, eudialyte, gadolinite or even steenstrupine and are often associated with the names of "exotic" rocks (ijolites, lujavrites, kakortokites, urtites or melteigites, etc).

Rifts in cratonic domains are zones with great economic potential, such as the Baltic shield, which includes the Kola Peninsula in Russia and Greenland (Goodenough et al. 2016). Only the loparite alkaline complexes of the Kola Peninsula are exploited today for their REEs and represent only 2% of global production. Although they are underused, alkaline complexes are the essence of new projects for exploring REEs. In 2015, Norra Kärr (Sweden) was the first deposit situated in the EU to certify resources and reserves evaluated at almost 55 Mt at 0.55% REO. Other deposits linked to alkaline intrusions show great potential. Moreover, the Khibina and Lovozero complexes in the Kola Peninsula include reserves estimated at more than 2 Gt at 0.6% REO. In southern Greenland, the magmatic province of Gardar is probably the most extraordinary. It is an ancient rift (1.35–1.12 Ga) over which a large number of alkaline complexes have been made (Grønnedal-Ika, Igaliko, Ilímaussaq, Nunarssuit or Tugtutôg) and has great potential since all have REE minerals, but also Nb and Zr ± U. Each intrusion contains a potentially world-class deposit such as the Ilímaussaq complex, including the Kvanefjeld and Kringlerne deposits. These alkaline complexes are mainly formed of nepheline syenite, associated locally with carbonatites. Concerning deposits associated with carbonatites, Norway, Finland and the Kola Peninsula in Russia have the greatest potential. The main REE-bearing minerals are apatite, bastnäsite, monazite and pyrochlore, and sometimes allanite, parisite and ancylite. The main targets are Fen in Norway (83.7 Mt at 1.08% REO), and there are many occurrences in the Kola Peninsula in Russia, such as Kovdor, which includes REO levels between 0.001 and 0.09% for an average tonnage of 500 Mt. In some deposits, substantial accumulations of apatite can form massive (apatitite) bodies within carbonatites. This is the case with the carbonatites at Siilinjärvi and Sokli (Finland), which, with 200 Mt of apatite reserves, could also form an important target for prospecting REEs. Indeed, Decrée et al. (2020) have recently demonstrated that the extraction of the rare earths contained in apatite from carbonatite at Siilinjärvi (Finland) would represent a tonnage of between 133,000 and 161,000 tons REO.

In less conventional form, the famous mine at Ytterby in Sweden (Romer and Smeds 1994) belongs to the NYF pegmatite model (Nb, Y, F). Similarly, it makes sense to distinguish deposits associated with skarns (e.g. Bastnäs).

1.4.2. *Secondary exogenous deposits*

Secondary deposits represented the major part of REE production before the 1970s, with the exploitation of monazite placers. Today, secondary deposits can be subdivided into two groups: deposits associated with surface processes and belonging to the regolith type and basin deposits associated with sedimentary environments (Figures 1.9 and 1.10).

Surface processes, including hydrolysis, oxidation, hydration, or indeed decarbonation reactions, trigger a chemical change in rocks and minerals in addition to physical phenomena. Thus, the soluble elements (Mg, Ca) are partly or sometimes entirely leached in very aggressive climates. The insoluble elements (Fe and Al, which form part of the REE group) remain in place to recombine into neoformed minerals, mainly clays, hydrophosphates or carbonates. In the context of laterite, the weathering of rocks initially rich in REEs (e.g. carbonatite) will therefore produce secondary, still richer REE deposits. This is the case with the REE laterite deposit at Mount Weld (Australia), which developed at the expense of a carbonatite deposit, the exploited content of which is higher than 8% (Lottermoser 1990). Ionic clay deposits (ion-adsorption laterite clays) are associated with the weathering of granites. REEs with weak solubility are adsorbed to the surface of neoformed clays (halloysite/kaolinite). These deposits are certainly numerous, but they are very small in size (some tens of thousands of tons) and are exploited despite their very low content levels. Southeastern Europe has many REE occurrences associated with weathering processes and linked to bauxites and laterites. These occurrences offer potential that remains to be ascertained in the Balkans and in Greece (Grebnik, Nazda-Vlasenica, Marmara).

Sedimentary basins, placer-type deposits, are accumulations of heavy minerals in sand and gravel separated by the process of gravity during their transportation by water or wind. Diagenesis consolidates these placers, transforming them into paleoplacers. The main REE-bearing minerals are monazite, xenotime, fergusonite, euxenite and allanite. These deposits represented the greater part of REE production before the 1970s, in particular through the exploitation of monazite placers. They are still exploited (2% of global production in 2018). Another accumulation process of microorganisms and algae in a marine setting at this time can, via diagenesis, produce concentrations of phosphates called phosphorites. These rocks, enriched with apatite can, in some settings, include REEs (e.g. Tåsjö in Norway). Finally, concerning authigenic deposits, the diagenesis of some silico-clastic rocks enriched with organic matter can in some cases, under the effect of a raised temperature, produce notable concentrations of REEs (Donnot et al. 1973; Pourret and Tuduri 2017).

In Europe, secondary paleoplacer-type deposits are generally of Cenozoic age, even though some date back to the Precambrian (e.g. the Nordkinn Peninsula in Norway). The main mineral encountered is monazite, sometimes with xenotime, allanite, apatite, euxenite, fergusonite, loparite and zircon. In Western Europe, these paleoplacers are mainly characterized by gray monazite enriched with intermediate REEs (Sm-Eu-Gd) and with very low concentrations of U and Th. These monazites result from the erosion of sedimentary rocks from the ancient basins of the Lower Paleozoic in Western Europe (France, Wales, Belgium, Czech Republic, Iberia). They remain small in size and without economic interest (Donnot et al. 1973; Tuduri et al. 2013). The phosphate sandstones of the Ordovician are also interesting targets, for example, Tåsjö (Sweden).

Like some deposits of sedimentary phosphates in the United States (Emsbo et al. 2015), deposits of oolitic iron and phosphate ooids in Norway, Sweden or Estonia (Sturesson 1995) and phosphate chalk at Ciply in the Mons basin in Belgium (Jacquemin 2020) may also form potential targets for the exploration of rare earths in Europe (see Decrée et al. (2017) and references therein).

1.5. Rare earths deposits in Europe

From analyzing the geographical distribution of REE deposits, it seems that they are present at all stages of the rock cycle. Indeed, REEs are often associated with metamorphism, plutonism, metasomatism, hydrothermal processes and sedimentary environments. Moreover, Fedele et al. (2008) argue that the large-scale distribution of REE deposits across Europe has a geogenic origin as their source. Goodenough et al. (2016) argue that the main economic REE deposits in Europe are linked to areas of intracontinental rifting. Their varied geological settings are the result of a region's evolution through geological time periods.

Thus, Europe and Greenland display extremely varied geological settings and include lithospheric blocks of varied nature that were assembled over the course of a geological history stretching back 3.8 billion years (see Choubert and Faure-Muret 1976; Ager 1980; Auboin 1980; Ziegler 1990; Asch 2003; Plant et al. 2005; Gee et al. 2008; Henriksen et al. 2009; Cloetingh et al. 2010 and references therein for a review).

The European continent has resulted from a long geological history over 3.6 billion years, with the assembly of many continental blocks. The European lithosphere can be divided into two distinct parts:

– an ancient cold craton (the East European craton, EEC) partially covered by a weakly deformed covering from the Phanerozoic and strewn with rifts from the Meso-Neoproterozoic in Eastern Europe;

– a thinner and cooler lithosphere dating mainly from the Phanerozoic and accreted to the craton during the course of the Paleozoic, including younger orogens in Western Europe (Figure 1.10) (Gee et al. 2008).

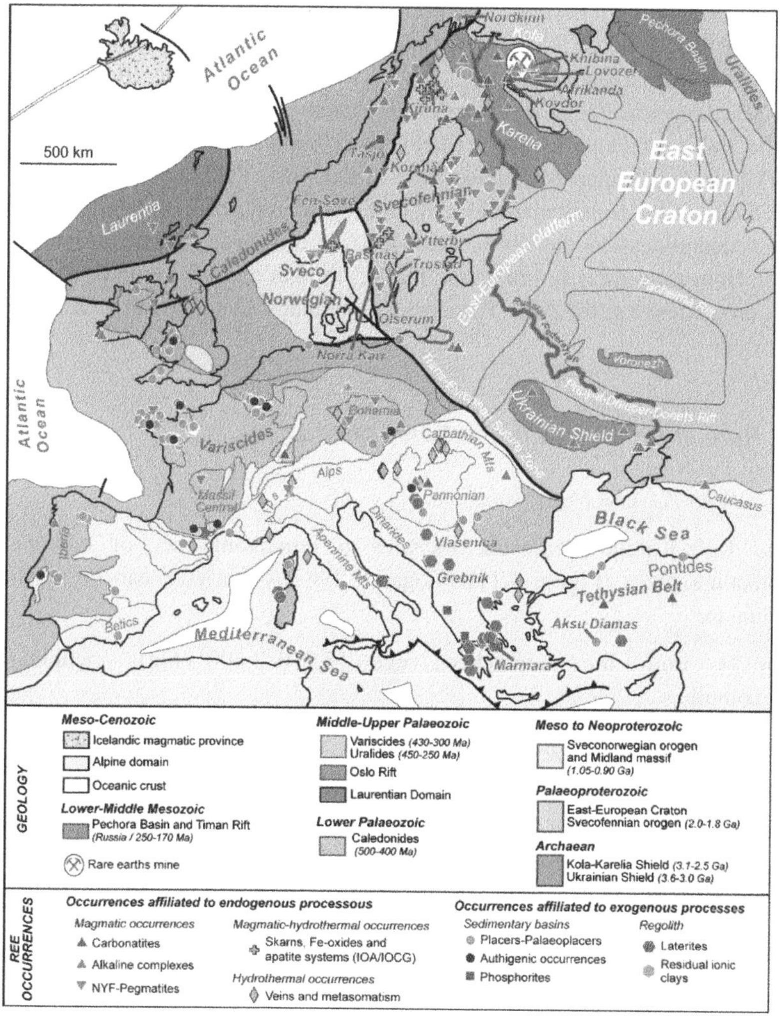

Figure 1.10. *Distribution of different occurrences of REEs affiliated with endogenous and exogenous processes in Europe (Charles et al. 2013; Goodenough et al. 2016; EURARE 2017; Tuduri et al. 2020)*

On the eastern flank of the craton, the Paleozoic orogeny of the Ural Mountains marks the boundary with Asia. To the north of Europe, the craton is surrounded by the Caledonids and the Timanids. Conversely, the southern border of the craton is less well defined since the evidence of accretions from the Neoproterozoic and the Paleozoic is partially obliterated by Alpine deformations and uplift, particularly well expressed in the Caucasus.

Finally, the western border of the craton is characterized by a major boundary of lithospheric scale, called a "Trans-European Suture Zone, TESZ" (Teisseyre 1903; Tornquist 1908; Thybo et al. 1999, 2002), stretching from the Black Sea to the North Sea. Thus, the TESZ separates the Precambrian lithosphere (EEC) to the east (e.g. the Baltic or Fennoscandian shield, the Ukrainian shield, and the Voronech massif) from the Phanerozoic lithosphere to the west. The latter is formed of land accreted during the Caledonian orogeny (e.g. Scandinavian, Irish-Scottish, German-Polish Caledonides) to the north, the Variscan orogeny (e.g. the French Massif Central, Bohemian massif, Armorican massif, Iberian massif) in the center, and the Alpine orogenies (e.g. the Alps, Pyrenees, Carpathians, Dinarides, Apennine, Betics) to the south.

1.5.1. *Rare earth indices in the Baltic shield*

1.5.1.1. *Kola-Karelia Peninsula*

The Kola-Karelia Peninsula includes a remarkable set of alkaline and hyperalkaline rocks (Figure 1.11) (Kogarko 1987). This area can be subdivided according to:

– rocks from the Neoarchean age (2,750–2,600 Ma), including the Keivy complex;

– rocks from the Paleoproterozoic age (1,900–1,600 Ma), including the Gremyakha-Vyrmes and Yeletozero intrusions (Arzamastsev et al. 2005);

– rocks from the Paleozoic age (370 Ma), such as the Khibina, Lovozero or Sokli massifs (Vartiainen and Woolley 1976; Kramm and Kogarko 1993; Arzamastsev et al. 2008).

All these massifs are composed of rocks that contain REE minerals (e.g. apatite, loparite, eudialyte and ancylite) and can constitute future deposits.

1.5.1.1.1. Neoarchean Keivy province (2,750–2,600 Ma)

The alkaline intrusions from the Neoarchean are situated in the western part of the Keivy greenstone belt. The alkaline province of Keivy is formed by six hyperalkaline granite laccolites that date back to 2,750 to 2,600 Ma for a surface of approximately 2,500 km² and two massifs of nepheline syenite injected along a fault on the border of the Keivy "terrane" (Mitrofanov et al. 2000; Zozulya et al. 2005, 2013).

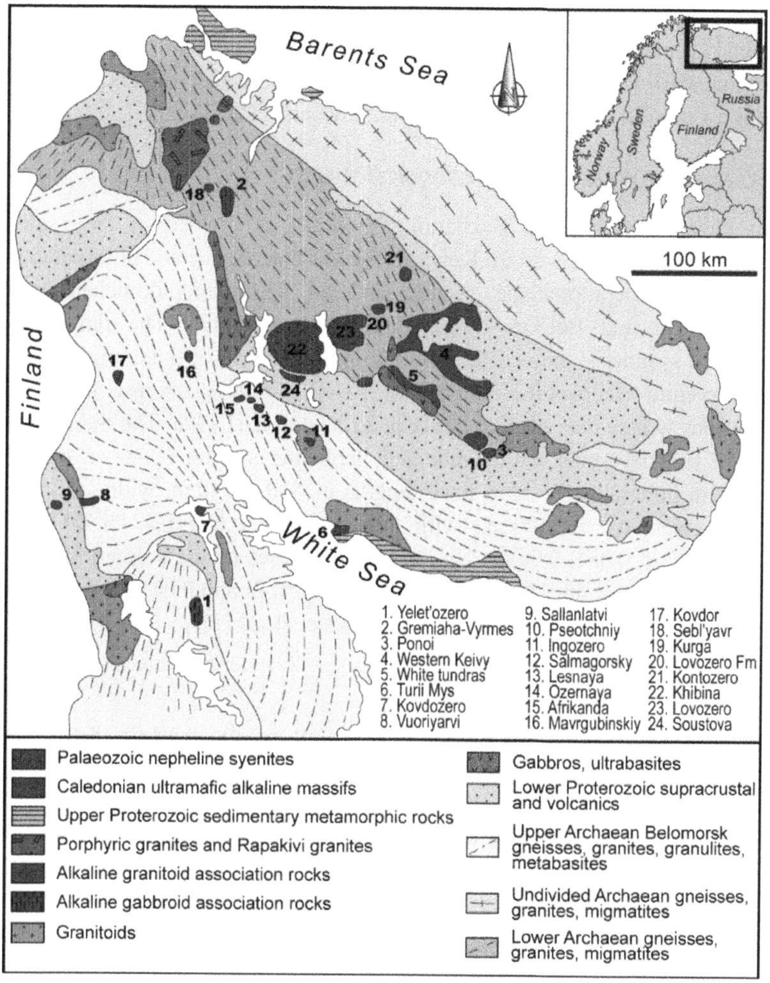

Figure 1.11. *The alkaline complexes of Kola-Karelia (Kogarko 1987)*

These massifs include six lithological groups: aegirine-arfvedsonite granites, ænigmatite-arfvedsonite granites, lepidomelane-arfvedsonite granites, lepidomelane granites, aegirine-magnetite granites and ferrohastingsite-lepidomelane-aegirine-augite syenogranites. The granites are enriched with Zr, REE, Nb and Ga, the main bearing minerals of which are zircon, monazite, britholite-(Y) and fergusonite. The known occurrences are pegmatitic veins (*elskoozerskoe*) present in microcline-albite-quartz mineralized zones within alkaline granites (*jumperuaiv*) and in albitized and zeolite zones in contact with nepheline syenites (Sakharjok 1984; Belolipetskyi et al. 1992; Korsakova et al. 2012; Zozulya et al. 2012). Little exploration work has yet been done, but – given the surface of this province – the potential for mining discovery is real (Korsakova et al. 2012).

Zr-REE alkaline complex at Sakharjok (Russia)

The mineralized zone at Sakharjok (5–6 km^2) is situated in the center of a hyperalkaline complex at western Keivy formed of a nepheline and phlogopite gabbro, a ferrohastingsite-lepidomelane syenite and an aegirine-lepidomelane nepheline syenite (Batieva and Bel'kov 1984; Zozulya et al. 2012). The ferrohastingsite-lepidomelane syenite is situated to the west and southwest of the complex, while the nepheline syenite occupies the eastern part. Zr-REE mineralization is linked to these two intrusions, which date back to between 2,682 and 2,613 Ma. The Sakharjok index consists of parallel veins within albitized nepheline syenite (Batieva and Bel'kov 1984). These mineralized bodies have lengths of 400–1,540 m, widths of 10–300 m and thicknesses of 15–100 m, with contents of 0.614–1.074% ZrO_2, 0.023–0.031% Y_2O_5 and 0.051–0.065% Nb_2O_5. The resources of the Sakharjok occurrence have been estimated at 35.8 Mt to 0.16% REO, 0.62% Zr and 0.041% Nb (Batieva and Bel'kov 1984; Pozhilenko et al. 2002; Korsakova et al. 2012).

Lesnoe alkaline intrusion (Russia)

The alkaline intrusion at Lesnoe forms part of the western Keivy alkaline complex, but the reading list for this area is both sparse and written in Russian (Korovkin et al. 2003). The Nb-REE-Zr Lesnoe occurrence is situated in a nepheline syenite, and the main REE-bearing minerals are fergusonite and britholite-(Y). The estimated resources range between 1.511 Mt and 0.22% REO, 1.21% Zr and 0.31% Nb (Korovkin et al. 2003; Korsakova et al. 2012).

1.5.1.1.2. Paleozoic alkaline province (370 Ma)

Khibiny and Lovozero Nb-P-REE alkaline intrusions (Russia)

The Khibiny and Lovozero massifs are the most substantial agpaitic nepheline syenite intrusions in the world.

The Khibiny intrusion (67°74′N; 33°72′E, Russia) covers a surface area of 1,330 km^2 and measures nearly 40 km in diameter. It is a massif with concentric intrusions of agpaitic nepheline syenite and alkaline and ultrabasic rocks (Figure 1.12) (Arzamastsev et al. 2008; Kalashnikov et al. 2016). The alkaline rocks of Khibiny have been dated back to 367.5 ± 5.5 Ma using the Rb–Sr method on whole rock (Kramm and Kogarko 1994). Although the size of this intrusion is incomparable, the alteration halo (fenitization) is nonetheless not very broad, not exceeding 50 m. Its main economic interest lies in the exploitation of phosphorus deposits from apatite, spatially associated with urtite and lujavrite. The most substantial deposits are situated along a narrow band to the south of the massif (e.g. Rasvumchorr, Kukisvum-chorr, Yuksporr, Koashva, Niorkpakhk, Oleny Ruchey, Partomchorr). Apatite, the main mineral of the exploited ore, contains 40–41 wt.% P_2O_5, 1.8–3.5 wt.% SrO, 9,000–11,000 ppm REE and 500–900 ppm Y. The resources measured for the deposits of the Khibiny intrusion are estimated at almost 3,200 Mt to 0.36% REO and 14.87% P_2O_5 (Eilu et al. 2013; Weihed et al. 2013; Kalashnikov et al. 2016). Many REE minerals have been identified: eudialyte, loparite, fluorapatite, ancylite-(Ce), belovite-(Ce), etc. (Arzamastsev et al. 2008).

The Lovozero intrusion (67°82′N; 34°75′E, Russia) covers a surface of 650 km^2 and measures approximately 25 km in diameter (Figure 1.13). The alkaline rocks of Lovozero have been dated back to 370.4 ± 6.7 Ma using the Rb–Sr method on whole rock (Kramm and Kogarko 1994). The massif intrudes garnet-biotite gneisses dating from the Archean. The Lovozero complex is often described as having four stages of intrusion (Eliseev and Fedorov 1953; Gerasimovsky et al. 1966; Kogarko et al. 1995; Kalashnikov et al. 2016):

– nepheline syenite with uniform grain, nosean nepheline syenite, metamorphozed nepheline syenite and poikilitic nosean syenite;

– embedded lujavrite-foyaite-urtite complex;

– layered urtite, lujavrite-eudialyte and foyaite-eudialyte complex;

– alkaline lamprophyre dykes (minchiquite, fourchite, tinguaite).

Moreover, the fenitized zone has a width of between 50 and 200 m. Many nepheline syenite and alkaline pegmatite veins intersect the host rocks and extend for more than 100 m around the contact. Surrounding the core of the massif, the lujavrite-foyaite-urtite layered sequence contains layers of loparite.

Loparite has been exploited episodically for 50 years for niobium and contains 30–36% REO. Loparite is present in submillimeter grains (0.2–0.6 mm) and more rarely in the form of large crystals (Vlasov et al. 1966; Smirnov 1977), with higher content levels in rocks enriched with nepheline. It is found mainly in porphyry rocks associated with concentrations of urtite and homologs containing aegirine in juvite and of malignite. The mineralization consists of banded horizons of some centimeters to some meters in thickness, with 2-3% loparite (Hedrick et al. 1997). The measured resources are estimated ca. 1,020 Mt to 0.75% REO (Kalashnikov et al. 2016).

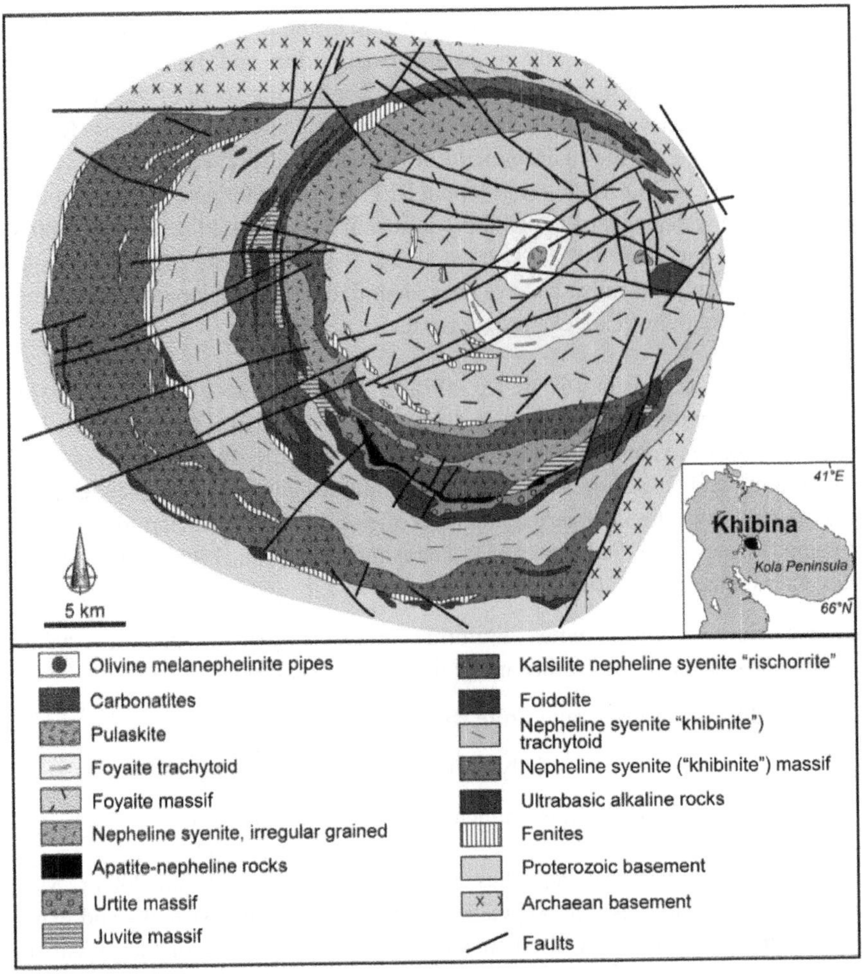

Figure 1.12. *Khibiny alkaline intrusion (Arzamastsev et al. 2008)*

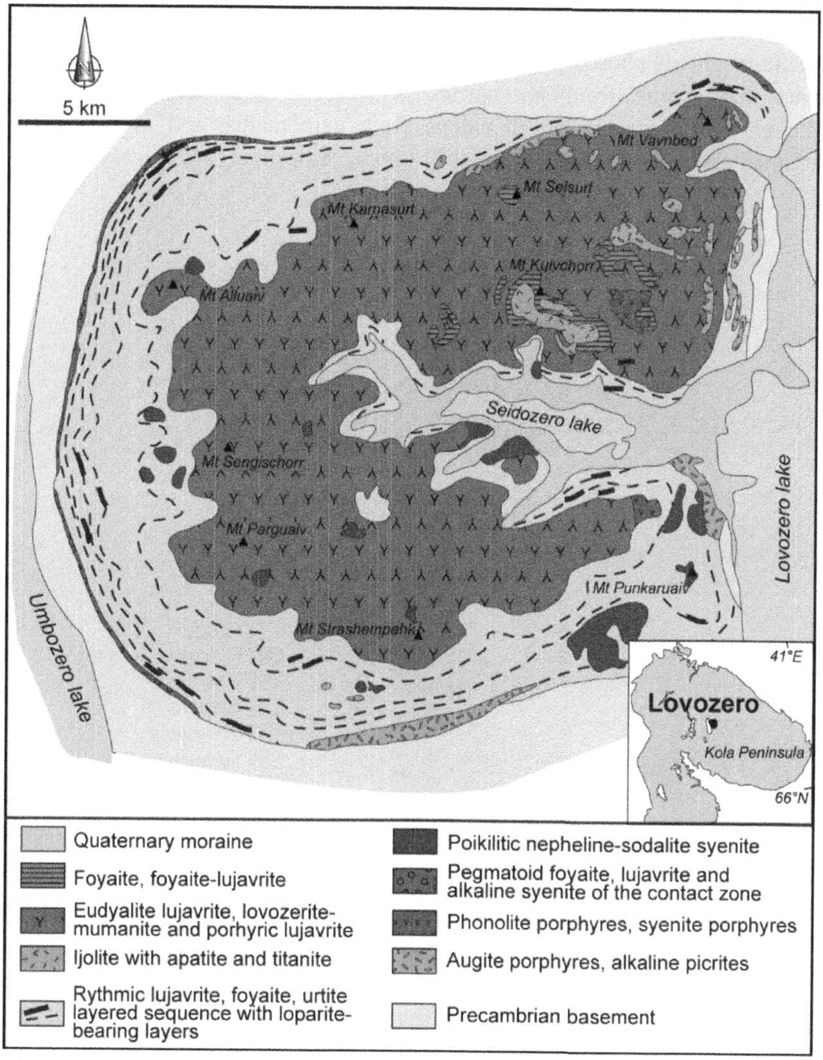

Figure 1.13. *Lovozero alkaline intrusion (Arzamastsev et al. 2008)*

Kovdor ultrabasic alkaline complex (Russia)

The Kovdor massif (Russia, 50 km SE of the Sokli carbonatite), with an area of 40 km^2, is an alkaline complex of multiple concentric intrusions (Kogarko et al. 1995; Kalashnikov et al. 2016) that has been dated back to 380 ± 4 Ma using the U/Pb method on baddeleyite (Bayanova et al. 1997). The formations, from the most ancient to the most recent, are ultrabasic rocks (dunite and clinopyroxenite), turjaites and melilitolites, an ijolite-melteigite series, a mineralized complex of apatite-forsterite-magnetite phoscorites, nepheline syenites and carbonatites (Figure 1.14). It appears that apatite from phoscorites is enriched with REEs (2,720 ppm). Moreover, REEs are present in calcite from carbonatites and in melilite from melteigites with levels of 886 ppm and 914 ppm, respectively (Verhulst et al. 2000).

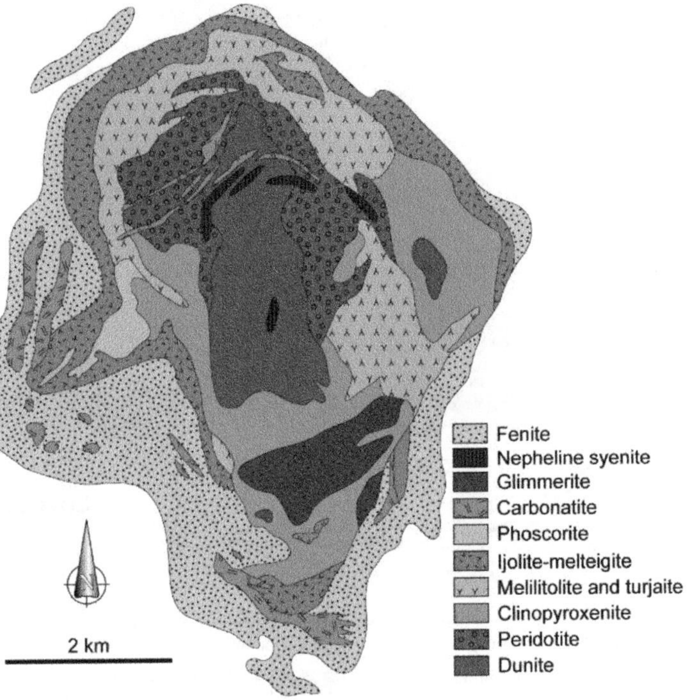

Figure 1.14. *Kovdor alkaline intrusion (Verhulst et al. 2000)*

Sokli Nb-REE carbonatite (Finland)

The Sokli carbonatite is situated to the northeast of Finland, 90 km from Savukoski, and forms part of the Kola alkaline province (Figure 1.15). This massif was discovered in 1967 during an airborne geophysics program (Paarma 1970) and described by Vartiainen and Woolley (1974, 1976), Vartiainen (1980, 2001), and Vartiainen and Vitikka (1994). This polyphase intrusion (~5 km in diameter) has a carbonatite core (phoscorites, sovite, silicosovite and rauhaugite) surrounded by a metacarbonatite and a broad fenite halo. The Sokli carbonatite dates back to the Devonian (360–380 Ma, (Vartiainen and Woolley 1974; Kramm et al. 1993)) and includes a deposit, which has still not been exploited, of P, Nb, Ta, Zr, REEs and U

(Korsakova et al. 2012; Sarapää et al. 2013). The most substantial REE potential occurs in the fenite halo and in the late carbonatite dykes intersecting the massif. In 2006, GTK (Finnish geological service) carried out a mineralogical and geochemical study on drill cores in carbonatites and reported a high total content (0.5–1.83% weight, including 0.11–1.81% weight LREEs and 0.01-0.041 wt. % HREEs; Kontio and Pankka 2006; Sarapää et al. 2013). The REE-bearing minerals are fluorapatite, apatite-Sr, monazite, bastnäsite-(Ce), ancylite-(Ce), strontianite, barite and brabantite (Al-Ani and Sarapää 2009, 2013). Finally, hydrothermal circulations are probably at the origin of the secondary monazite alteration. The Sokli phosphorus deposit displays substantial resources, with 190 Mt of ore in altered rock and 12,200 Mt of non-weathered rock ore (GTK).

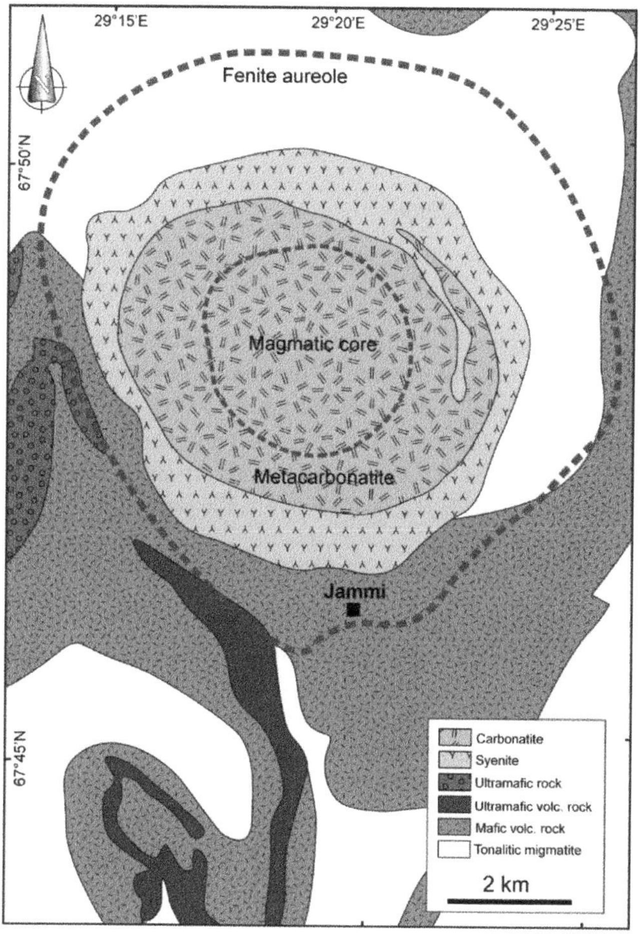

Figure 1.15. *Sokli massif in Finland (modified from Browne (2008))*

Iivaara alkaline complex (Finland)

The Iivaara massif is an alkaline complex situated to the northeast of Finland and forms one of the most western intrusions of the Devonian Kola alkaline province (Kramm et al. 1993; Sarapää et al. 2013). This complex is an oval intrusion (> 10 km^2) that can be subdivided into three parts, from the center to the rims: alkaline rocks (nepheline and clinopyroxene urtite, ijolite, melteigite), a transition zone, and a fenitized zone up to 1,000 m in width (Lehijärvi 1960). The Iivaara alkaline complex has been dated back to between 367.4 ± 4.9 and 373.3 ± 8.1 Ma using the Rb–Sr method on whole rock (Kramm et al. 1993). Note that Iivaara is the type locality of ijolite, a fairly widespread rock in alkaline complexes. A mineralogical study of GTK has demonstrated the presence of REE-bearing minerals within nepheline syenites, mainly with apatite and allanite to a lesser degree. The samples analyzed contain <200 ppm REE and 1–5% P_2O_5, leading to a lack of economic interest where REEs are concerned (Sarapää et al. 2013). However, the fenitized zone of the Iivaara complex is still being studied, and the most optimistic results remain possible in terms of REE content.

1.5.1.2. *Svecofennian domain*

1.5.1.2.1. The Appinitic intrusions at Suhuvaara, Lehmikari and Vanttaus (Finland)

In the Lapland granulitic belt, to the north of Finland, the Suhuvaara massif is an intrusion formed mainly of syenitic to dioritic rocks (Sarapää et al. 2013). Similar appinitic intrusions have been identified further to the south in the central Lapland granitic complex with the Lehmikari and Vanttaus massifs (Mutanen 2011; Sarapää et al. 2013). High Ba, Sr, Zr, P and REE levels have been detected in these rocks. For syenitic rocks, the REE levels are above 1,300 ppm.

1.5.1.2.2. REE-Nb-Zr sheared gneiss at Otanmäki (Finland)

In central Finland, to the south of Lake Oulujärvi, the Otanmäki-Katajakangas region contains REE-Nb-Zr within an alkaline granitic gneiss (Figure 1.16) (Hugg 1985; Hugg and Heiskanen 1986; Sarapää et al. 2013 and references therein). The region is formed mainly of gneiss of Archean age, intruded by gneissic granites and gabbro-anothorsites dating back to approximately 2.50 Ga. The Otanmäki area was exploited in the 1950s and 1980s by the Otanmäki Oy company for its deposits of Fe-Ti-V gabbro-anorthosites. Since 2010, the region has been explored by Tasman Metals for REEs through two licenses, those of Katajakangas and Kontioaho.

REE-Nb-Zr mineralization at Katajakangas was discovered by GTK in 1982. It is located at the contact between gneiss and metasediments (turbidites) in the form of intrusive bodies of some meters in thickness (Kärenlampi et al. 2020). This main lens is greatly deformed. It is formed of a fine-grained gneiss (mylonite), reddish to gray, with quartz-feldspar, riebeckite and alkaline pyroxene. The main REE-bearing minerals are as follows: fergusonite, allanite, bastnäsite, and to a lesser extent columbite, thorite and xenotime (Marmo et al. 1966; Hugg and Keiskanen 1986; Äikäs 1990; Tasman Metals Ltd. 2010a). In 1985, an initial estimate of resources for Katajakangas yielded 0.46 Mt to 2.71% REO, 0.76% Nb_2O_5 and 1.13% ZrO_2 (Hugg 1985). This estimate was based on an analysis of 14 core drillings carried out over seven profiles with a total length of 850 m and was calculated for a vertical depth of 150 m. More recently, Sarapää et al. (2013) again analyzed some samples of alkaline gneiss from cores from the 1980s. Analyses have demonstrated high REE contents (2.4 wt. % at most), with LREE/HREE ratios ranging from 8.82 to 9.77.

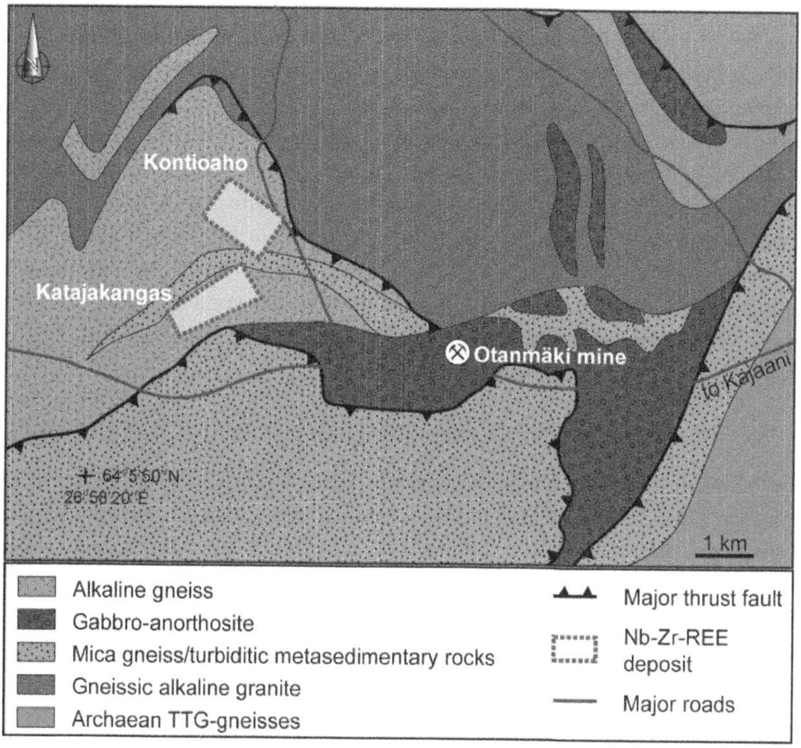

Figure 1.16. *Shear gneiss with REE-Nb-Zr at Otanmäki-Katajakangas, Finland (Sarapää et al. 2013)*

The permit at Kontioaho is situated 1.3 km to the northeast of the Katajakangas permit. REE mineralization was discovered within an altered area formed of intrusive bodies 7–12 m in width, oriented northwest, and formed of quartz, microcline, plagioclase, zircon and magnetite within quartzo-feldspathic schists. The mineralized horizon is characterized by a strong magnetic anomaly oriented NW–SE over more than 400 m. The main REE-bearing minerals are allanite, fergusonite and xenotime (Tasman Metals Ltd. 2010a). The REO total content was estimated, on the basis of new drill core data, to be 0.59%.

The intrusive bodies of the permits at Kontioaho and Katajakangas carry a complex mineralization of Ti, Zr, Nb, REE, Th and U. This alkaline magmatism has been dated back to 2.04-2.05 Ga and then deformed over the course of the Svecofennian orogen approximately 1.9–1.8 Ga (Kärenlampi et al. 2020).

1.5.1.2.3. Albite and felsite Sc-REE-U at Biggejav'ri (Norway)

The Sc-REE-U albitite at Biggejav'ri is situated in the volcano-sedimentary belt at Kautokeino, of Early Proterozoic age, in the extreme north of the county of Finnmark, in Norway. The rocks date back to 1,890–1,750 Ma (Statherian to Orosirian) and are mainly represented by Sc-U-REE-enriched albite felsites (Olerud 1985, 1988; NGU 2013). Felsite contains calcite, quartz, muscovite and rutile, but also minerals rich in Cr and V, such as davidite-loveringite and chromite. Additional minerals include La-Ce-bearing carbonates, monazite, orthite, thortveitite, coffinite, uranophane, brannerite, zircon, pyrite, sphalerite, galena, chromium chlorite and an unidentified REE-bearing mineral. The rock's total Sc content, and perhaps REE content, could present economic benefit (e.g. 122 ppm Sc, 1 027 ppm La, 605 ppm Ce, 248 ppm Y) (Olerud 1988).

1.5.1.2.4. Kiruna-type iron ores with magnetite and apatite REEs (Sweden)

To the north of Sweden, close to Kiruna (67.85°N–20.22°E, Norrbotten region), Kiruna type P-Fe-apatite-REE iron ores form part of a group of occurrences formed by a magnetite-hematite-apatite assemblage, with rocks from the Paleoproterozoic from the Fennoscandian shield (Geijer 1931; Frietsch 1977; Frietsch and Perdahl 1995) (Figure 1.17). These surrounding rocks are alkaline rhyolites, trachytes and trachyandesitic lava flows with intrusive rocks and a sedimentary sequence (Geijer 1931; Geijer and Ödman 1974).

It is important to note that the deposits are directly associated with sodium and potassium alterations (Frietsch et al. 1997; Martinsson 1997). The ore is formed of a magnetite-apatite assemblage that contains nearly 30% apatite (mainly fluorapatite) and some additional minerals (i.e. actinolite, calcite, quartz, sphene, talc, albite). The mineralized bodies are massifs, forming stockwerks and irregular, bulbous, but sometimes also concordant masses. The mineralization dates back to between 1.88 and 1.90 Ga (Cliff et al. 1990). Discovered in the 17th century, the apatite iron ores of Kiruna are the subject of intense discussion (Cliff et al. 1990; Hitzman et al. 1992; Evans 2000; Harlov et al. 2002). Two main hypotheses are still discussed to explain the origin of the mineralizations:

– immiscibility between an iron-rich magmatic fluid and hydrothermal remobilization (Bottke 1981; Harlov et al. 2002);

– or an exhalative hydrothermal process (Cliff et al. 1990; Evans 2000).

More broadly, according to Hitzman et al. (1992) and Barton and Johnson (1996), Kiruna-type mineralization forms a subclass of IOCG (iron-oxide copper gold) type deposits from the Proterozoic: IOA (iron-oxide apatite) type deposits.

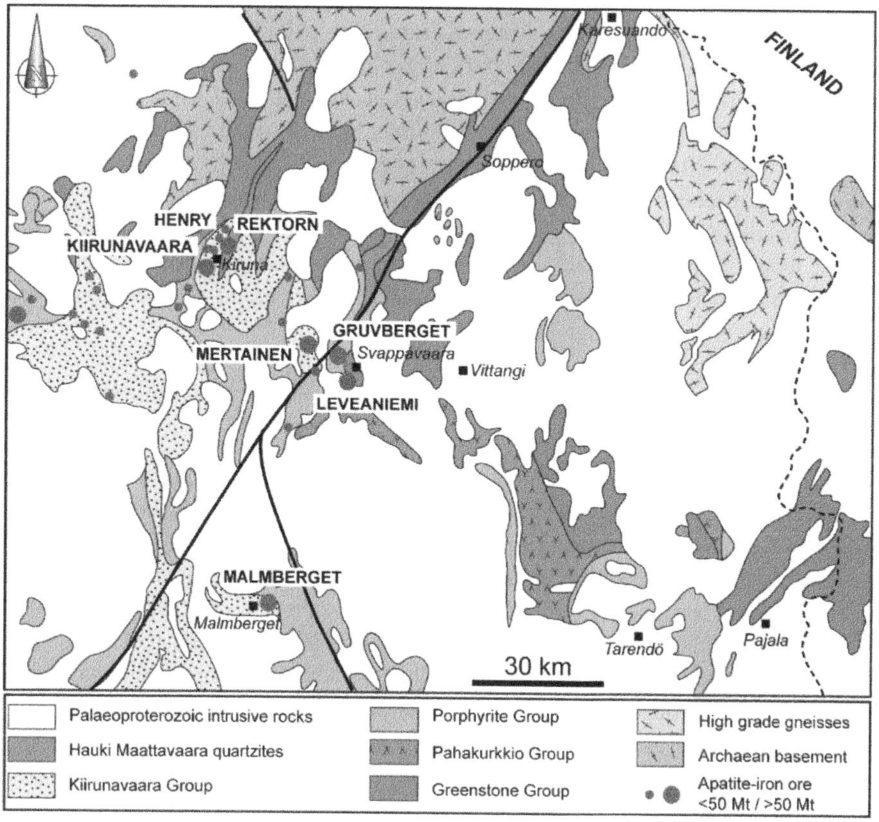

Figure 1.17. *REE magnetite-apatite iron ores of the type found in Kiruna (Sweden), modified from Martinssen (2007)*

The peculiarity of Kiruna-type iron mineralizations is their relatively high REE content (Parák 1973, 1975, 1985; Frietsch and Perdahl 1995). These significant REE contents in ores were first mentioned by Geiser (1931). The district of the first Kiruna-type iron deposits has an average content of 0.7% REO, mainly carried by apatite. The REEs are contained in apatite and to a lesser extent in monazite, which is not homogeneously present in the mineralizations. Moreover, the hematite ores of Kiruna have an average content of 0.5% REO and are still mainly contained in apatite. In a more precise fashion, Parák (1973) analyzed the REE content of apatite iron ores from several deposits from the Kiruna district: Kiirunavaara (6,560 ppm), Malmberget (5,760 ppm), Leveaniemi (1,545 ppm), Rektorn (3,275 ppm), Haukivaara (5,295 ppm), Henry (5,210 ppm), Nuku-tusvaara (6,730 ppm) and Lappmalmen (5,890 ppm).

The Swedish LKAB (Luossavaara-Kiirunavaara AB) company has extracted more than 950 Mt of ore since the start of exploitation a century ago, and only a third of the mineralized body was initially exploited. Recently, LKAB has sought to extract REE from the apatite present in tailings because of the ReeMAP research project (LKAB 2018). Tailing ponds that contain apatite represent a substantial volume with a relatively low REE content. These tailing ponds at Kiruna represent approximately 8 Mt/year and contain 4–8% apatite and 1,200–1,500 ppm REEs, which would represent 10–15 kt REEs per year (Renberg 2014). Industrial production of REE concentrate is predicted for 2021 and would represent more than 500% of Sweden's demand for REEs.

1.5.1.2.5. Albitites at Uuniniemi, Honkilehto and Palkiskuru-Palovaara (Finland)

In the north of Finland, three small areas of hydrothermal alteration are presented as showing REE anomalies: Uuniniemi, Honkilehto and Palkiskuru-Palovaara (Sarapää et al. 2013). The Uuniniemi area (near Kuusamo, 680 km to the north of Helsinki) is formed of albitites, syenites and carbonatite veins (Mutanen 2011). The carbonatite has been sampled and analyzed, revealing levels of 2.80% P_2O_5, 0.43% REE and 256 ppm Nb (Sarapää et al. 2013). Moreover, the REE content is 0.1% in the albitite and carbonates and albite rocks. REE-bearing minerals have been identified through analyses using electron probes (monazite-Ce, colombite, ferro-columbite, euxenite-Y, zircon, Fe-thorite, britholite and bastnäsite in minority quantities).

In the central part of the Kuusamo belt (Vanhanen 2001), Au-Co mineralization at Honkilehto, surrounded by sericite and sulfide schists, also includes uranium minerals (davidite), Pb-U and U-Si minerals. Bastnäsite and allanite are always associated with these anomalies. Analyses of drill core samples (e.g. DH398, 81–93 m intersection) by Sarapää et al. (2013) have revealed the following average levels: 1,552 ppm REE (maximum 3,442 ppm), 1,472 ppm Co (maximum 3,410 ppm), 175 ppm U (maximum 430 ppm) and 6.1% S (maximum 13.9%).

Finally, to the northeast of Finland, the Palkiskuru-Palovaara area shows albitites formed of REE-bearing minerals with bastnäsite, monazite, allanite and xenotime (Sarapää et al. 2013). The uranium minerals are davidite, masuyite and sayrite. The highest REE content is 0.4%, with levels of 209 ppm Th and 3,570 ppm U.

1.5.1.2.6. Kaolinic weathering profiles at Virtasalmi and Mäkärä-Vaulo (Finland)

Laterization processes lead to the partial or total leaching of metals such as Al, Mg, Fe, Cu, Ni and REE from the bedrock and are then accumulated within a weathering profile between laterite and saprolite (Morteani and Preinfalk 1996).

To the southeast of Finland, the kaolin deposits of the Virtasalmi region (from 0.5 to 2 km in length and 50–400 m in depth, such as those of Litmanen, Eteläkylä, Vuorijoki and Montola) are covered by 20–30 m of formations from the Quaternary. Kaolin presents over an average thickness of 30–40 m, indeed up to 100 m, and contains 40–75% weight kaolinite, 20–30% quartz and some grains of potassium feldspar and muscovite. An area of transition is observed between the kaolin and the surrounding rocks (gneiss and tonalite), which present well-preserved structures. The kaolin deposits at Virtasalmi have been dated back to 1,180 Ma using the K/Ar method on authigenic illites from the Litmanen deposit (Sarapää 1996). Sarapää et al. (2013) selected some samples of drill cores to carry out mineralogical and chemical studies to estimate their REE potential. The kaolins show REE enrichment with LREE/HREE ratios between 6 and 19, (La/Yb)n between 6 and 66, and Ce/Ce* equal to 1 (Al-Ani and Sarapää 2009). At the base of the kaolinic weathering profiles of the deposits at Litmanen and Eteläkylä, the total REE content may reach at most 0.1–0.2% (Sarapää et al. 2013).

To the north of Finland, near Sodankylä (Laponie), the region of Mäkärä-Vaulo forms part of the Tana belt (the southern part of the Lapland granulitic belt). It has been subject to exploration for gold, which had previously been detected through geochemical anomalies in REEs (anomalies strongly positive in La and Y). The Tana belt is a strongly tectonized area formed of amphibolites, from gneiss to garnet-biotite and arkosic gneiss. The whole area overlaps the central Lapland greenstone belt at approximately 1.91 Ga (Tuisku and Huhma 2006). In the areas of Mäkärä and Vaulo, a thick kaolin profile is developed at the expense of basement rocks, the whole covered by glacial clay deposits (tills) rich in La and Y (Sarapää and Sarala 2011, 2013).

The Mäkärä profile shows levels of approximately 0.05% REE (maximum 0.1%), LREE/HREE ratios between 2.7 and 15.1 and Ce/Ce* between 0.7 and 0.9. The profile of Vaulo shows a level of 0.40 to 0.45% REEs in the fine fraction of glacial deposits. The REE-bearing minerals are monazite, rhabdophane and xenotime. The REE potential of the Tana belt could be interesting since only the Mäkärä-Vaulo region has been explored (Sarapää et al. 2013).

1.5.1.2.7. REE mineral pegmatites (Sweden)

The Svecokarelian orogeny includes some pegmatites that contain uranium- and REE-bearing minerals (Sadeghi et al. 2013). Allanite is the main REE-bearing mineral in pegmatites with high LREE/HREE ratios (Castor and Hedrick 2006). In Sweden, pegmatites have been classified into two categories following the study of additional minerals carried out by Smeds (1990):

– tourmaline pegmatite ± Li-bearing mineral (rocks with low to moderate metamorphic facies);

– pegmatites with REE-Y minerals (various geological settings).

Unfortunately, the pegmatites are generally too small or present contents too low to be of economic interest. Nevertheless, for pegmatites already exploited for feldspar and mica, the exploitation of the REEs contained in other minerals would be possible.

1.5.1.2.8. P-REE and carbonatite complex at Siilinjärvi (Finland)

The carbonatitecomplex at Siilinjärvi is situated in eastern Finland (Figure 1.18), near Kuopio (380 km to the north of Helsinki). It is a steeply dipping lenticular body approximately 16 km long and 1.5 km broad (Pusstinen 1971). The calcite, dolomite and phlogopite rocks of this complex form a tabular body 600 to 700 m in width surrounded by a fenitized area. White to greenish medium-grained carbonate rocks are rarely present at Siilinjärvi, and carbonatite is not very widespread. The majority of the central body are rocks rich in phlogopite, ranging from pure glimmerite to silicocarbonatite, followed by some carbonatites. Dating using the U/Pb method on zircon has given an age of 2 610 ± 10 Ma (Tichomirowa et al. 2013), which makes the carbonatite from Siilinjärvi one of the oldest in the world.

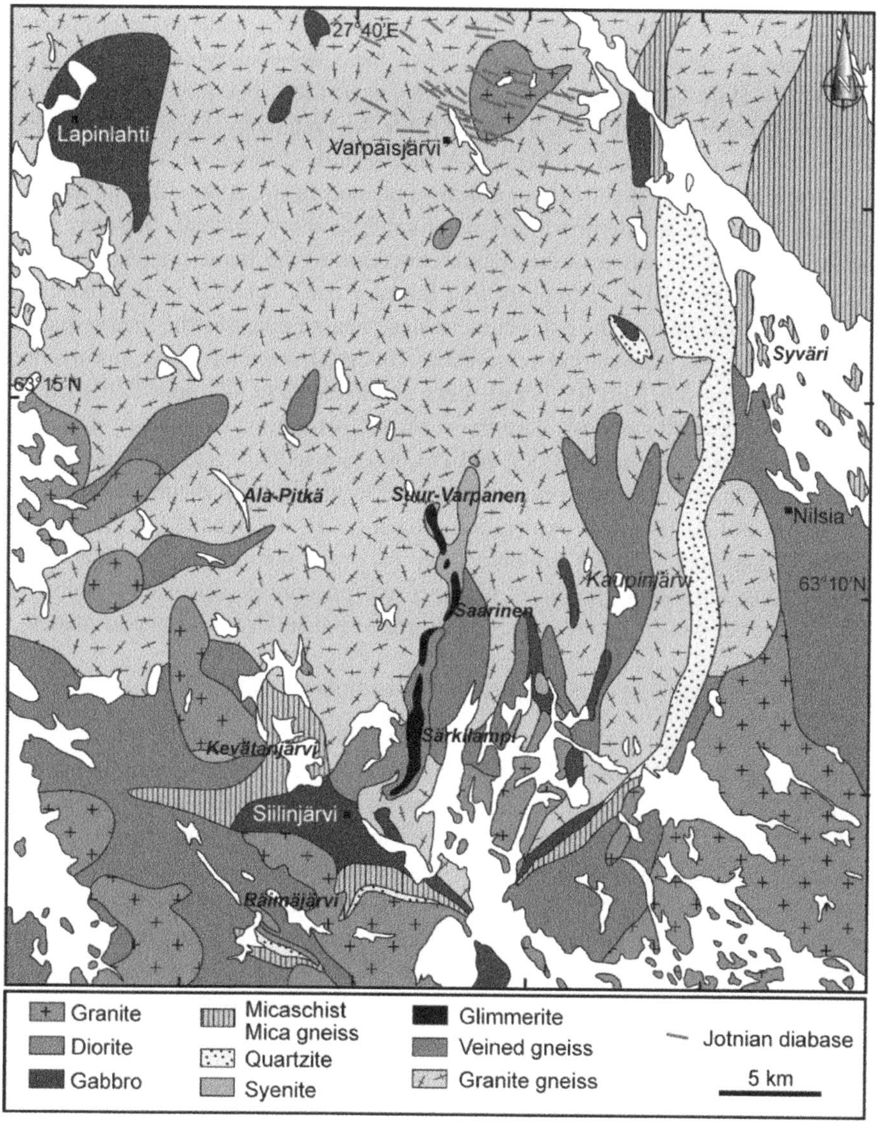

Figure 1.18. *P-REE and carbonatite complex in Siilinjärvi, Finland (Pusstinen 1971)*

Apatite is situated in the central part of the magmatic body in the form of gray to greenish-yellow prisms and irregular grains in glimmerite-carbonatite-type rocks. Apatite shows no inclusion in the thin sheets. It is fluorapatite with an average content of 1.2% Sr and 0.4% RE_2O_3 (Pusstinen 1971). The minerals in glimmerite are phlogopite, apatite and carbonates. They have been analyzed for their REE potential by Hornig-Kjarsgaard (1998). Samples of total rock have revealed REE contents between 499 and 1,629 ppm. Apatite has relatively high REE contents varying between 2,986 and 3,820 ppm. A recent study carried out by Decrée et al. (2020) revealed the content of rare earths contained in the apatite of this alkaline complex. Thus, the apatite present in the glimmerite shows contents between 3,195 and 6,665 ppm REE, whereas the apatite contained in the fenite contains between 663 and 4,567 ppm REE. It appears that the rare earth contents in glimmerite or fenite are proportional to the P_2O_5 content since apatite encloses the greater part of the rare earths. By considering reserves of 200 Mt @ 4 wt. % P_2O_5, and average contents between 576 ppm REEs (all rocks combined) and 697 ppm REEs (in the glimmerite-carbonatite series), Decrée et al. (2020) estimated that it would be possible to extract between 133,000 and 161,000 tons REO. It must be remembered that in 2012, an exploration permit was obtained for the Siilinjärvi complex (Tasman Metals Ltd. 2012). This REE exploration project covers 450 ha and lies close to the phosphate (apatite) mine of Siilinjärvi (by the Yara company), 30 km north of Kuopio. This site mainly produces fertilizer and phosphoric acid and remains the sole producer of phosphate ore in western Europe (800,000 ton/year of phosphates).

1.5.1.2.9. Carbonatites at Kortejärvi and Laivajoki (Finland)

In central-northern Finland, the carbonatite massifs of Kortejärvi and Laivajoki intrude basic volcanic rocks dating back to the lower Paleoproterozoic along a crustal-scale fault and form a substantial part of Finland's REE potential (Sarapää et al. 2013). The Kortejärvi carbonatite is an intrusive massif approximately 2 km long and 60 m thick, while the Laivajoki massif measures 4 km in length and 20 m thick (Nykänen et al. 1997). The two intrusions are formed above all calcite carbonatite and dolomite, even though the intrusion at Laivajoki also contains silico-carbonatite. Moreover, glimmerite and olivine-magnetite rocks are also visible in the Kortejärvi intrusion. Within the Kortejärvi carbonatite, a study has been carried out on cores from older drillings and has revealed the abundant presence of apatite with monazite and allanite (Al-Ani et al. 2011). For the Laivajoki carbonatite, some samples analyzed have shown high calcite, dolomite, phlogopite, magnetite and apatite contents. The main REE-bearing minerals in the two massifs are allanite and monazite. Unfortunately, the REE contents in the Laivajoki (210–1,644 ppm) and Kortejärvi massifs (443–892 ppm) are relatively low. Further studies are needed to better define the REE potential of the Kortejärvi and Laivajoki carbonatites.

1.5.1.2.10. The REE-Zr nepheline syenite of Norra Kärr (Sweden)

The nepheline syenite of Norra Kärr was discovered in 1906 by Törnebohm in central southern Sweden, 300 km from Stockholm and 15 km to the NNE of Gränna, near the eastern bank of lake Vättern. The intrusive complex is elongated N–S and is nearly 1.3 km long and 460 m wide, covering a surface of approximately 380,000 m^2 (Figure 1.19). This complex intrudes into the gneiss and granites of the Paleoproterozoic (1,791 ± 8 Ma; Johnsen and Grice 1999), belonging to the Småland-Värmland granitoids of the "Växjö" type, which form part of the Trans-Scandinavian igneous belt (1.85–1.65 Ga; Högdahl et al. 2004). The granite of Växjö is a red biotite granite, generally of coarse and isotropic grain, although a cataclastic fabric oriented N–S is visible along the lake Vättern (Adamsson 1944). Contacts between the Norra Kärr nepheline syenite and the surrounding Växjö granite dip approximately 45° toward the west. Until recently, the age of the Norra Kärr intrusion was wrongly established at approximately 1,545 ± 61 Ma by Rb–Sr (Blaxland 1977; Welin 1980). Indeed, more recent dating using the U/Pb method on zircon revealed the metasomatic alteration of the surrounding rocks (fenitization) and hence the magmatic activity linked to the intrusion dating back to 1,489 ± 8 Ma (Christensson 2013; Sjöqvist et al. 2013). The emplacement of the Norra Kärr complex is still under discussion, although it already seems that this intrusion is considered to be a metamorphosed agpaitic complex (Schilling et al. 2011; Sjöqvist et al. 2013 and references therein).

The Norra Kärr complex seems to be concentrically zoned and is formed of four main lithologies: "grenaite", "kaxtorpite", "pulaskite" and "lakarpite" (Figure 1.19). Grenaite is a gray–green rock, partially recrystallized and fine-grained with alkaline feldspars, nepheline, aegirine, eudialyte and catapleiite. The central part of the massif shows pegmatite zones and schlieren surrounding a recrystallized to migmatized unit of grenaite. There are pulaskite (alkaline rock of medium to thick grain) outcrops to the west of the massif, and they are formed of albite, microcline, aegirine, sodium amphibole, biotite and nepheline as minor phases. The additional minerals are rosenbuschite, apatite, titanite and fluorine (Sjöqvist et al. 2013; Gates et al. for Tasman Metals Ltd. 2012, 2013). Outcropping mainly in the central part of the intrusion, Kaxtorpite generally consists of black alkaline rocks that are coarse-grained and have low Zr contents. Its constitutive minerals are microcline, which is present within a matrix formed of dark alkaline amphibole, aegirine, eckermannite, pectolite and nepheline. Finally, lakarpite is a medium-grained rock rich in albite-arfvedsonite-nepheline with small amounts of microcline and rosenbuschite and titanite, apatite and fluorine as minor phases.

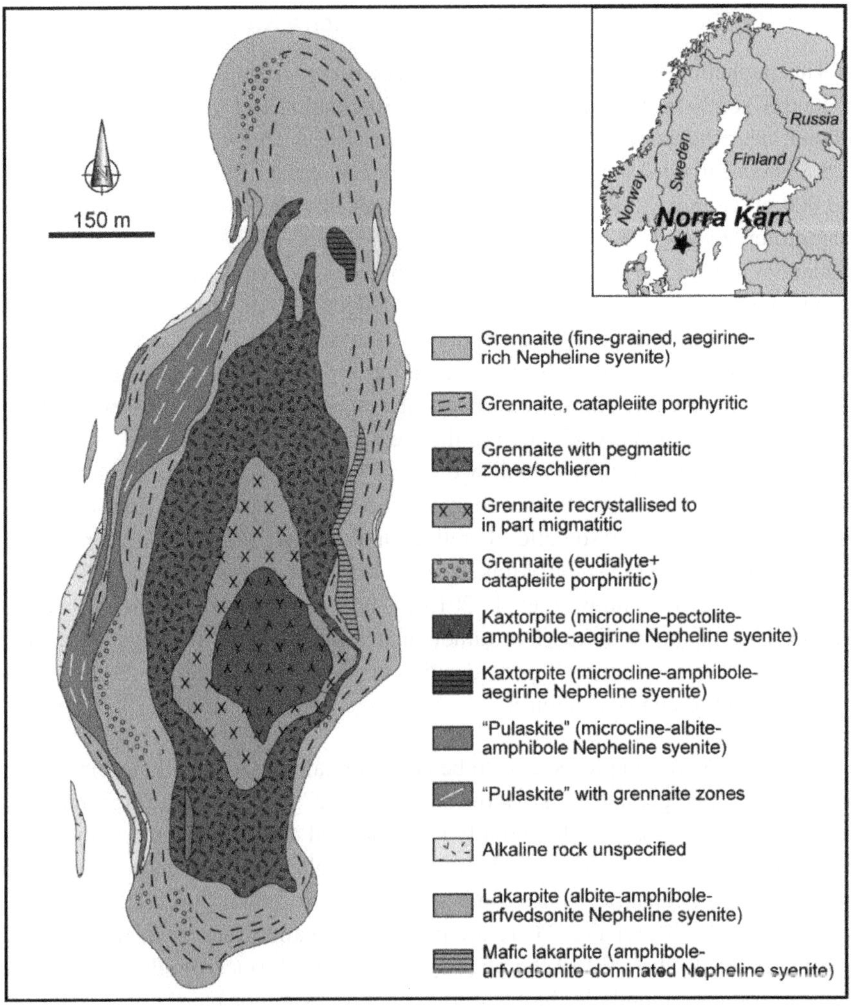

Figure 1.19. *REE-Zr nepheline syenite at Norra Kärr, Sweden (Gates et al. 2013)*

Chemical analyses have been carried out on each of these lithologies and are presented in Gates et al. (2013). Thus, grenaite contains between 0.14 and 0.67% REO, with maximum levels in the pegmatitic facies. Pulaskite contains between 0.09 and 0.24% REO. Kastorpite shows values of 0.20% REO. Finally, lakarpite includes between 0.42 and 0.54% REO. Whole-rock analyses are also presented in the study by Sjöqvist et al. (2013) for each of the lithologies of the Norra Kärr complex. Thus, grenaite with catapleiite has levels of 0.28% REO+Y_2O_3, while the grenaite associated with the pegmatites and the deformed grenaite (recrystallized texture) have levels of 1.19 and 0.47% REO+Y_2O_3, respectively. The levels of kaxtorpite and lakarpite are 0.08% and 0.62% TREO+Y_2O_3, respectively.

Previous mineralogical studies have allowed the reporting of REE-bearing minerals such as eudialyte, rinkite-mosandrite, britholite-(Y), lessingite, apatite and tritomite-(Ce), as well as a rosenbuschite REE-bearing assembly. In a recent mineralogical study on the eudialyte group from Norra Kärr, Sjöqvist et al. (2013) distinguished three distinct groups:

– classic pink eudialyte enriched in Fe and low in REEs in the lakarpite;

– eudialyte rich in HREEs and Fe-Mn in grenaite associated with the pegmatites;

– eudialyte rich in Mn and LREE in the "migmatitic" grenaite.

Chemical analyses carried out by Tasman Metals on eudialyte concentrates show values ranging from 3.91 to 4.63% REO, with high contents of eudialyte collected in pegmatitic grenaite (Gates et al. 2013). In 2011, an initial estimate of resources for the REE-Zr deposit at Norra Kärr was published according to standard NI43-101. With a cutoff grade of 0.4% REO, the Norra Kärr deposit contains 60.5 Mt @ 0.54% REO and 1.72% ZrO_2. In 2012 and 2013, the resources indicated were 41.6 Mt @ 0.57% REO, corresponding to 237,120 t REO, and the resources inferred from 16.5 Mt @ 0.64% TREO, corresponding to 105,600 t REO (Gates et al. 2013). In July 2013, a preliminary economic study estimated production capacity at 6,800 ton/year for a lifespan for the mine of 40 years, although the deposit was still exploited underground.

1.5.1.2.11. Phosphorite at Påläng (Sweden)

The phosphorite at Påläng is an occurrence situated 50 km to the northwest of Luleå. The surrounding rocks are formed of metasediments and volcanic rocks dating back to the Paleoproterozoic (2.1–2.3 Ga), which belong to the Karelia supergroup (Notholt et al. 1989; Frietsch and Perdahl 1995): micaschists with quartzite, phyllite and limestone-dolomite interbeds. The limestone and dolomite contain levels of apatite, or even fragments of apatite associated with a breccia formed during regional metamorphism. The phosphate rocks are situated in two steeply dipping areas at levels of 1 m thickness showing contents of 10% P_2O_5. The Påläng deposit has an average content of 2% P_2O_5 and 0.03% U for 6 Mt of ore and a thickness estimated at 100 m (Gustafsson 1979). Frietsch and Perdahl (1995) analyzed the REE content in apatite and showed a low splitting $(La/Yb)_{cn} = 4$) and REE levels 50–400 times higher than chondritic values (i.e. ~730 ppm REE). It should be noted that 100 km to the north of the Påläng zone, a similar deposit is present in Mustamaa (Finland), where phosphorite and dolomite outcrop in basic volcanic rocks, the phyllites and schists of the Karelia supergroup.

1.5.1.2.12. REE-Pb carbonatite at Korsnäs-Svartören (Finland)

The Korsnäs deposit is a historic REE-Pb mine situated in central-western Finland, 350 km north of Helsinki and only 25 km southwest of the port of Vaasa on the Baltic Sea. The Korsnäs mine was exploited by the Outokumpu Oy company, above and below ground from 1959 to 1972. The Korsnäs deposit was identified in 1950 through the discovery of erratic blocks rich in lead sulfides. Afterward, the ore was intersected by a substantial drilling campaign carried out by GTK in 1955 (see the references in Sarapää et al. (2013)). The deposit was formed of a network of carbonatite dykes, the most important of which is the one at Svartönen, which is home to the main mineralized body. This dyke, the establishment of which has been dated back to 1,830 Ma using the U/Pb on titanite method (Papunen 1986), measures 5–30 m in width for 1.5 km in length and continues at nearly 350 m underground. The network of dykes intrudes migmatitic gneisses that date back to 1.9 Ga and belong to the south-ostrobothnian schist belt (Lehtonen et al. 2005). The deposit is situated in an intensely deformed zone and oriented N–S with a 40–60° dip toward the east. Immediately next to the mineralized body, a chlorite alteration zone is weakly expressed in the surrounding rocks. The deposit appears as a heterogeneous body with mineralizations in the pegmatites and carbonatites or in the form of a scapolite-diopside-barite skarn in the surrounding rocks.

The main REE-bearing minerals are apatite, monazite and allanite, although the site is also known for its many other exotic REE minerals. More precisely, monazite is found in the altered zones and presents in the form of small crystals or in diffuse inclusions in apatite (Isokangas 1975). The apatite has REE contents of more than 6%, with a slight excess of HREEs compared to monazite and allanite (Papunen and Lindsjö 1972). On the basis of a sample – certainly fragmented – of REE minerals, the total level of these elements is mainly represented by Ce, La and Nd for 80–90% (Papunen 1986). Moreover, more than 50 different minerals have been identified in the deposit, including rare phases such as carbonates and phosphates, and famous harmotome crystals (hydrated barium silicate) and apophyllite crystals within the small vacuoles (Isokangas 1975). Historically, the Korsnäs mine provided a total of 0.87 Mt of ore with 3.6% Pb and 0.83% REE between 1959 and 1972 (Himmi 1975; Tasman Metals Ltd. 2010c). Past production is evaluated at a total of 45,000 tons of Pb and 36,000 tons of La between 1961 and 1972 (Himmi 1975). Other figures have been published and give an average ore grade of 3.57% Pb and 0.91% RE_2O_3 (Papunen and Lindsjö 1972). Since then, the GTK has resumed exploratory work on REEs and discovered erratic blocks formed of Svartören-type rocks at approximately 1 km to the southwest of the Korsnäs mine (Kärkkäinen and Huhta

1993). These blocks seem to come from a source that has still not been discovered and that is similar to the Svartören dyke. Moreover, some samples taken from drill cores in the 1950s have been analyzed again and have revealed REE contents ranging from 0.7 to 2.2% with a fraction rich in LREE. More specifically, the Eu content is high (between 66 and 242 ppm), and the Th content varies between 107 and 604 ppm. Finally, weak negative gravimetric anomalies have been measured to the south of the Korsnäs zone and could be the marker of a potential zone for identifying REE-Pb weathered carbonatite dykes. The Tasman Metals company has acquired the old Korsnäs REE-Pb mine (Tasman Metals Ltd. 2010c).

1.5.1.2.13. The metamorphozed paleoplacer and REE hydrothermal alteration at Olserum (Sweden)

The Olserum occurrence is situated near Gamleby, 220 km southwest of Stockholm. This occurrence was discovered in the 1950s during prospection for uranium and was studied for its REE content in the 1990–2000s. In 2013, an exploration campaign was carried out by the Tasman Metals company (IGE Nordic 2007; Reed 2013; Tasman Metals Ltd. 2013).

The Olserum zone forms part of the Västervik formation, which is a metasedimentary succession formed between 1.88 and 1.85 Ga in the southernmost part of the Svecofennian domain (Kleinhanns et al. 2012). This psammitic to pelitic formation has been intruded by granitoids dating back to 1.85–1.80 Ga and linked to the Småland granite, which belongs to the Trans-Scandinavian igneous belt. HT-BP metamorphism affected the sediments transformed into mica quartzites, gneiss and metapelites. In this metasedimentary sequence, intercalations of black sandy horizons rich in iron oxides (i.e. magnetite) are present in abundance and contain the main REE mineralization. However, metasomatism contemporary with metamorphism is well developed (Andersson et al. 2018b). It is responsible for the hydrothermal alteration of the black sandy horizons and leads to a remobilization of the REE-bearing minerals. The authors consider the Olserum deposit to be a paleoplacer that has been metamorphosed and weathered by the concomitant hydrothermal episode (Reed 2013; Tasman Metals Ltd. 2013).

The most abundant REE-bearing minerals are monazite and xenotime. Moreover, apatite is abundant, albeit with a low REE content. These three REE-bearing phosphates seem to be of metamorphic origin and associated with the hydrothermal episode, although a primary sedimentary (detritic) origin may also be possible (Reed 2013). Monazite and xenotime are present mainly in the form of inclusions in apatite and biotite, suggesting that apatite is the old, main REE-bearing mineral, which was then leached during metamorphism. The highest REE contents are associated with the layers rich in magnetite and the quartzite to biotite-amphibole layers, while the lowest REE contents are associated with the levels of gneiss and quartzites.

In 2007, the IGE Nordic AB company estimated the resources of the Olserum deposit to be 2.8 Mt @ 0.83% REO (0.53% LREOs + 0.30% HREOs) or 23,226 t REO (IGE Nordic 2007). In 2013, an estimate following standard NI43-101 was published by the Tasman Metals company, indicating resources of 4.5 Mt @ 0.6% REO (33.9% HREOs from REO) or 27,260 t REO and inferred resources of 3.3 Mt @ 0.63% REO (33.7% HREOs from REO) or 20,770 t REO (Reed 2013; Tasman Metals Ltd. 2013). The cutoff grade was fixed at 0.4% REO.

1.5.1.2.14. Nb-Zr-REE alkaline complex at Alnö (Sweden)

The alkaline complex at Alnö is situated in the middle of eastern Sweden and outcrops to the north of the isle of Alnö. The main intrusion is formed of alkaline silicate rocks and carbonatites that can be divided into four series (Morogan 1988; Hode Vuorinen and Hålenius 2005): pyroxenite, ijolite (melteigite-ijolite-urtite), nepheline syenite and calcium carbonatite, laid down in this order (Figure 1.20). The magmatic rocks of this complex are surrounded by a fenitized zone 500-600 m broad within the surrounding rocks (Morogan and Woolley 1988). The complex has been dated back to 553 ± 6 Ma using the Rb–Sr method on whole rock (Brueckner and Rex 1980). A more recent and reliable dating using the Ar-Ar method has given an age of 584 ± 7 Ma (Ediacarian; (Meert et al. 2007)). This age is interpreted as that of the crystallization of the alkaline complex (Andersen 1996). The Alnö complex intrudes the Proterozoic basement, mainly the migmatites belonging to the Svecofennian domain.

Additional minerals such as titanite and apatite are present in the ijolite series and incorporate non-negligible contents of Nb, Zr and LREE (Vuorinen and Hålenius 2005). Indeed, the titanite shows significant contents of Nb (0.4-7.3 wt.% Nb_2O_5) within the melteigite and to a lesser extent in the ijolite with higher contents (6.5 wt.% Nb_2O_5). Contents of a similar order of magnitude are sometimes measured in the titanite present in the nepheline syenite (up to 7 wt.% Nb_2O_5). For zirconium, the highest grades are found in the borders of the titanite present in the melteigite and reach more than 2.8 wt.% ZrO_2. Finally, LREE contents (i.e. $\sum La_2O_3 + Ce_2O_3 + Pr_2O_3 + Nd_2O_3$) are significant in the titanite present in the melteigite and can reach 1 wt.% with maximum values of up to 3.9 wt.%.

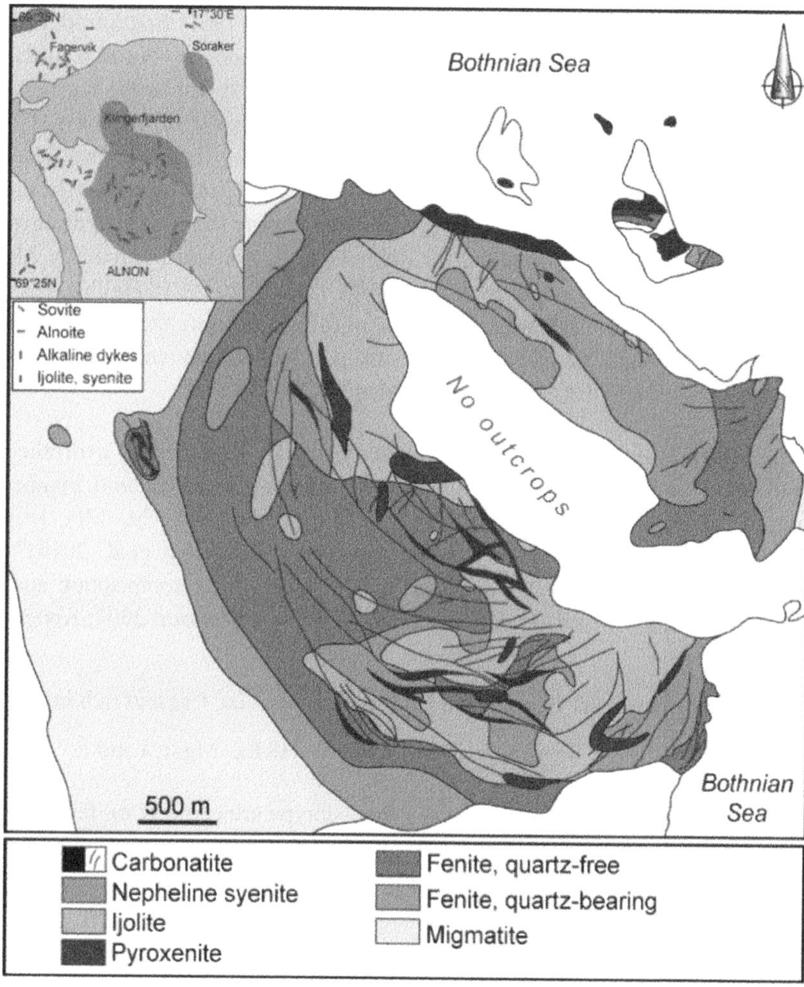

Figure 1.20. *Nb-Zr-REE alkaline complex in Alnö (Sweden), modified from Vuorinen and Hålenius (2005)*

1.5.1.2.15. Bastnäs type Fe-REE-(Cu) skarns (Sweden)

The Fe-REE-(Cu) deposit at Bastnäs is situated in the Svecofennian domain, close to Skinnskatteberg, in the Bergslagen region in the center of southern Sweden (Figure 1.21). This deposit forms part of the mining district of Riddarhyttan and is a historical site for the discovery of REEs, with unpublished descriptions of many REE-bearing minerals (e.g. bastnäsite, cerite, törnebohmite). The Bastnäs deposit was first exploited for iron and copper, and approximately 160 tons of REEs were extracted between 1860 and 1919 (Carlborg 1923). Geijer (1961) first introduced the notion of a "Bastnäs type" deposit in reference to iron deposits very rich in REEs. These deposits are situated near the districts of Norberg (30 km to the northeast of Bastnäs) and Rödbergsgruvan-Nora (50 km to the southwest of Bastnäs; (Geijer 1961; Holtstam and Andersson 2007, 2013 and references therein). The Bastnäs-type deposits are surrounded by felsic metavolcanic rocks and marbles from the lower Svecofennian (1.91–1.88 Ga), which have undergone significant hydrothermal alteration with metasomatism leading to richness in K, Na and particularly Mg in areas close to the mineralizations (Geijer 1923; Trägårdh 1991; Hallberg 2003). Afterwards, these rocks were metamorphosed in the amphibolite facies and intruded by at least two generations of plutonic rocks (i.e. 1.90–1.86 Ga and 1.81–1.75 Ga) (Lindh and Persson 1990; Andersson and Öhlander 2004).

Thus, Bastnäs-type deposits are characterized by skarn-type hydrothermal mineralization and a complex assemblage of iron oxides, calcic and magnesian silicates, REE silicates, REE fluorocarbonates, sulfides and Cu, Mo, Bi, Co sulfo-salts and Au-Ag minor alloys (Holtstam 2004; Sahlström et al. 2019). The deposits can be divided into two subtypes depending on their composition and the paragenesis of the REE-bearing minerals (Holtstam and Andersson 2007; Andersson et al. 2013):

- the LREE deposits at Bastnäs and Rödbergsgruvan (La, Ce) and rich in Fe;
- the Norberg district deposits, rich in LREE, Y+HREE, Mg, Ca and F.

The main REE-bearing minerals of the first subtype are cerite-(Ce), ferriallanite-(Ce), törnebohmite-(Ce) and bastnäsite-(Ce), while those of the second subtype are fluorobritholite-(Ce), västmanlandite-(Ce), dollaseite-(Ce), gadolinite-(Ce) and gadolinite-(Y). The precipitation of REE silicates in these deposits (i.e. cerite and fluorobritholite) results from the reaction between the relatively acidic fluids carrying the majority of the REEs complexified by ligands (F and Si) with the surrounding dolomitic rocks (Holtstam and Andersson 2007). Bastnäs-type deposits are almost unique, for example, an exoskarn with carbonated protolith of sedimentary origin. Recent datings using the Re–Os method on molybdenite have given crystallization ages for the sulfides associated with REE-bearing minerals between 1.90 and 1.84 Ga (Andersson et al. 2013). In detail, a space-time evolution is observed between the Rödbergsgruvan deposit, which is considered the most ancient in the southwest, and the most recent deposits from the Norberg district in the northeast.

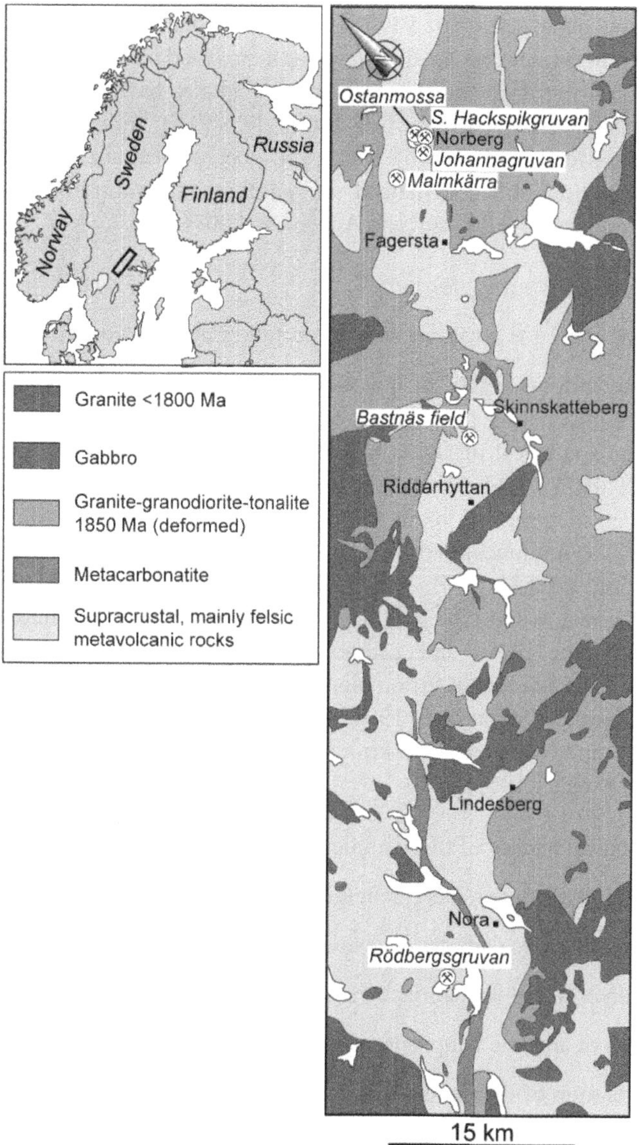

Figure 1.21. *Bastnäs type Fe-REE-(Cu) skarns, Sweden (Holtstam and Andersson 2007)*

In 2010, a joint venture was signed between a Swedish consultancy group and the Tasman Metals company to carry out an exploration campaign for REEs in the

Bastnäs region on a permit of 915 ha (Tasman Metals Ltd. 2010b). A series of samples taken in former mining sites have been analyzed by the Swedish Geological Survey (SGU) using XRF and ICP techniques. Samples have provided variable but high contents comprising between 1.82 and 17.47% REO, with an average of 6.97% REO (Tasman Metals Ltd. 2010b). Moreover, geochemical analyses have revealed U and Th depletion, still unexplained, in gadolinite, fluorobritholite and minerals from the allanite group (Holtstam and Andersson 2007).

1.5.1.3. *Sveconorwegian domain*

1.5.1.3.1. Nb-REE-Fe and carbonatite complex at Fen-Søve (Norway)

The carbonatite complex at Fen-Søve (Figure 1.22) is established within Telemark gneisses dating from the Lower Proterozoic. It is situated near Ulefoss, 120 km to the southwest of Oslo, in Norway (Barth and Ramberg 1966; Mitchell and Brunfelt 1975; Andersen 1986, 1988; Kresten and Morogan 1986). This complex covers a surface of 9 km^2 and has already been exploited for iron between 1657 and 1927 (Fen mines) and for niobium from 1953 to 1965 (Søve mine), with 1 Mt of raw ore @ 0.31–0.57% Nb_2O_5 (Dahlgren 2005). Moreover, the Fen-Søve complex is known as the type locality for søvite and fenite. The massif is formed mainly of alkaline rocks and carbonatites dating back to the Ediacaran, or approximately 565 Ma (age K/Ar on biotite) (Faul et al. 1959) and 583 ± 15 Ma (age $^{40}Ar/^{39}Ar$ on biotite) (Meert et al. 1998). The erosion level makes it possible to have a section at approximately 1–3 km depth compared to the complex's initial emplacement (Sæther 1957).

The geological history of this complex can be broken down as follows:

– intrusion of alkaline rocks (melteigite, ijolite, urtite series);

– intrusion of calcic carbonatites (søvite and silico-søvite);

– intrusion of dolomite carbonatites (magmatic rauhaugite I);

– intrusion of alkaline lamprophyres (damkjernite);

– late alteration of the damkjernite, forming metasomatic rauhaugite II (Kresten and Morogan 1986; Andersen 1988; Dietzel et al. 2019 and references therein);

– a fenite aureole surrounds the complex at the level of the western and southern borders.

In 2011, the Danish exploration company 21st North obtained a permit for REEs covering the external carbonatite part of the Fen-Søve complex. Indeed, the eastern parts of the complex are richer in REEs, especially in ankerite-dolomite carbonatite (rauhaugite) and hematite carbonatite (rødberg) (Lie and Østergaard 2011).

However, large parts of the complex are still little known, given the well-developed surficial cover. Before 21st North's recent works, the Geological Survey of Norway (NGU) carried out a detailed study on REE potential between 1967 and 1970, focusing on the northeastern part of the complex. The most promising REE contents are situated in hematite carbonatite (rødberg) with approximately 2.8% REO (8 elements + Y), and in ankerite-dolomite carbonatite (rauhaugite) with 1.51% REO (Lie and Østergaard 2011). Historical mineralogical and geochemical studies in the Fen complex have shown that the abundance of REEs increased in the following order: ijolite-damkjernite < søvite-rauhaugite < rødberg, reflecting the order of differentiation of the magma with the highest REE contents within the last rocks formed (Mitchell and Brunfelt 1975). REEs are mainly carried by monazite, bastnäsite, synchysite and parisite.

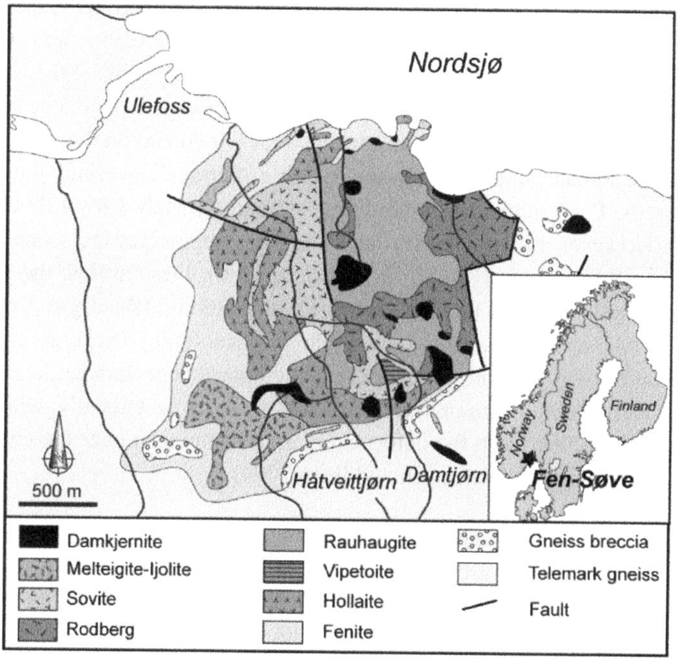

Figure 1.22. *Nb-REE-Fe and carbonatite complex at Fen-Søve (Norway), modified from Lie and Østergaard (2011)*

In 2011, the 21st North company collected 58 samples to measure the REE content and identify the minerals within each type of REE mineralized rock. Ankerite-dolomite carbonatite contains approximately 1.32% REO carried by bastnäsite, synchysite and parisite (in the form of aggregates with grains from 0.1 to

1.0 mm). Hematite carbonatite contains approximately 2.32% REO carried by the finer monazite and complex phases of Nb-Fe-Ti-Y-REE, rich in HREEs. Moreover, apatite concentrates extracted from calcic carbonatites contain 3,752–5,220 ppm REE and 164–288 ppm Y, and those extracted from ankerite-dolomite carbonatites contain 5,143 ppm REE and 199 ppm Y (Hornig-Kjarsgaard 1998; Ihlen et al. 2013). To the south and east of the massif, an envelope has been estimated with inferred resources of 84 Mt @ 1.08% REO (0.8% REO in cutoff grade), but the resources of the entire complex are probably greater (Lie and Østergaard 2014).

1.5.1.3.2. Pegmatitic Ta-Nb-REE fields at Evje-Iveland, Froland and Glamsland (Norway)

To the south of Norway, at the end of the Sveconorwegian orogeny, the NYF-type pegmatite fields outcrop (Figure 1.23): Evje-Iveland, Froland and Glamsland ((Andersen 1926; Bjørlykke 1937; Larsen 2002; Müller et al. 2012, 2018) and references therein). These pegmatites form the south Norwegian pegmatite belt formed over the course of the Sveconorwegian orogeny (1.14–0.90 Ga) on the western margin of the Fennoscandian Shield, when the Bamble "terrane" overlaid the Telemark terrane along the Porsgrunn-Kristiansand fault at approximately 1.09–1.08 Ga (Bingen et al. 2008). However, the datings available for these pegmatites are scarce and often based on obsolete methods (Rb–Sr on simple mineral). Despite this, the pegmatites date back to $1{,}060^{+8}/_{-6}$ Ma at Froland (age U/Pb on euxenite) (Baadsgaard et al. 1984) and 910 ± 14 Ma at Evje-Iveland (age U/Pb on gadolinite) (Scherer et al. 2001), leading the authors to consider a post-orogenic emplacement. Indeed, later structural studies have shown that the pegmatites were laid down syntectonically, which implies a space-time relation between the deformation and magmatism over the course of the Sveconorwegian orogeny (Hendersen and Ihlen 2004).

Most of the pegmatites are laid down in basic rocks (amphibolite, norite and basic gneisses). The veins rarely exceed 20 m in width in the form of subhorizontal lenses that very often intersect the foliation of the surrounding rocks. In the Froland field, the pegmatites belong, according to the classification by Černý and Ercit (2005), to the HREE abyssal pegmatites and do not seem to be linked to granitic plutons such as that of Herefoss (in the southwest of the zone). The pegmatites contain, in varying proportions, quartz, alkaline feldspar, plagioclase, biotite and a small amount of white mica, and REE minerals and other additional minerals are rare (Ihlen et al. 2001, 2002). Nevertheless, the only pegmatites to present a significant REE mineral content are those at Bjortjørn and Gloserheia in Froland (Müller et al. 2005); this field has approximately 100 pegmatite veins. In the Evje-Iveland field, the pegmatites are classed as REE pegmatites and muscovite and REE pegmatites (Černý and Ercit 2005; Müller et al. 2012). The genetic relationship between the pegmatites and the granitic pluton of Høvringsvatnet ($971^{+63}/_{-34}$ Ma) (Andersen et al. 2002) is still under debate. These pegmatites contain mainly K-feldspar, plagioclase, quartz, biotite and muscovite, magnetite and garnet as minors, and variable proportions of REE minerals and rare metals (beryl, allanite, monazite, euxenite, aeschinite, gadolinite or columbite). Approximately 350 pegmatites crop out in the Evje-Iveland field, including the Steli pegmatite and some other lithium pegmatites that include REEs (monazite and euxenite). The Slobrekka pegmatite is highly fractionated and rich in Y-REE minerals (gadolinite-(Y), allanite-(Ce), aeschinite-(Y) and euxenite-(Y)).

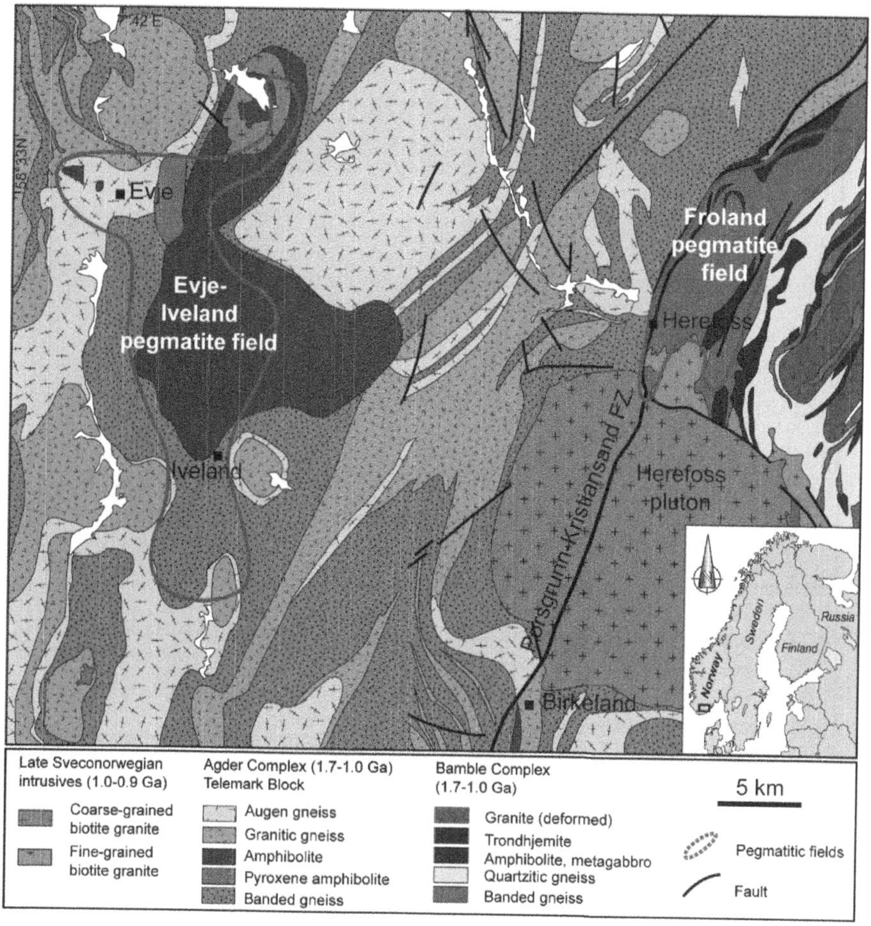

Figure 1.23. *Ta-Nb-REE pegmatite fields at Evje-Iveland, Froland and Glamsland, Norway (Muller et al. 2012)*

1.5.2. *The rare earths indices of the Caledonides*

1.5.2.1. *Scandinavian Caledonides*

1.5.2.1.1. U-REE-P phosphorite at Tåsjö (Sweden)

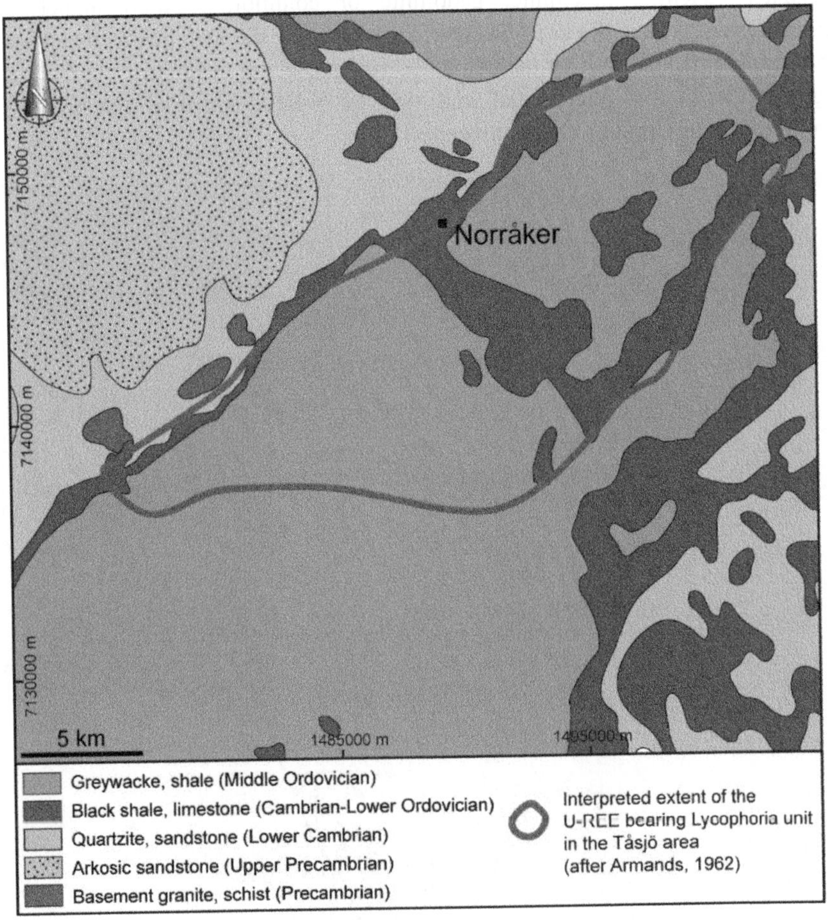

Figure 1.24. *U-REE-P phosphorite at Tåsjö (Sweden)*

The Tåsjö occurrence of U-REE-P is situated in Sweden (Figure 1.24), 200 km north of Ostersund, between the cities of Dorotea and Strömsund (in the counties of Västerbottens and Jämtlands län). This zone was prospected for uranium in 1957 by the Swedish atomic energy company. The Tåsjö occurrence is situated on the Caledonian front in the lower native unit, and mineralization is encased in Lycophoria shales of the Lower Ordovician and aluniferous shales from the

Cambrian-Ordovician (see Browne (2008) and references therein; Sadeghi et al. (2013)). The thickness of the mineralized series reaches 15–20 m in the aluniferous shales and 3–10 m in the Lycophoria shales. Moreover, this zone is greatly deformed by the caledonian tectonic, with overlaps, sheets and folds that led to a thickening of the series (more than 400 m in thickness for the aluniferous shales).

Aluniferous shales are black, foliated pyrite shales, with variable proportions of glauconite and apatite, while Lycophoria shales are formed by siltstone layers and pelitic layers with phosphate nodules. The peculiarity of Tåsjö is the thick phosphate unit (lycophoria shale) with high concentrations of uranium (associated with apatite, which forms 9 to 20% of the rock) and REEs. Estimates show that the uranium content in apatite is 0.16%, while the content of the Lycophoria shale (approximately 75–150 Mt) is 0.03–0.07% U_3O_8, 0.11–0.24% REE and 3.75–7.5% P_2O_5 (Armands 1964, 1970). Therefore, the total tonnage of this occurrence has been estimated at 47–53 Mt U_3O_8, 165,000–180,000 tons REE and 5.63 Mt P_2O_5. The Tåsjö zones continued to be explored for uranium and REEs by the Mawson Resources company (Browne 2008).

1.5.2.1.2. Pyroxenite and P-REE complex at Misværdalen (Norway)

Discovered by Farrow in 1974, the Misværdalen complex contains several intrusions of clinopyroxenite dating from the middle Ordovician to the upper Devonian to the north of Norway, 40 km east of Bodø (Figure 1.25). These pyroxenite bodies occupy approximately 8 km^2 and show richness in apatite. In 2006, the Nordland mineral company and the NGU considered the latter to present mineral potential (Ihlen et al. 2013). Pyroxenite is formed in variable proportions from biotite, alkaline feldspar and apatite at more than 70 vol.%. The P_2O_5 content of the clinopyroxenites is estimated at approximately 2.4 wt.% (see Ihlen et al. (2013) and references therein) with higher values in the coarse-grained pyroxenites (2.5 wt.% P_2O_5 or approximately 6 wt.% apatite). Chemical analyses of the apatite grains extracted from the coarse-grained pyroxenites present REE contents between 1,243 and 11,180 ppm, with averages varying between 4,150 and 6,363 ppm (Ihlen et al. 2013). This great variation in REE levels is a priori linked to the retrograde formation of allanite, which is observed along microfractures within the apatite grains. Moreover, chemical analyses on the apatite concentrates from the fine-grained and coarse-grained pyroxenites show values between 4,276 ppm REE and 3,883–4,075 ppm REE, respectively. Indeed, the relatively high Th content detected in some samples (as high as 100 ppm) does not allow the use of apatite in fertilizers.

1.5.2.2. *British-Irish Caledonides (United Kingdom)*

Several alkaline intrusions crop out to the northeast of the Caledonian belt to the north of the Highlands (Figure 1.26), dated using U/Pb on zircon to the Silurian and Ordovician (Halliday et al. 1987; Thirlwall 1988). The magmatic source seems to have been linked to the limited partial melting of the lithospheric mantle in line with the Caledonian orogeny (Halliday et al. 1987). The rocks are formed of differentiated series ranging from melanocratic syenites to more evolved leucosyenites, with the local presence of ultrabasic rocks such as pyroxenites. The British Geological Survey (BGS) has carried out a study on these alkaline intrusions, especially on their REE potential (Shaw and Gunn 1993; Walters and Lusty 2011; Walters et al. 2012).

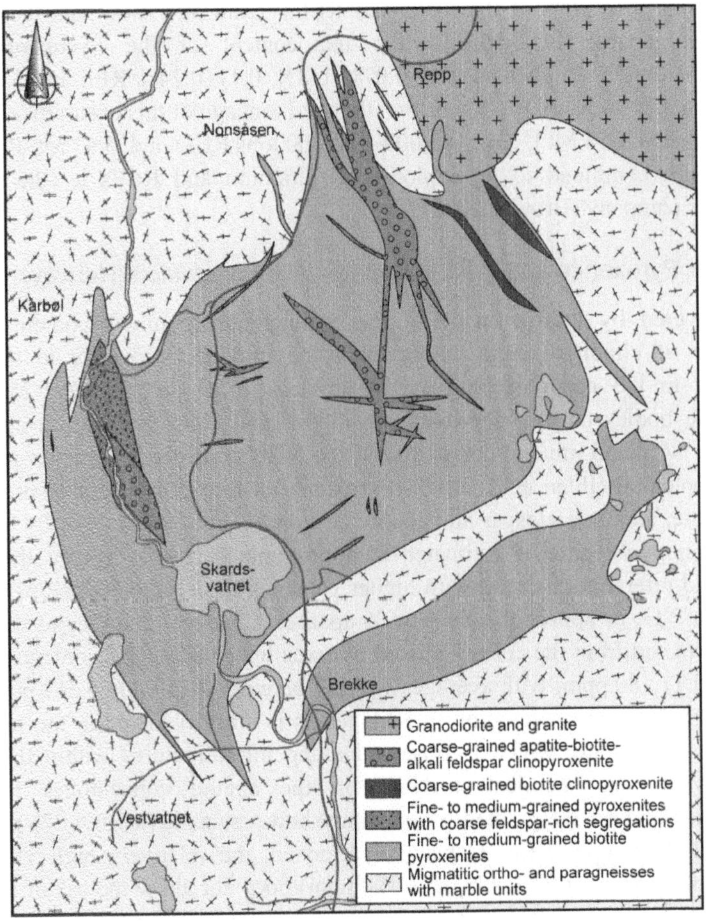

Figure 1.25. *Pyroxenite and P-REE complex at Misværdalen, Norway (Ilhen et al. 2013)*

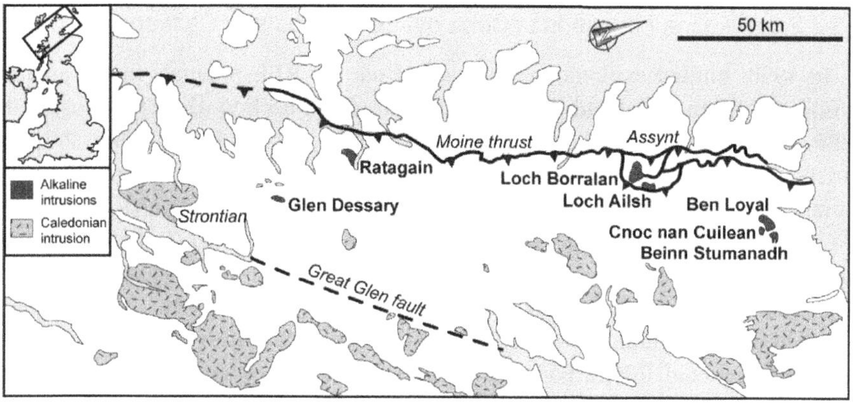

Figure 1.26. *Caledonian magmatism and associated alkaline intrusions (Halliday et al. 1987)*

1.5.2.2.1. Alkaline intrusion at Loch Borralan

The intrusion at Loch Borralan has been dated back to 426 ± 9 and 430 ± 4 Ma using the U/Pb method on zircon. From 1988–1989, geochemical analyses were carried out on samples from 37 drillings performed in apatite-biotite-melanite-phlogopite pyroxenites, with the aim of exploring elements of the platinum group. These 37 drillings had been made 10 years earlier for an exploration of phosphates. Apatite is present in the form of euhedral crystals (of some millimeters). REEs have been detected in apatite thanks to measurements taken using X-ray fluorescence spectrometry. In 1990, an additional 37 drillings in the pyroxenites and syenites revealed REE contents of up to 739 ppm La, 1,764 ppm Ce and 986 ppm Y. Apatite contents reach up to 2% and make this mineral the main REE carrier. Unfortunately, the alkaline intrusion at Loch Borralan is not an economic target for REE exploitation.

1.5.2.2.2. Alkaline intrusion at Loch Ailsh

Following the discovery of levels rich in platinoids in the Loch Borralan intrusion, additional studies were carried out in the Loch Ailsh alkaline intrusion in 1989 and 1990. This alkaline intrusion has been dated back to 430 ± 4 and 439 ± 4 Ma using the U/Pb method on zircon. The average contents of La and Ce are similar to those found in the Loch Borralan intrusion. Nevertheless, the Allt Cathair Bhan zone presents significant enrichment, with maximum concentrations of 3,239 ppm La and 3,956 ppm Ce in syenite and 840 ppm La and 1,456 ppm Ce in pyroxenite. An enrichment in U (maximum 65 ppm) and in Th (maximum 244 ppm) has also been measured. Finally, allanite is another REE-bearing mineral in addition to apatite.

1.5.2.2.3. Alkaline intrusions at Ben Loyal and Cnoc nan Cuilean

In 1990, limited exploration was carried out for REE minerals in the alkaline intrusions of Ben Loyal and Cnoc nan Cuilean (dated back to 426 ± 9 Ma using the U/Pb method on zircon). Two samples of Cnoc nan Cuilean syenite show an enrichment of REEs with a maximum of 5,667 ppm La and 19,785 ppm Ce and an average value of 3,800 ppm REO (Shaw and Gunn 1993; Walters and Lusty 2011; Walters et al. 2012). Many REE minerals have been recognized, and allanite is the main REE mineral. Three generations of allanite have been identified: a late generation in a "crown" around the apatite, a generation of crystals in microveins intersecting the syenite, and a generation of crystals within biotite and magnetite veins (Walters et al. 2012). For all allanite generations together, the average content is 22% REO. Finally, it should be noted that the alkaline intrusion of Ben Loyal presents no major enrichment of REEs among the samples analyzed to date.

1.5.3. *Rare earth occurrences in the Variscan belt*

1.5.3.1. *The domains of western and central Europe (Gondwana-Avalonia-Armorica)*

1.5.3.1.1. The Armorican massif – gray monazite shales from the Ordovician in Brittany (France)

Since the 1960s, the Armorican massif has been known for black shales from the Paleozoic, including nodular gray monazites, to heights of 50–200 g/t (Donnot et al. 1973). The monazite grains (up to 2 mm in diameter, flattened ellipses) are characterized by many mineral inclusions (phyllites, quartz, rutile, graphite, iron oxides), the absence of an inherited core, a notable enrichment of intermediate rare earths (Sm, Eu, Gd) and very low concentrations of Th and U. The core of the grains presents up to 10 wt.% Sm_2O_3, 1.3 wt.% Eu_2O_3 and 5 wt.% Gd_2O_3, whereas their rims are rich in LREEs. The crystallization of monazites is authigenic and occurs under conditions of deep diagenesis at the limits of very low-grade metamorphism (Tuduri et al. 2013). The erosion of these gray monazite shales gave rise to the formation of placers similar to those in Iberia, the Belgian Ardenne or the Welsh basin.

1.5.3.1.2. Wales and the southwest of England – nodular gray monazites of the Ordovician-Silurian of the Welsh basin (United Kingdom)

In the United Kingdom, nodular gray monazite occurrences have been listed in the Welsh basin due to pan concentrates collected over the course of the inventory by the BGS mineral exploration programme (Cooper et al. 1983, 1984; Smith et al.

1994) in Wales, Exmoor and southwestern England. The concentrates presented high levels of REE with, notably, more than 5,000 ppm of Ce (Figure 1.27).

These anomalies have been attributed to the presence of nodular gray monazite. The highest Ce contents have been measured in rocks from the Silurian and Ordovician, which are detrital sediments, including black shale layers intercalated in clays, siltstones and sandstones with turbidite facies. The monazite grains are on the millimeter scale (0.05–2 mm), ovoid and have marked zoning between LREE and HREE between the core and the rims. The monazites also present low-grade metamorphic fabrics, similar to the surrounding rocks, and a low Th content (Read et al. 1987). Milodowski and Zalasiewicz (1991) and Evans and Zalasiewicz (1996) proposed that the nodular monazites were formed by the migration of REEs from the clay layers to layers rich in organic matter where grains could crystallize over the course of late diagenesis.

These gray monazite occurrences are not isolated in the Welsh basin and are also found in Brittany (Donnot et al. 1973), the Ardenne massif (Nonnon 1984; Limbourg 1986; Burnotte et al. 1989) and northwestern Spain (Vaquero 1979; Windle and Nesbitt 1993). These nodular monazites, present in sediments from the Lower Paleozoic, have – on a European scale – similar chemical and mineralogical properties (Smith et al. 1994).

In the United Kingdom, an estimate of the economic potential of these occurrences has been carried out by Smith et al. (1994). Although the nodules present low contents of Th and interesting contents of Eu, the surrounding rocks do not contain sufficient monazites to be exploited economically. Moreover, the study recommends studying placers resulting from the erosion of shales from the Ordovician-Silurian more precisely. These are the source of the monazites and so of the REEs.

1.5.3.1.3. The Bohemian massif – NYF Pegmatites from the Třebíč pluton (Czech Republic)

The Třebíč pluton is situated in western Moravia (Czech Republic) and is formed of rocks running from melasyenite to melagranite (durbachite) corresponding to type I granitoids (Foley et al. 1987; Finger et al. 1997). The rocks are generally amphibole, biotite and quartz porphyries. They are locally foliated, especially at the edges. The additional minerals are fluorapatite in abundance, zircon and titanite. Phlogopite is associated with trace minerals such as allanite-(Ce), thorite, thorianite, monazite-(Ce), xenotime-(Y), cheralite and sulfides (Sulovský 2000). The granite has been dated back to 342 ± 2 Ma (Kusiak et al. 2010) using the U/Pb method on zircon.

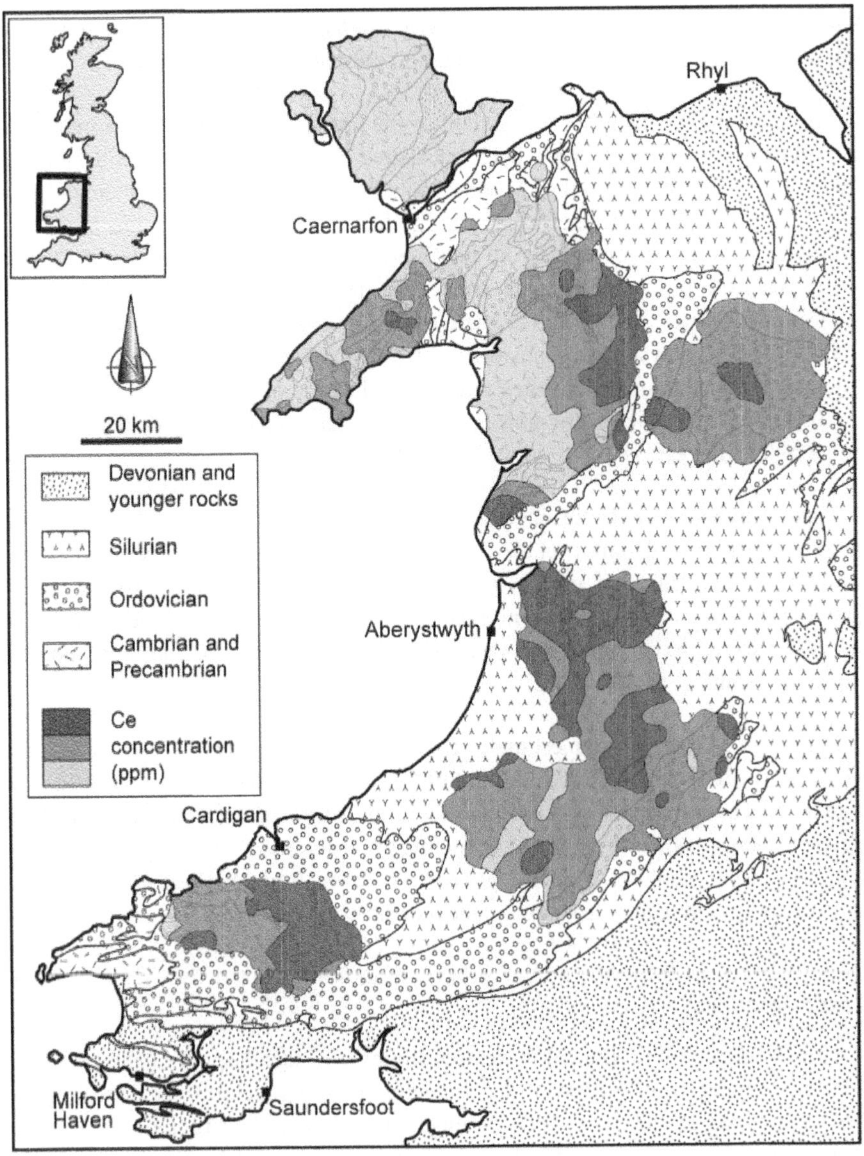

Figure 1.27. *Welsh basin and Ce anomalies associated with gray monazite occurrences (Smith et al. 1994)*

The Třebíč pluton also includes pegmatites distributed over two zones (Škoda et al. 2006; Škoda and Novák 2007): to the south (Třebíč-Vladislav) and the north

(Velké Meziříčí-Bochovice). On the basis of the internal structure, geochemical fractionation and mineralogical composition, two varieties of pegmatites (allanite subtype and euxenite subtype) are distinguished and are of the NYF type (Černý and Ercit 2005; Škoda et al. 2006). The pegmatites are characterized by K-feldspar, phlogopite and the absence of muscovite, titaniferous minerals (titanite, ilmenite) and REE minerals (allanite-(Ce), aeschinite and euxenite) (Škoda et al. 2006; Škoda and Novák 2007). Despite these mineralogical specifics, the Třebíč pluton is not an economic target.

1.5.3.1.4. The Bohemian massif – Ti-Zr-Th-REE paleoplacer from the Ordovician at Brtná (Czech Republic)

On the basis of radioactive sites listed by the Czech uranium industry (Krištiak and Záliš 1994), Goliáš (2001) published a summary of Th occurrences in the Czech Republic. Where REEs are concerned, the Brtná region (in the west of the country, near Dolní Žandov) presents radiometric anomalies associated with sericite quartzite layers 50–80 m thick within the Dyleò micaschists (Saxothuringian).

The Th concentration varies between 126 and 542 ppm. This rock contains zircon, rutile of metamorphic origin, hematite and xenotime. Th is situated within the monazite and the cheralite of metamorphic origin. The Th occurrences, although they underexplored, are interpreted as belonging to a Ti-Zr-Th-REE paleoplacer dating back to the Ordovician and metamorphosed (Goliáš 2001).

1.5.3.1.5. The Ardenne massif – gray monazite shales from the Ordovician in the Ardenne (Belgium)

The first nodular gray monazite occurrences in the Belgian Ardenne were described in shales from the Upper Cambrian ("Revinian") in the Rocroi massif by Nonnon (1984) from the mineralogical pan bottom study. Other occurrences have been discovered in metasediments from the Upper Devonian (Siegenian) in the syncline at Neuf-château (Limbourg 1986), shales from the Cambrian-Ordovician in the Stavelot massif (Burnotte et al. 1989) and the Brabant massif (Nonnon 1989) (Figure 1.28). In the Stavelot massif, the gray monazites are present in black shales intercalated into a microconglomerate with phosphate nodules (Burnotte et al. 1989). The monazite grains rarely exceed 700 µm, with an average size of 300 µm. The grains are formed of polycrystalline aggregates, with the exception of a few single crystals or, more rarely, pairs. The chemical and mineralogical characteristics are similar to the gray monazites of the Welsh basin (Smith et al. 1994) and Brittany (Donnot et al. 1973).

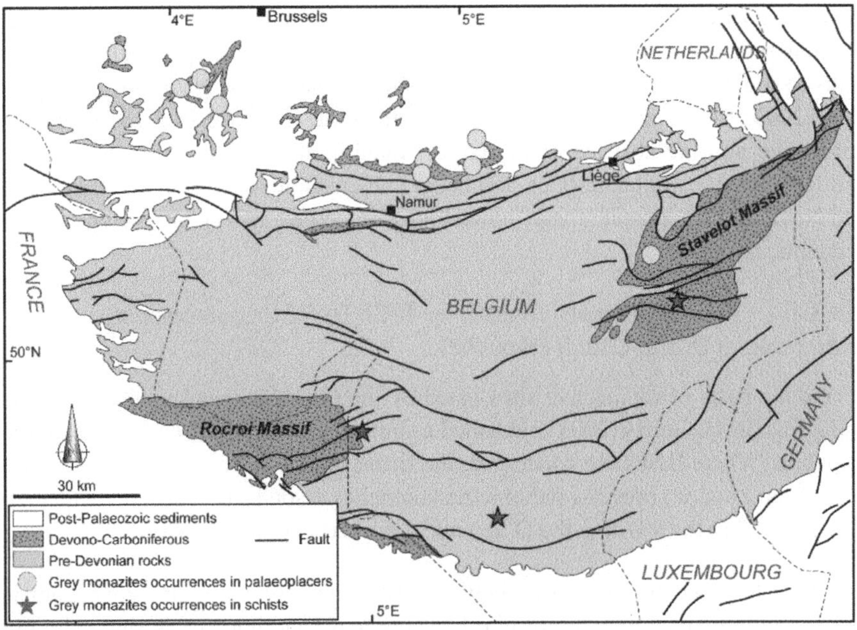

Figure 1.28. *Gray monazite shales from the Ordovician in the Ardenne, Belgium (Burnotte et al. 1989; Nonnon 1989)*

1.5.3.1.6. French Massif Central – the phosphorites of the Cévennes and the Montagne Noire (France)

To the south of the French Massif Central, the Cévennes and the Montagne Noire include dolomites and limestone with phosphate nodules that show high radioactive anomalies (2,000 ppm U and 1,500 ppm Th) (Laval et al. 1990). In the Cévennes, close to Le Vigan, phosphate levels of 2–3 m thickness and limited lateral extension have been sampled with dolomites from the Cambrian to estimate their REE content. The nodules are mainly formed of microcrystalline apatite associated with quartz and white mica geodes. The nodules sampled are rich in U (1,000–2,200 ppm), Th (1,000–1,800 ppm) and REO (1,700–2,500 ppm, with wealth in HREEs). Unfortunately, the low REO content (0.2%) is not sufficient to excite economic interest (Laval et al. 1990). In the Montagne Noire, the phosphate levels are localized in carbonates from the Cambrian on the northern flank of the massif. The phosphate nodules are on the centimeter scale, gray–black and form levels of 20–30 cm thickness broadly parallel to the bedding of the carbonates. The nodules are formed of quartz, microcrystalline apatite and micas (Laval et al. 1990; Clausen and Alvaro 2007). The REO content varies between 300 and 2,100 ppm and is associated with low U concentrations (10–40 ppm) and Th concentrations (5–20 ppm) (Laval et al. 1990). The negative Ce anomaly led Laval et al. (1990) to consider a marine origin for the phosphate nodules. The phosphate nodules of Montagne Noire no longer present any economic interest for REEs.

1.5.3.1.7. Iberia – rocks with nodular gray monazite (Spain and Portugal)

In Portugal, REEs were prospected for the first time by the IGM (Institute of Geology and Minerals) in the regions of Beira Baixa and to the north of Alentejo. In the region north of Alentejo, gray monazites resulting from the weathering of quartzites from the Silurian-Ordovician were identified on the southwest flank of the Portalegre syncline (IGM 1998; Rosa et al. 2010; Salgueiro et al. 2020). An enrichment of REEs was measured within the quartzites intercalated into the Alegrete schists (Portalegre) at the level of the regions of Vale de Cavalos and Penha Garcia.

In Spain, at Becerreá in Galicia, the Aqueira Formation is considered one of the potential sources of gray monazites that form the region's paleoplacers (Vaquero 1979; Rosenblum and Mosier 1983). This formation is formed of layers of clay, silt and sandstone affected by low-grade regional metamorphism and contact metamorphism (Marcos et al. 1980).

1.5.3.1.8. Iberia – hyperalkaline complex at Galiñeiro (Spain)

The hyperalkaline complex at Galiñeiro is a circular massif of approximately 10 km^2 close to Vigo in northwestern Spain. This complex forms one of the many orthogneisses from the Lower Paleozoic in the Variscan belt (the Armorican massif, Portugal, northwest and southwest Spain). These complexes mainly outcrop in the central part of the orogen and form calc-alkaline, alkaline and sometimes hyperalkaline rocks. The Galiñeiro complex has been dated back to 484 ± 5 Ma (Kusiak et al. 2010) using the U/Pb method on zircon (Montero et al. 2009). It is formed of two lithologies: aegirine-riebeckite gneiss and amphibole-biotite gneiss. Within the massif, veins and small hydrothermally altered clusters contain additional mineral phases rich in Th (thorite, thorite enriched in P), as well as REE-bearing minerals: bastnäsite, REE silicates (allanite, thalenite-yttrialite), REE niobiotantalates (aeschinite, fergusonite-formannite, samarskite, pyrochlore-betafite) and REE phosphates (monazite, xenotime) (Montero et al. 1998; Floor et al. 2007). The magmatic fractioning is initiated from an alkaline basaltic mantle magma, which has resulted in differentiated siliceous and hyperalkaline rocks with residual fluids extremely rich in fluorine and other ligands that have made it possible to complex the REEs. The decrease in fluorine activity with the drop in temperature and the crystallization of fluorine minerals remobilized the complexes created previously, leading to the formation of minerals rich in REEs. Moreover, the Galiñeiro complex after alteration and erosion would be the source of REE minerals that form the alluvial and coastal placers and the Ría Vigo valley (Marmolejo-Rodríguez et al. 2008).

1.5.3.2. *The Oslo Rift (Norway)*

1.5.3.2.1. Pegmatite field at Langesundsfjord-Larvik

The pegmatite field at Langesundsfjord-Larvik is situated to the south of the Oslo rift, at approximately 100 km from Oslo (Larsen et al. 2008). This field forms part of the plutonic complex at Larvik, known for nearly 200 years. A book published by Larsen (2010) retraces the geology, mineralogy and history of the district. The Larvik complex (50 km long and 20 km large) is formed of monzonite and larvikite, and integrates lardalite and foyaite in the Lågendalen-Farris region, as well as the nepheline syenites around Langesundsfjord (Larsen 2010). In the Langesundsfjord region, larvikite intrudes sediments from the Silurian and Carboniferous as well as basalt flows. Larvikite and lardalite have been dated back to 298.6 ± 1.4 Ma and 292.1 ± 0.8 Ma, respectively, using the U/Pb method on zircon and baddeleyite (Dalghren et al. 1996). Moreover, the larvikite and the pegmatites near Larvik and Tveladen were laid down at approximately 290 Ma.

Within the Larvik complex, a great variety of pegmatites have been described according to the mineral textures and parageneses: Stålaker type (coarse-grained alkaline feldspar, dendritic crystals of amphibole and biotite, thin prisms of apatite and zircolinite), Stavern type (prismatic zircon of up to 5–6 cm, alkaline feldspar of 0.3–0.5 m in length), Tveladen type (microcline, nepheline, aegirine, ferro-edenite, magnetite and biotite with many REE minerals), Bratthagen type (feldspar alkaline, nepheline, aegirine prisms and additional minerals such as loparite) and the nepheline syenite type from Langesunfjord (agpaitic pegmatites). The known pegmatite sites in the Larvik complex (quarries, mainly) are Låven, the southern part of Stokkøya, the Arøy isles, Eikaholmen, Skutesundskjærene, Sagåsen, Tvedalen, Jahren, Bratthagen, Ula and Vøra (Larsen 2010; Piilonen et al. 2018 and references therein). The main REE minerals described in the pegmatites are allanite-(Ce), ancylite-(Ce), bastnäsite-(Ce), britholite-(Ce), cerite, chevkinite-(Ce), eudialyte, fergusonite-(Y), gadolinite-(Ce), hingganite, kainosite-(Y), loparite-(Ce), monazite-(Ce), parisite-(Ce), perrierite-(La), thorite, tritonite-(Ce), xenotime-(Y), zirconolite and zirsilite-(Ce).

1.5.3.2.2. Fe-Ti-REE apatite pyroxenite ores at Kodal

The Kodal deposit is situated in the Oslo rift (Permian), 100 km south of Oslo, and belongs to the plutonic complex at Larvik (Bergstøl 1972; Lindberg 1985; Korneliussen et al. 2000; Ihlen et al. 2013) (Figure 1.29).

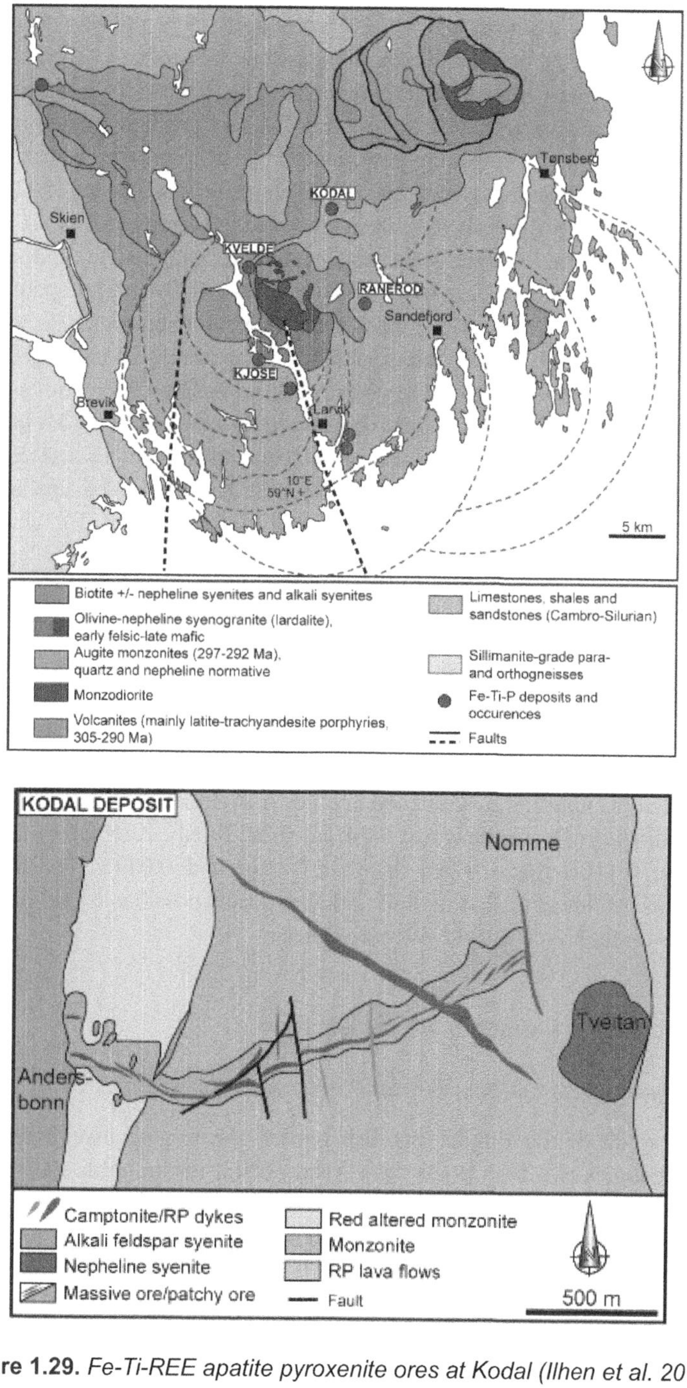

Figure 1.29. *Fe-Ti-REE apatite pyroxenite ores at Kodal (Ilhen et al. 2013)*

This deposit was explored by Norsk Hydro from 1959 to 1984 for its apatite potential, with ilmenite and magnetite as possible by-products. A recent survey campaign was carried out by Kodal minerals. The deposit is a set of massive pyroxenite lenses within nepheline monzonite. The ores are fine-grained and contain clinopyroxene (20–40% diopside) surrounded by an ilmenite matrix (5–15%), magnetite (25–60% ilmenomagnetite and titanomagnetite) and apatite (15–24% and an average of 17%). Other minerals have been identified: phlogopite (3–10%), magnesio-hastingsite hornblende, olivine, feldspar, nepheline, fluorapatite and carbonates (Bergstøl 1972; Lindberg 1985; Andersen and Seiersten 1994). The apatite contains fluid inclusions (H_2O-CO_2) and crystals of calcium amphibole, titanite, calcite and REE carbonates (Andersen and Seiersten 1994). According to Lindberg (1985), the reserves are 70 Mt @ 4.9 wt.% P_2O_5, considering a transition zone with surrounding rocks. Moreover, the ore massif would represent deep reserves of 35 Mt @ 6.8 wt.% P_2O_5. The Kodal apatite presents acceptable contents of Th for use as fertilizer (26–71 ppm and an average of 40 ppm) (Ihlen et al. 2013). The apatite contains approximately 1 wt.% REE (between 7,588 and 10,990 ppm REE), which would represent a by-product of value.

1.5.3.2.3. Zr-Nb-REE alkaline volcanic rocks at Sæteråsen

The Sæteråsen occurrence is southwest of Oslo in the eponymous rift. This Zr-Nb-REE index is formed of two aphyric trachyte flows within a broader volcanic sequence (the Vestfold Plateau). Petrographic and mineralogical studies have shown the presence of minerals bearing Nb, Ce, La and Y within lavas: euxenite, pyrochlore, chevkinite, fergusonite and apatite. A preliminary estimate on the basis of four drillings gives a tonnage of 8 Mt @ 0.245% Nb; 0.35% REE; 2.25% Zr; 0.180% Ce; 0.110% La; 0.075% Y; 0.069% Nd and 0.049% Th (Ihlen 1983; Ryghaug 1983; Ihlen et al. 2013; NGU 2013). It seems possible to find other similar occurrences on the vast Vestfold volcanic plateau.

1.5.4. *Rare earths indices from the Mesozoic*

1.5.4.1. *The fluorine district in Pennines (United Kingdom)*

To the south of the English-Scottish border, the mining district north of the Pennines occupies two high points, the Alston block to the north and the Askrigg block to the south (see Bouch et al. (2008) for details). This district is reputed for its Mississippi-Valley type (MVT) Pb–Zn-F-Ba-Ag deposits. Its past production is estimated at approximately 4.0 Mt Pb; 0.3 Mt Zn; 2.1 Mt of fluorine and 1.5 Mt of barite (Bouch et al. 2008). This region (Figure 1.30) is formed of volcanic and sedimentary rocks from the Lower Paleozoic intruded by Caledonian granitoids (Tynehead, Scordale, Corsnay and Weardale plutons) (Kimbell et al. 2010), since the whole is covered by Carboniferous sediments (limestone, sandstone, pelites with coal veins). The mineralizations (lenses and veins) were established at the end of the Permian in three stages:

– dolomitization and ankeritization of the surrounding carbonate rocks;

– the main sulfide mineralization;

– barite mineralization (Cann and Banks 2001; Bouch et al. 2008).

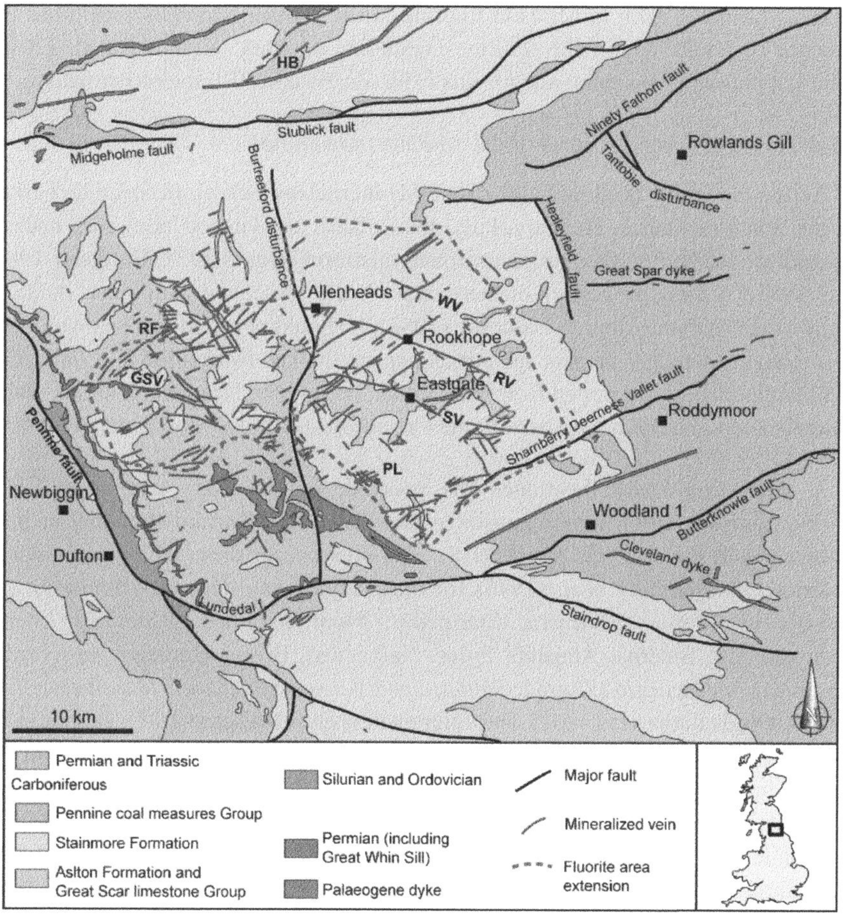

Figure 1.30. *Fluorine district in the Pennines (United Kingdom). GSV: Great Sulfur Vein; HB: Haydon Bridge mining area; PL: Pike Law; RF: Rotherhope Fell; RV: Rookhope Red Veins; WV: Hunstanworth White Vein (Kimbell et al. 2010)*

A quantity of synchysite and other REE-bearing minerals has been identified in these deposits (Shepherd et al. 1982; Ixer et al. 1996; Plant and Jones 1999; Walters 2011). Near the Stanhope and Rookhope deposits, synchysite has been described in quartz and bismuthinite veins in the form of intergrowth between crystals in the bismuthinite (Ixer et al. 1996). To a lesser extent, similar crystals of monazite, xenotime and adular have also been observed. Synchysite has also been observed in the Tynebottom mine at Garrigill. Some studies have shown that fluorine in Pennines is rich in REEs (e.g. > 1,888 ppm Ce + La + Y) (Smith 1974; Shepherd et al. 1982; Plant and Jones 1999; Walters 2011; Kraemer et al. 2019), the origin of which is debated. Indeed, an initial hypothesis recorded the presence of microinclusions of REE minerals in fluorine, while another hypothesis mentions the presence of REEs within the fluorine crystalline network. Kraemer et al. (2019) suggest a Permian (Asselian) age for this F-Pb–Zn-Ba and REE hydrothermalism.

1.5.4.2. *Bauxite and Ni karst of the Mediterranean belt*

Across the whole northern edge of the Mediterranean belt, from Spain to Turkey via the South of France, Hungary, Italy, the Balkans and Greece, numerous bauxite deposits are located on the Alpine carbonate platform (Agard 1975; Bárdossy 1982; Özlü 1983). A large proportion of these deposits are associated with karsts defining a type of deposit according to Bárdossy (1982). It should be noted that some of these karst bauxite deposits present an enrichment in nickel (Maksimović 1987). The weathering profiles developed from the Middle Trias to the Neogene, with the major mineralization episode dating back to the Cretaceous (Bárdossy and Dercourt 1990).

A pioneering study mentioned the presence of secondary authigenic REE-bearing minerals within these deposits, especially at San Giovani Rotondo, in Italy (Bárdossy and Pantó 1973; Sinisi 2018). The presence of these REE minerals was confirmed by Bárdossy et al. (1976) for the Nagyharsány deposits (Hungary), the deposits in Greece (Marmara, Marmeiko, Mont Parnasos, Panormos, Stenis Arahovis, Nisi, Smerna, Atalandi, Pylos, Nafpactos), Bosnia-Herzegovina (Nazda-Vlasenica), Montenegro (Zagrad, Štitovo), and Kosovo (Grebnik) (Maksimović and Pantó 1996; Kalatha et al. 2017 and references therein) (Figure 1.31). It should be noted that Mondillo et al. (2019) analyzed the REE potential of substantial bauxite deposits in southern France. Average REE concentrations in French bauxites are approximately 700 ppm (Mondillo et al. 2019). REEs, such as Ni, Co, Mn and Zn, are mobile during the bauxitization process, and the base of a weathering profile is generally rich in these elements (Maksimović 1976; Maksimović and Roaldset 1976). From a clayey material initially rich in REEs within karst cavities, bauxitization leads to the formation of authigenic REE minerals where there is contact between the bauxite and the carbonate hanging wall of the surrounding rock. Moreover, ultrabasic rocks may also be weathered in some places, leading to enrichment in nickel at the clayey levels within the weathering profile. In this region, it is thus possible to observe a transition between the bauxite karst deposits and those with nickel with the unusual association of REE and Ni minerals (takovite, brindleyite, nepouite). This unusual mineralogy is doubtlessly linked to the multitude of rocks leached over the course of bauxitization. The main deposits illustrating this transition are Marmara, Nisi and Marmeiko (Greece), and Grebnik in Kosovo.

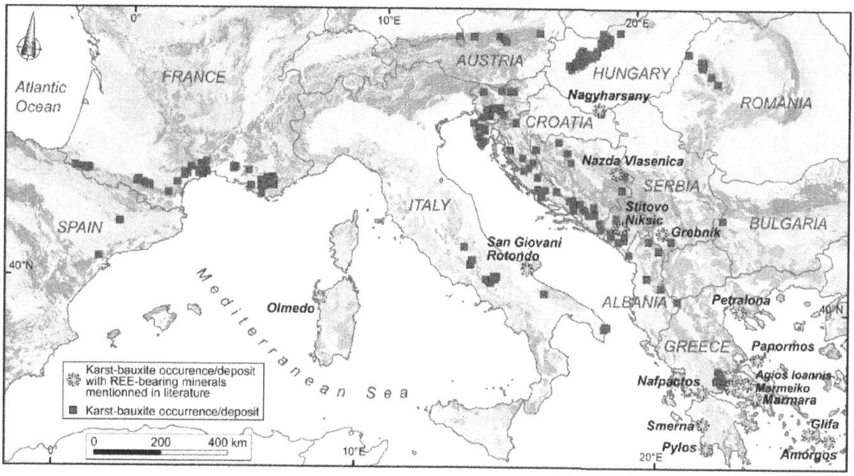

Figure 1.31. *Bauxite and Ni karsts of the Mediterranean belt (Promine and EURARE bases). References for bauxite deposits and EE minerals (Bárdossy and Pantó 1973; Maksimović and Pantó 1991, 1996; Jones et al. 1996)*

REE minerals are only present in deposits where intense leaching occurs in situ. In the Mediterranean belt, the majority of authigenic REE minerals were formed during the Cretaceous period. The most abundant minerals are hydroxylbastnäsite, followed by synchysite-(Nd), bastnäsite-(Ce) and bastnäsite-(Nd). To a lesser extent, the following are also found: monazite-(Nd), monazite-(La), goyazite-Nd, florencite and crandallite (Maksimović and Pantó 1996; Gamaletsos et al. 2017, 2019). The maximum REE levels are situated at the bottom of the profile, as in Montenegro (6.6 wt.% Nd + La + Pr). However, the basal part of the karstic nickel deposit at Aghios Ioannis (Greece) has only 180 ppm REEs. In fact, weathering profiles developed on basic rocks are naturally poor in REEs since the initial rocks were also poor in REEs. The Ni deposit at Marmeiko (Greece) is an exception. This shows richness in REEs with authigenic REE minerals from two initial rocks (an ultrabasic Cr-Ni rock and siliceous rock rich in REEs). In contact with the surrounding limestone, which acts as a geochemical barrier, the REE content reaches nearly

1,140 ppm at Nazda-Vlasenica (Bosnia-Herzegovina), 1,153 ppm at Marmeiko (Greece), 1,320 ppm at Štitovo (Montenegro), 1,450 ppm at Grebnik (Kosovo), 1,582 ppm at Marmara (Greece), and 4,576 ppm at the base of the Nisi deposit (Greece) (Maksimović and Pantó 1996). The REE content also depends on the intensity of the leaching from the rocks by the meteoric waters circulating in the karstic network.

1.5.4.3. *Authigenic black clays and gray marls from the Southeast basin (France)*

The Southeast basin (France) includes thick sedimentary series of gray marls, black clays, limestones and detritic interlayers (more than 9 km thickness) deposited during the Mesozoic (Debrand-Passard et al. 1984; Lemoine 1984). This basin was then affected at the end of the Cretaceous and Cenozoic by the Pyrenean and Alpine orogenies, leading to its uplift, erosion and folding (Roure et al. 1992). During the Jurassic, especially during the Callovian-Oxfordian, thick layers (several hundreds of meters) of black clays with carbonated interbeds filled the basin and are locally called "black earths". In 1990, a study was carried out on these rocks' REE level and level of organic matter (Laval and Dromart 1990). It appears that these black clays have a low REE content (158 ppm REO), which seems disconnected from the organic matter content. Conversely, the gray marls dating back to the Cretaceous (the Albian especially) contain phosphate nodules rich in REEs and present an average content of 2,924 ppm REO. Finally, the study revealed that the LREEs are contained in the marls, whereas the HREEs are concentrated in the phosphate nodules.

1.5.5. *Rare earth occurrences from the Cenozoic and Quaternary*

1.5.5.1. *Paleoplacers (from the Neogene to the Quaternary)*

1.5.5.1.1. Gray monazite paleoplacers of the Armorican massif (France)

Since the 1960s, the Armorican massif has been known for its monazite paleoplacers (Figure 1.32), including gray monazite placers resulting from the erosion of black shales from the Paleozoic (Donnot et al. 1973). Two old mining projects have made it possible to estimate the resources for two occurrences: the Ti-Zr-REE-Th paleoplacer at Bailleul (3.5 Mt @ 0.45% REO) (Lemarchand 1970) and the gray monazite placer at La Monnerie (0.08 Mt @ 0.08% REO) (Cornet and Sapinart 1967).

The gray monazite is present in the form of small ellipsoidal lenses, the lengths of which measure 0.1–0.8 mm along the main axis. Nodules are often formed by polycrystalline aggregates. The monazites present the following REE contents:

23.5% Ce_2O_3; 8.4% La_2O_3; 3.1% Pr_2O_3; 13.4% Nd_2O_3; 2.9% Sm_2O_3; 2.2% Gd_2O_3; 0.7% Y_2O_3; and 0.5% Eu_2O_3 (Donnot et al. 1973). The monazites have similar properties to those of the Welsh basin, the Belgian Ardenne or Iberia (Burnotte et al. 1989; Ferrero et al. 1989; Smith et al. 1994).

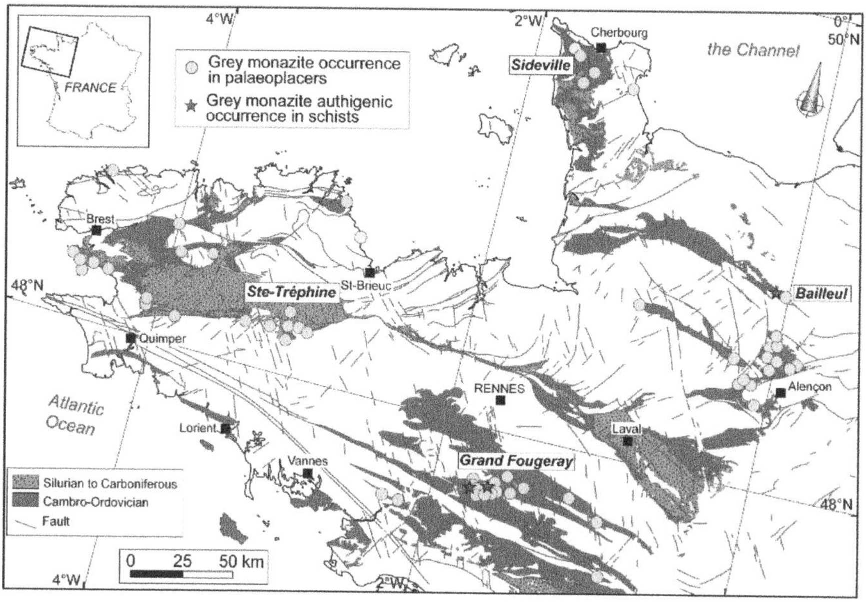

Figure 1.32. *Gray monazite paleoplacers and black shales in the Armorican massif (France) (Donnot et al. 1973)*

1.5.5.1.2. Zr-REE paleoplacers of the Ardenne massif (Belgium)

The Belgian Ardenne massif includes several gray monazite paleoplacers, especially in the Rocroi massif, the Neufchâteau syncline, and the Stavelot and Brabant massifs (Nonnon 1984; Limbourg 1986; Burnotte et al. 1989). The source of the monazites has been identified as being the Upper Cambrian schists ("Revinian") of the Rocroi massif, the metasediments of the Lower Devonian ("Siegenian") of the Neufchâteau syncline, and the rocks from the Cambrian-Ordovician of the Stavelot and Brabant massifs. The monazite grains (polycrystalline aggregates) measure as much as 700 μm for an average of 300 μm. The chemical and mineralogical characteristics of the monazites are similar to those of the Welsh basin (Smith et al. 1994), Brittany (Donnot et al. 1973) and Iberia (Ferrero et al. (1989) and references therein). The monazites contain on average 14.7% La_2O_3, 32.4% Ce_2O_3, 14.6% Nd_2O_3, 0.4% Eu_2O_3, 1.3% Gd_2O_3 and 77.0 ppm Yb_2O_3 (Burnotte et al. 1989).

1.5.5.1.3. The gray monazite placers of Spain

In Galicia, Calderón (1910) was the first to identify the presence of monazite in alluvial and coastal placers. Later, Pardillo and Soriano (1929) mentioned these placers with monazite concentrations of up to 1.25% around Vigo. Monazite placers are also described at the Insuela beach and in the Ría de Arosa valley by Parga Pondal (1935), with concentrations that can reach as high as 30%.

Both gray and yellow monazites exist. The gray monazites are similar to those described in Brittany (Donnot et al. 1973) and contain between 0.3 and 1.0% Eu. The yellow monazites are rich in LREEs and are mainly present in the coastal placers (along the coast north of Ría Arose and Ría Vigo) and the placers developed on granitic terrains (Ferrero et al. (1989) and references therein). The gray monazite placers developed thanks to the alluvial network with sediments of the Middle Ordovician from the Agüeira and Pizarras de Luarca formations (Ferrero et al. 1989).

Placers generally date back to the Cenozoic and Quaternary and contain xenotime, as well as Ti-Zr-Sn-(W)-Nb-Ta minerals (ilmenite, zircon, rutile, columbo-tantalite, wolframite and scheelite). Recent analyses have shown interesting concentrations of Y (0.3–23.1 ppm, an average of 14.5 ± 3.5 ppm) with the Ría Vigo sediments (Marmolejo-Rodríguez et al. 2008; Prego et al. 2009). The origin of the yttrium would be linked to the Galiñeiro complex. To the north of Galicia, Prego et al. (2012) measured the REE content within alluvium from the Ortigueira, Barqueiro and Viveiro valleys, as well as in the Mera, Sor and Landro estuaries. The highest REE contents are found in sediments from the heart of the valleys (233 ppm), while most of the alluvium has between 11 and 70 ppm REEs.

1.5.5.1.4. The gray monazite placers of Portugal

In the center-east of Portugal, near Bisa, Santo António das Arcias and Marvão, black sands from pan concentrates were collected by the Geological and Mining Institute (IGM) in 2005-2007 to identify the REE-bearing minerals in the placers (Inverno et al. 2007). Rosa et al. (2010) have shown that these sands contain xenotime in the form of round grains of approximately 250 μm (between 100 and 375 μm) and nodular monazite, as well as the following minerals: ilmenite, cassiterite, scheelite and wolframite. The xenotime shows concentrations of 39.7% Y_2O_3, 3.5% Gd_2O_3, 51% Dy_2O_3, 4.3% UO_2 and 0.31% ThO_2. The direct association of xenotime, monazite, tourmaline, rutile and zircon shows that the source should be linked to the granitic rocks of the Nisa and Penamacor massifs. However, it should also be noted that the

Ordovician quartzites near Salvador, Penha Garcia, Monfortinho and Portalegre are rich in nodular gray monazite and in REEs (Inverno et al. 1998; Salgueiro et al. 2020).

1.5.5.1.5. Ti-Zr-Th-REE-Nb placer at Lužnice (Czech Republic)

On the basis of the radioactive sites listed by the Czech uranium industry (Krištiak and Záliš 1994), Goliáš (2001) published a summary of the Th occurrences in the Czech Republic. Where REEs are concerned, the alluvium of the river Lužnice (in the south of the country, near Austria) presents gamma spectrometric anomalies linked to the contrast between sediments, resulting from the erosion of granitoids from the moldanubian and Cretaceous rocks from the bedrock of the river (Goliáš 2001). The main anomaly is a recent alluvial silty horizon described by Maòour et al. (1992). The thickness has been measured to be between 0.8 and 1.1 m, and the Th and U contents reach 47 ppm and 12 ppm, respectively. A zircon-ilmenite concentrate with nearly 9 wt.% monazite and 0.36 wt.% Th has been obtained. Th and U contents have been measured in zircon (0.022 wt.% Th and 0.046 wt.% U) and monazite (5.8 wt.% Th and 0.29 wt.% U). Brabantite forms the most radioactive grains. The recent placers of the river Lužnice show a Ti-Zr-REE-Th-Nb mineral association that could be exploited as a by-product of the exploitation of feldspathic gravels (Maòour et al. 1992).

1.5.5.1.6. The paleoplacers of the Welsh basin (United Kingdom)

In the United Kingdom, nodular gray monazite paleoplacers have been reported in the Welsh basin (draining the shales from the Ordovician-Silurian), thanks to pan concentrates collected over the course of the mineral inventory by the BGS (Cooper et al. 1983, 1984; Smith et al. 1994) in Wales, Exmoor and southwestern England. The study by Smith et al. (1994) on the potential of these paleoplacers concluded that there was no economic benefit, even though more in-depth studies would be desirable.

1.5.5.1.7. The paleoplacers and gray monazite placers of the Arize massif (France)

The Arize massif, in the French Pyrenees, 80 km south of Toulouse, revealed the presence of alluvial deposits containing gray monazites (Lacomme et al. 1993). In the Pyrenees, the gray monazite was discovered for the first time to the north of the belt within the alluvial deposits of the Pliocene-Quaternary associated with cassiterite, scheelite and gold (Lacomme and Fontan 1971). It seems that this alluvium resulted from the erosion of, among others, metasediments from the Ordovician and Silurian that contain gray monazites. Moreover, the gray monazite alluvial occurrences of the Arize are almost exclusively found near the Ordovician sandstones and metapelites (Lacomme et al. 1993). Unfortunately, the torrential character of the water courses limits the accumulation of these monazite alluviums, excluding any economic benefit. The monazites measure between 80 and 500 μm. Analyses carried out on the Ordovician gray rocks give concentrations between 65 and 67% REO, and an average level of 0.5–1.0% ThO_2.

1.5.5.2. *The Alpine and Pyrenean orogenies*

1.5.5.2.1. Carbonatite and lamprophyre complex at Delitzsch (Germany)

In eastern Germany, the carbonatite and lamprophyre complex at Delitzsch lies under a cover of sediments from the Cenozoic, approximately 100 m thick (Seifert et al. (2000) and references therein). This complex was discovered during uranium prospecting from 1960 to 1970. It is formed of dykes and diatremes established during the Campanian (75–71 Ma) (Krüger et al. 2013) within Upper Paleozoic volcano-sedimentary rocks. REEs are known in carbonatites in the form of apatite, pyrochlore and bastnäsite. Within the Storkwitz diatreme, the carbonatites have been explored from surveys, and the Deutsche-Rohstoff company published an estimate of resources in 2013, compatible with the JORC code. The total resources are estimated at 4.4 Mt @ 0.45% REO (part indicated, part inferred), which is 20,100 tons of REEs.

1.5.5.2.2. REE-Sc and lazulite-quartz Alpine veins (Austria)

There are many Alpine veins in the belt (Luckscheiter and Morteani 1980; Mullis 1996; Wagner et al. 2010; Rauchenstein-Martinek et al. 2014). These veins measure some meters in length for some tens of centimeters in width and are associated with shear zones at the brittle–ductile boundary. The Alpine veins are present in the form of veins alone or in an echelon formation and contain euhedral crystals of quartz and other minerals that depend on the surrounding lithology (adularia, albite, muscovite and quartz in the granitoids; calcite, ankerite, rutile and quartz in the carbonates; albite, chlorite, epidote, actinolite, titanite and quartz in the basic rocks). In southeastern Austria (47.5°N-15.6°E near Mürzzuschlag), the Pretul site in the Styria region is known for its Alpine lazulite-quartz veins that contain REE minerals and Sc (apatite, florencite-(Ce), goyazite, xenotime-(Y), monazite-(Ce), and pretulite; Cornelius 1931; Berl 1996; Bernhard et al. 1998). These occurrences present no economic benefit for REEs.

1.5.5.2.3. Talc deposit at Trimouns-Luzenac (France)

In the French Pyrenees, the Trimouns-Luzenac quarry is the largest talc exploitation in the world, with more than 400,000 tons/year (Fortuné 1971; Fortuné et al. 1980; Moine et al. 1989; de Parseval et al. 1993). The deposit is situated on the eastern flank of the Saint-Barthélémy massif, 100 km south of Toulouse, at an altitude of 1,700 m. The talc and chlorite are formed by the hydrothermal alteration of Paleozoic dolomites, micaschists and pegmatites along a shear zone (Fortuné et al. 1980; Moine et al. 1989; de Parseval 1992). Thanks to exploitation, centimeter-scale geodes within dolomites containing REE minerals have been identified. The minerals reported are allanite, bastnäsite, parisite, synchysite, gadolinite, monazite, trimounsite-(Y), hellandite-(Y), imoriite-(Y), hingganite-(Y), aeschynite-(Y), gatelite-(Ce), dissakisite-(Ce) and xenotime (Moëlo et al. 1974; Piret et al. 1990; Grew et al. 1991; Favreau 1994; de Parseval et al. 1997; Bonazzi et al. 2003). The formation of geodes seems to be linked to metasomatism along the shear zone (300–350°C/0.2–0.3 GPa). The monazite and xenotime of these geodes have moreover allowed the deposit to be dated back to between 112 and 97 Ma using the U/Pb method (Schärer et al. 1999).

1.5.5.3. *The British part of the Cenozoic North Atlantic igneous province*

Some REE mineral occurrences are described in the granitic massifs from the Cenozoic British igneous province (Exley 1980; Harding et al. 1982; Hyslop et al. 1999; Green et al. 2005; Green and McCallum 2005) (Figure 1.33). This magmatic province is situated in the northwest of the British Isles and forms essential evidence of magmatic activity linked to the North Atlantic in the Paleocene and Lower Eocene (Emeleus and Gyopari (1992) and references therein). The main intrusions are the plutons of the Isle of Skye, Saint-Kilda, Arran and the Mourne Mountains.

1.5.5.3.1. Granitic intrusion on the Isle of Skye

The Isle of Skye (northwestern Scotland) is formed of four granitic complexes dating back to between 55 and 59 Ma (Harker 1904; Brown 1963; Thorpe et al. 1977): Cuilin, Strath na Creiteach, Western Red Hills (57.0-58.5 Ma using the U/Pb method on zircon) (Hamilton et al. 1999) and Eastern Hills (55.7-55.9 Ma using the U/Pb method on zircon) (Hamilton et al. 1999). These complexes integrate hyperaluminous, metaluminous and hyperalkaline granites. Allanite and sphene are the main additional minerals and are particularly rich in REEs (between 25 and 45% RE_2O_3) (Exley 1980). Moreover, secondary allanite has been identified. It formed over the course of a late hydrothermal episode and during the stages of meteoric alteration. Finally, miarolitic cavities have been identified as containing REE minerals (Green and McCallum 2005): gadolinite and kainosite.

1.5.5.3.2. The Arran granitic pluton

The Arran granitic pluton is situated on the eponymous island to the southwest of Scotland and has been dated back to 60.3 ± 1.0 Ma using the U/Pb method on zircon (Dickin et al. 1981) and to 58.5 ± 0.8 Ma by $^{40}Ar/^{39}Ar$ (Mussett et al. 1987). To the north of the pluton, REE minerals have been mentioned: fergusonite, gadolinite and allanite (Hyslop et al. 1999). The miarolitic cavities also include fergusonite and gadolinite crystals associated with albite and orthoclase.

1.5.5.3.3. Saint-Kilda granitic complex

The granitic complex at Saint-Kilda is situated on an island to the extreme northwest of the Hebrides in Scotland. It is formed of the granites of the Conachair plutons (55 ± 1 Ma, using the Rb/Sr method) (Harding et al. 1984), Glen Bay and h-Heagen and gabbroids. The granites contain quartz, alkaline feldspar, sodium plagioclase, chlorite and amphibole (Cockburn 1935; Harding et al. 1966). Three REE minerals have been identified by order of abundance: chevkinite, allanite and zirkelite (Harding et al. 1982). They represent less than 1% of the rock, and chevkinite contains more than 41% of all the REEs.

1.5.5.3.4. The granitic pluton of the Mourne Mountains

The Mourne Mountains are situated in Northern Ireland and are formed of a granitic pluton that has been dated back to between 53.3 and 55.0 Ma using the $^{40}Ar/^{39}Ar$ method (Thompson et al. 1987). These rocks contain REE minerals, gadolinite and honey-colored monazite within the miarolitic cavities (Hyslop et al. 1999; Green et al. 2005).

1.5.5.4. *The European Cenozoic Rift System – the Kaiserstuhl Nb-Ta-REE alkaline volcanic complex (Germany)*

The Kaisertuhl massif is situated in the center-south of the Rhine Graben, to the southwest of Germany, 15 km from Fribourg. It is a Miocene alkaline volcanic complex (15-18 Ma) that is 16 × 12 km (Van Wambeke et al. 1964; Wimmenauer 1974; Keller 1981; Schleicher et al. 1990; Weisenberger et al. 2014). The predominantly volcanic to subvolcanic rocks have compositions of tephrite and phonolite, while the carbonatites were formed toward the end of the emplacement. The main types of volcanic rock are as follows:

- lava flows (tephrite);
- pyroclastic rocks (tephrite, phonolite and carbonatite) (Keller 1981);
- a porphyric dyke with essexite, phonolite and carbonatite;
- fine-grained subvolcanic intrusions of essexite, phonolite and carbonatite;
- volcanic breccias (Figure 1.34).

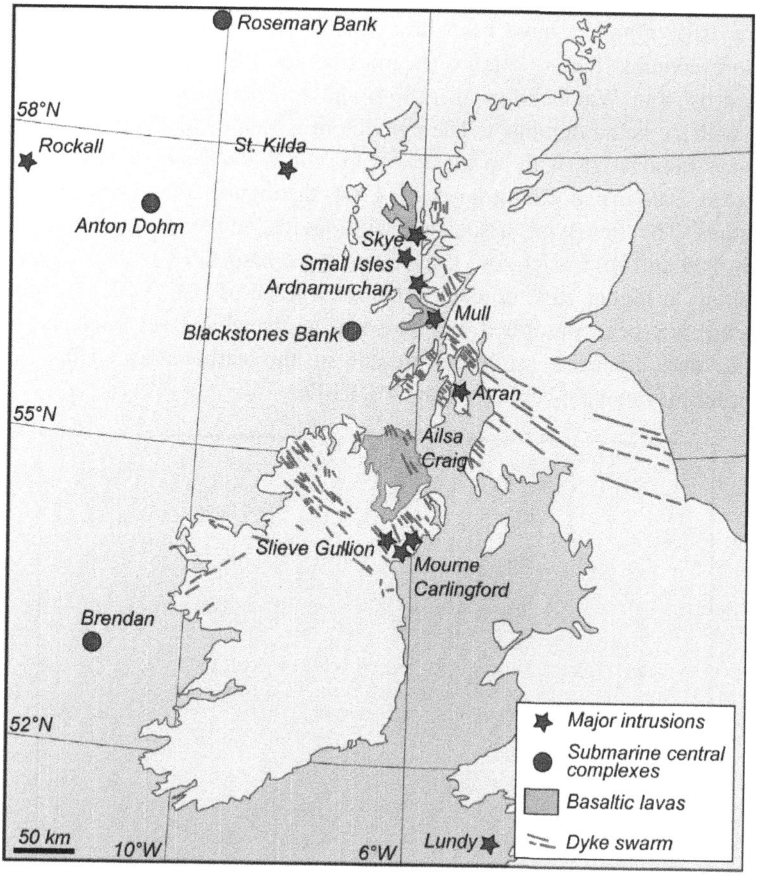

Figure 1.33. *British part of the North Atlantic Cenozoic igneous province*

Carbonatites are present at the core of the complex in the form of an intrusion of approximately 1 km^2. This mainly involves coarse-grained søvite mainly formed of calcite, then apatite and magnetite, with the following additional minerals: pyrochlore, Nb perovskite (dysanalite), bastnäsite and monazite (Van Wambeke et al. 1964; Keller 1981). A second type of carbonatite is presented in the form of hundreds of small, fine-grained veins and a width varying between 1 cm and 1 m. Calcite forms 90–95% of its volume, followed by magnetite and apatite. The additional phases are pyroxene, melanite and micas. Pyrochlore and Nb-perovskite are rarer. The veins are concentrated at the core of the massif and are the most recent rocks. Thus, the Kaisertuhl carbonatites are for the most part calcitic, and the ankerite-dolomite carbonatites seem restricted to late veins associated with a few søvite intrusions (Wimmenauer 1963).

Some REE minerals have been described in the complex's carbonatites. The pyrochlore contains REEs, koppite (Brandenberger 1931), and chemical analyses published by Van Wambeke et al. (1964) show variation between 4.3 and 6.4% RE_2O_3, with Ce being dominant. The pyrochlore is rich in LREE (up to 19% REO) and always radioactive, with up to 11% U_3O_8 and 6.5% ThO_2 (Kirchheimer 1957; Walter et al. 2018). Van Wambeke et al. (1964) distinguished two types of calcite in carbonatites. The first type, associated with søvite, contains variable REE levels between 0.09 and 0.15% RE_2O_3. The second type, associated with late carbonatite veins, shows a higher REE content, with an average of 0.22% RE_2O_3. Moreover, fluorapatite has been described and presents appreciable REE contents. Finally, bastnäsite and monazite are only present in the carbonatites. Thus, the late carbonatites may contain between 1 and 4% REEs.

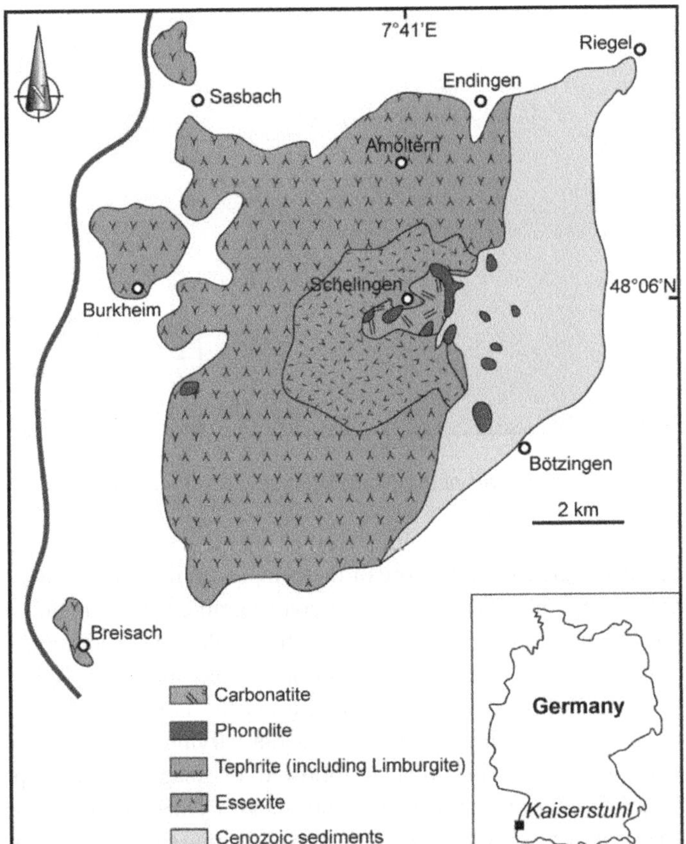

Figure 1.34. *Kaiserstuhl Nb-Ta-REE alkaline volcanic complex, Germany (Weisenberger et al. 2014)*

1.5.5.5. *Coastal placers*

The different REE occurrences of the Quaternary are heavy mineral placers mainly integrating monazite, zircon and ilmenite.

1.5.5.5.1. The coastal placers of Chalkidiki and Kavala (Greece)

In northeastern Greece and north of the Aegean Sea, geochemical studies have been carried out along the coast for 240 km between the Chalkidiki and Kavala Peninsulas, revealing the presence of REEs in the coastal sediments (Perissoratis et al. 1988; Pergamalis et al. 2001a, 2001b; Melfos and Voudouris 2012). Indeed, geochemical anomalies have been revealed in placers, such as that of Nea Peramos, with levels of up to 8,000 ppm REE (including 93.5% LREE with mainly Ce, La, Nd and Pr). The main carrier is allanite. Recent studies mention the magmatic massifs of the Pangaion Mountains near Kavala as the primary source of REEs. Indeed, analyses carried out on these mountains show concentrations of 6,171 ppm La, 10,883 ppm Ce, 1,062 ppm Pr, 3,249 ppm Nd, 334 ppm Sm and 214 ppm Eu.

1.5.5.5.2. The coastal placer at Hanstholm (Denmark)

The placer at Hanstholm is situated in northern Denmark in the province of North Jutland. This heavy mineral placer contains ilmenite, rutile, zircon and monazite sands (Hedrick and Templeton 1991), and has been the subject of exploration by the Morstal Minerals company (Orris and Grauch 2002).

1.5.5.5.3. The coastal placer at Cuxhaven (Germany)

The placer at Cuxhaven is situated on the North Sea coast, 100 km west of Hamburg, and contains heavy mineral sands. The resources are estimated at 10 Mt of heavy minerals, including ilmenite, rutile, zircon and monazite (Hedrick 1985).

1.5.5.5.4. Radioactive anomaly on Espiguette beach (France)

In the South of France, near Grau-du-Roi, the beach at Pointe de l'Espiguette presents radioactive anomalies, with average values between 600 and 850 c/s and a maximum of 1,050 c/s (Vazquez-Lopez and Jézéquel 2002). This anomaly extends over 1,500 m in length and 10–50 m in width, and is situated above the intertidal zone. The heavy mineral sands contain crandallite (the florencite type) and zircon, which are probably responsible for the radioactive anomaly.

1.6. The rare earths deposits of Greenland

In response to increasing demand for mineral resources, Greenland is today generating international interest in the discovery of new deposits (Stensgaard et al. 2016). Greenland benefits, in particular, from varied geological settings favorable to the presence of REE deposits attractive to exploration companies (GEUS 2010; Sørensen and Kalvig 2011; Steenfelt 2012; Stendal 2014; Goodenough et al. 2016; EURARE 2017; Andersson et al. 2018a). The main REE occurrences in Greenland are situated in carbonatites, alkaline intrusions, pegmatites and paleoplacers (Figure 1.35).

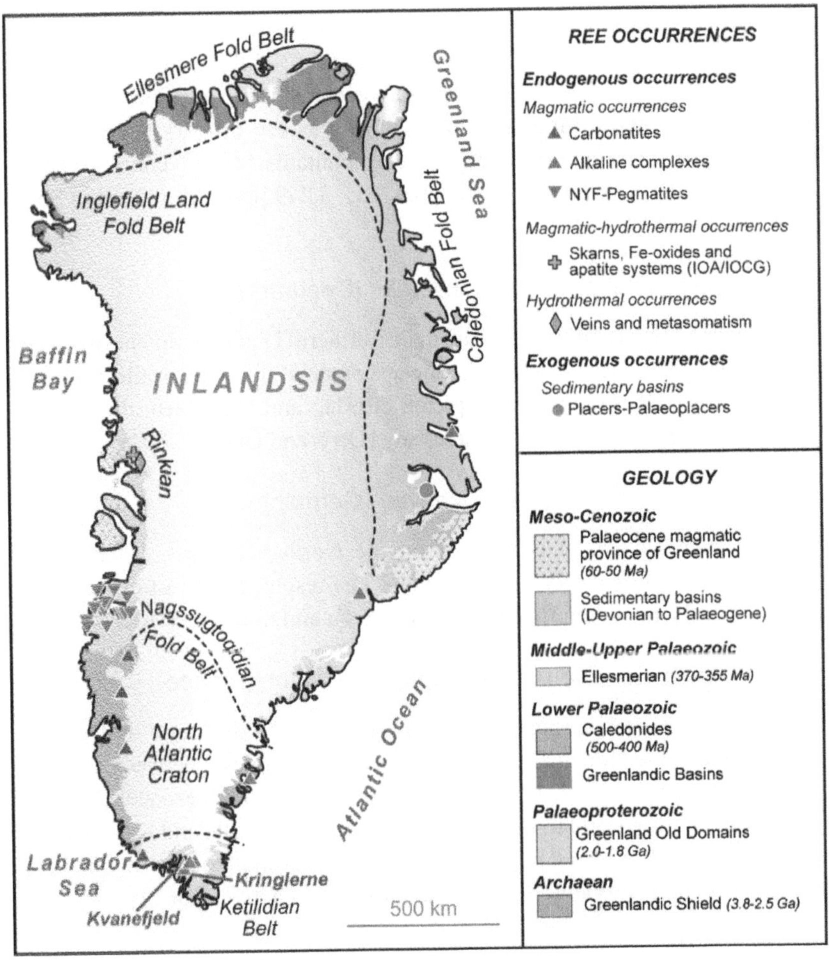

Figure 1.35. *REE occurrences in Greenland (Steensgaard et al. 2010; Charles et al. 2013; Tuduri et al. 2020)*

1.6.1. *The alkaline provinces of the Archean (2,750–2,700 Ma)*

The alkaline rocks of the Archean are rare globally and only outcrop in the Baltic shield (e.g. Keivy), the Canadian shield, Australia (Yilgarn), Greenland (e.g. Skjoldungen) and South Africa (Fitton and Upton 1987; Blichert-Toft et al. 1995; Zozulya et al. 2007). These geological features form REE potential, such as the Siilinjärvi carbonatite (Finland) or the Sakharjok alkaline complex (Russia). In Greenland, two regions have Archean alkaline rocks:

– the Tupertalik carbonatite in the west;

– the Skjoldungen alkaline province in the southeast, with the Singertât carbonatite and the Timmiarmiut alkaline complex (Larsen et al. 1983; Fitton and Upton 1987; Nielsen and Rosing 1990; Blichert-Toft et al. 1995; Steensgaard et al. 2010; Berger et al. 2014).

1.6.1.1. *Sr-Zr-REE carbonatite at Tupertalik (western Greenland)*

In western Greenland, the Tupertalik carbonatite massif (65.50°N–51.85°W) extends over 500 × 200 m and is situated only 11 km north of the Upper Jurassic carbonatite at Qaqqarsuk, near the Sønder Isortoq fjord (Larsen et al. 1983). The carbonatite is present in the form of a blade, with a thickness of 10 m within orthogneiss, having undergone metamorphism in the granulite facies during the Archean, without undergoing a major retrograde stage afterwards. The high concentrations of Sr, Zr and REE are comparable to those of other carbonatites. Archean carbonatite dykes with a similar geochemical signature have been discovered at Fiskenæsset, 310 km south of Tupertalik (Bollingberg et al. 1976). The Tupertalik carbonatite has been dated back to 2,620 ± 80 Ma and 2,691 ± 80 Ma using the K/Ar method on phologopite (Larsen et al. 1983). This age has been re-evaluated and established at 3,007 ± 2 Ma using the U/Pb method on baddeleyite (Bizzarro et al. 2002). Further studies are needed to better estimate the real REE potential of the older, land carbonatite massif.

1.6.1.2. *Alkaline province from the Neoarchean at Skjoldungen (southeast Greenland)*

In southeastern Greenland, the alkaline province of Skjoldungen (63°N–43°W) extends along a NW–SE axis over a width of 30 km, with a variety of late to post-tectonic alkaline intrusions established between 2,700 and 2,665 Ma (Rosing et al. 1992; Blichert-Toft et al. 1995). The alkaline rocks intrude a craton formed of migmatitic orthogneiss and supracrustal rocks (e.g. amphibolite, metapelite), all affected by metamorphism in the granulite facies (2,750–2,740 Ma) during the Skjoldungen orogeny (Kolb et al. 2013; Berger et al. 2014).

1.6.1.2.1. Singertât Sr-REE carbonatite

The Singertât carbonatite-ijolite complex (63.23°N–42°W) intrudes fenitized gneisses south of the Skjoldungen province, near the Kattertooq fjord (Nielsen and Rosing 1990; Blichert-Toft et al. 1995). The core of the massif is formed by bedded sills of coarse-grained melteigite, ijolite and urtite containing aegirine-augite and/or alkaline amphibole, nepheline and biotite in variable proportions. The additional phases are magnetite, apatite, sphene, alkaline feldspar, calcite and zircon. The cancrinite (feldspathoid) often presents as a reaction rim between calcite and nepheline. Dykes and veins (<1 m) of nepheline syenite and carbonatite intersect the massif. The nepheline syenite dykes are dominated by perthite with variable levels of nepheline, aegirine-augite and/or alkaline amphibole, biotite, apatite, calcite and Fe-Ti oxides. The carbonatite dykes are formed of apatite-biotite søvite. The peripheral alteration (fenite) zone reaches approximately 2 km in width. The Singertât carbonatite dates back to 2,664 ± 3 Ma (U/Pb age on zircon) (Blichert-Toft et al. 1995) and is rich in Sr (5,260 ppm) and REE (2,400 ppm with 1,122 ppm Ce, 464 ppm La, 524 ppm Nd). Further studies are needed to estimate the REE potential of this massif.

1.6.1.2.2. Timmiarmiut Sr-REE alkaline complex

Approximately 100 km south of the Singertât carbonatite, the Timmiarmiut alkaline complex (62.37°N–42.28°W) includes alkaline carbonate rocks rich in Sr (2,000 ppm) and REE (2,300 ppm REE with 1,100 ppm Ce and 120 ppm Y) (Steensgaard et al. 2010). This zone remains largely underexplored, although it presents good REE potential.

1.6.2. *Paleoproterozoic deposits at Karrat (NIAQ and UMIA)*

In central-west Greenland, in the Rinkian fold thrust belt, the Karrat project (71.67°N–52.77°W) includes two potential deposits (Figure 1.36): Niaqornakassak (NIAQ) and Umiammakku Nunnaa (UMIA), discovered by Avannaa Resources in 2007–2009 (Avannaa Resources 2010; Sørensen and Kalvig 2011; Stendal 2014). The sites are located on two peninsulas 7 km apart, and the two deposits are grouped under the name "Karrat". Exploratory works are focused on NIAQ, where the geometry of the mineralized body has been defined by eight drilling campaigns (1,320 m in total). It is a layered horizon of metacarbonates parallel to the regional foliation (30° toward the south) of the surrounding rock (amphibolite from the Karrat group dated back to the Paleoproterozoic) (Garde 1978; Henderson and Pulvertaft 1987; Kalsbeek et al. 1998; Mott et al. 2013).

The mineralized body is at least 1.5 km long, while the extent of its extremities is not known. Its thickness varies between 10.3 and 32.5 m for depths between 56 and 168 m above sea level. The main REE minerals are as follows: bastnäsite (39%), allanite (26%), monazite (22%), monazite-(Th) (4%), fergusonite-(Y) and fluorides-Y (4%), and chevkinite and diversilite (Fe-Ti-Ba-REE silicates) (4%). Avannaa Resources proposes a tonnage and grades: 26.3 Mt @ 0.996% REO+Y_2O_3 and 0.13% HREO+Y_2O_3 in an optimistic scenario and 19.5 Mt @ 1.02% REO+Y_2O_3 and 0.13% HREO+Y_2O_3 in a more limited scenario. The average Th content varies between 400 and 500 ppm (huttonite), and uranium seems to be absent from the deposit NIAQ (< 5 ppm). We should note the low contents of Be, Zr, Sc or Ta, with Nb contents varying between 300 and 500 ppm. The origin of the REEs would be linked to the metasomatic fluids affiliated to a ferrocarbonatite (Mott et al. 2013). Indeed, the mineralized body is not considered a carbonatite, given its SiO_2 content of 15–25% (< 5% for a carbonatite) and CaO content of 18–20% (> 30% in a carbonatite).

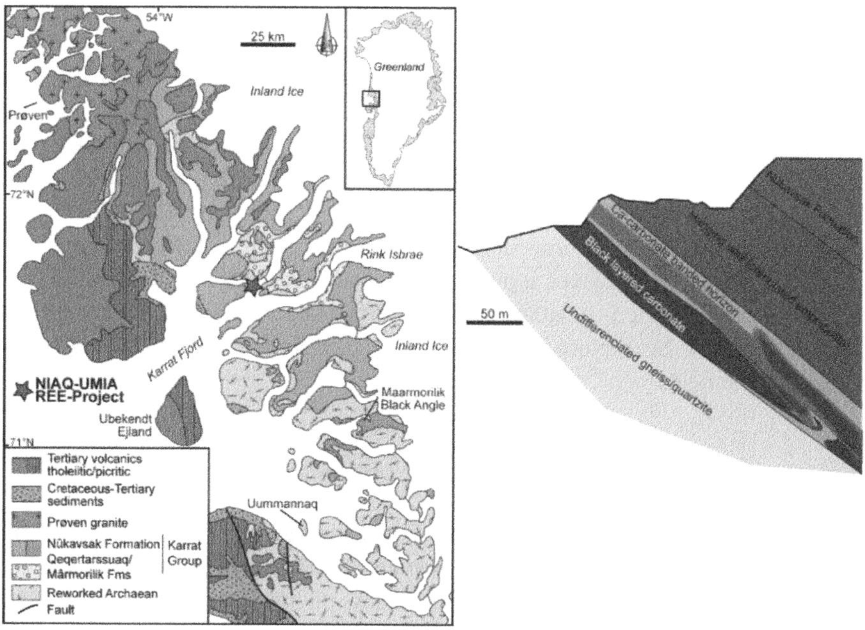

Figure 1.36. *Deposits from the Paleoproterozoic at Karrat (NIAQ and UMIA) (Avannaa Resources 2010)*

Although exploration is limited at the UMIA site, two drillings (420 m in total) and 347 samples have provided an estimate of the REE content, reaching up to

1.98% REO+Y. The real thickness and global geometry are still not defined due to a lack of sufficient exploratory work.

1.6.3. *Pegmatites from the late Paleoproterozoic (1,800–1,785 Ma)*

To the southwest of Greenland, to the north of the Nassugtoqidian orogeny, three pegmatite fields are present within the Archean base. Two pegmatite fields (Nassuttutata tasia and Nordre Isortoq) are situated in the regions of Nassuttooq and Nordre Strømfjord, respectively (Secher 1976, 1980). These are white biotite and plagioclase pegmatites, concordant with the foliation of the surrounding rock (Archean gneisses affected by metamorphism in the granulite and amphibolite facies and supracrustal rocks from the Paleoproterozoic) (Secher 1980; GEUS 2014) and dating back to approximately 1,800 Ma. The width of the pegmatite bodies varies from 5 to 10 m, with lengths of 50–200 m. The main REE mineral is monazite in the form of orange euhedral crystals of 0.5–5 mm. The monazite contains Nd (5–7%) and 1–2% La, 2.5–4.0% Ce, 500 ppm Y, and 5–10% Th (GEUS 2014).

The third pegmatite field is situated in the Attu region, where the pegmatites intrude the Archean gneisses and amphibolites. The pegmatites, dated back to 1 785 ± 5 Ma, are concordant or discordant and present in the form of veins up to 5 m in width over lengths of several hundred meters. The main REE mineral is allanite (crystals of up to 5 cm), accompanied by sphene, apatite, magnetite (centimetric aggregates with allanite) and iron sulfides in a gangue where K-feldspar, quartz and biotite are associated. Moreover, a pyrite and magnetite alteration zone is observed at the contact between the pegmatites and the surrounding amphibolite, over nearly a meter in thickness. No exploration campaign has been carried out until now in these three pegmatite fields.

1.6.4. *Gardar Mesoproterozoic alkaline province*

The province of Gardar (Figure 1.37) is a rift zone formed at the end of the Mesoproterozoic south of Greenland (Upton and Emeleus 1987; Upton et al. 2003; Upton 2013). Associated with sediments, many plutonic rocks were laid down between 1,280 and 1,120 Ma. These alkaline intrusions are formed of syenites, nepheline syenites, quartz syenites and granites associated with a swarm of giant basic veins. They form the main REE and U potential in Greenland. The intrusion at Ilímaussaq (1,160 ± 5 Ma; Krumrei et al. 2006) includes the two world-class deposits at Kringlerne and Kvanefjeld, and the Igaliko intrusion includes the Motzfeldt Sø deposit.

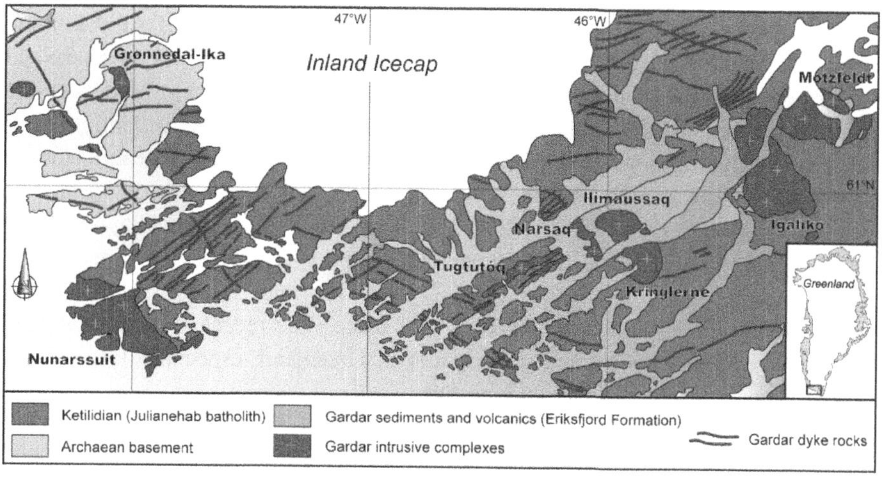

Figure 1.37. *Alkaline province from the Mesoproterozoic at Gardar*

1.6.4.1. *Kvanefjeld U-REE-Zn-F deposit*

To the south of Greenland, the U-REE-Zn deposit at Kvanefjeld (Figure 1.38) is situated a few kilometers from Narsaq and occupies the northeastern part of the Ilímaussaq intrusion (Ferguson 1964; Larsen and Sørensen 1987; Sørensen et al. 2011). This alkaline complex consists of four successive intrusions:

– an augite syenite in the form of a discontinuous band on the edge of the intrusion;

– a hyperalkaline quartz syenite and an alkaline granite at the top of the complex and in the form of enclaves in the third intrusion;

– a layered series hyperalkaline nepheline syenite (agpaite);

– cumulates of kakortokite, aegirine lujavrite and arfvedsonite lujavrite (Sørensen 1969; Bohse et al. 1971; Nielsen and Steenfelt 1979; Sørensen et al. 2011 and references therein).

These rocks (especially sodalite foyaite, naujaïte, kakortokite and lujavrite) are interpreted as resulting from highly fractionated phonolitic magma rich in iron and incompatible elements (Zr, Hf, Nb, REE, U). The Ilímaussaq complex (1,160 ± 5 Ma; Krumrei et al. 2006) is an igneous body, unique with regard to its extreme enrichment of many rare elements in more than 200 minerals.

The world-class deposit at Kvanefjeld occupies a plateau of 2.1 × 0.9 km and forms part of the 4.5 km lujavrite belt in the Ilímaussaq complex. The northern and western edges of the plateau correspond to the northern and western borders of the alkaline complex. To the northeast, the lujavrite level at Kvanefjeld is continuous as far as the Steenstrup fjord. On the western part, the lujavrite of the plateau forms the base of the augite syenite and the naujaite. Finally, the eastern part is formed of medium- to coarse-grained lujavrite ("M-C lujavrite"), fine-grained lujavrite and then naujaite. In the 1950s, the Ilímaussaq complex was widely prospected for uranium, then for REEs, as well as Zr, Be, Nb and Ta. For political and socioeconomic reasons, no development took place. The recent back and forth on legislative relaxation on mineral exploration in Greenland, especially for uranium and oil, is not conducive to foreign investors.

The Kvanefjeld deposit is being explored by the Greenland Minerals company for U, REE, F and Zn. It appears that U, Th and REEs are concentrated in the upper parts – that is, the most recent parts – of the complex. Thus, hyperagpaitic lujavrite and agpaitic nepheline syenite are the main ores. The upper units of the lujavrite are very rich in REEs, U and Zn, with REE contents of over 1.5% and 400 ppm U_3O_8. At depth, these concentrations decrease. The mineralized areas may exceed 250 m in width within dome-shape structures. The mineralized lujavrite mainly outcrops in the northern half of the Ilímaussaq complex. Kvanefjeld is the main deposit, and satellite deposits have recently been discovered at Sørensen (formerly zone 2) and in zone 3 (GME 2014). The main REE-U-Zn minerals are steenstrupine, britholite, monazite, vitusite, xenotime and various Na-Zr silicates, such as eudialyte and lovozerite. In 2015, an estimate of NI43-101 or JORC compatible resources showed a total resource of 673 Mt @ 1.09% REO and 248 ppm U_3O_8 (7.34 Mt REO including 0.27 Mt HREO). The indicated resources represent 308 Mt (3.42 Mt REO including 0.13 Mt HREO) and the inferred resources are 222 Mt (2.22 Mt REO including 0.08 Mt HREO), with a cutoff grade of 150 ppm U_3O_8. The total project (Kvanefjeld, Sørensen and zone 3) represents 1,010 Mt @ 11,000 ppm REO, including 9,700 ppm LREO, 399 ppm HREO and 266 ppm U_3O_8 (11.14 Mt REO, including 0.40 Mt HREO). Moreover, tests for extraction using froth flotation and open air leaching are conclusive (GME 2012). In 2014, 30 tons of ore was used for the pilot plant in the EURARE project (GME 2014; EURARE 2017).

1.6.4.2. *Kringlerne Ta-Nb-REE-Zr deposit*

The world-class polymetallic deposit (Ta-Nb-REE-Zr, hence the name Tanbreez) at Kringlerne is situated in southern Greenland, 15 km southeast of the Kvanefjeld deposit to the south of the Ilímaussaq intrusion (Bohse et al. 1971; Bohse and Andersen 1981; Sørensen 2001; GEUS 2014). This deposit (Figure 1.38) is situated at the base of the alkaline complex in the lower levels of layered agpaitic nepheline syenite cumulates called kakortokite. Kringlerne is situated in the Kangerluarssuk region, and the kakortokite cumulates are formed of 29 regular layers over a total thickness of 200 m. These layers are a repetition of banding of black syenite (dominated by arfvedsonite), red syenite (dominated by eudialyte) and white syenite (dominated by feldspar). Each banding measures between 1 and 20 m in width. The origin of this banding is still being debated and results from minerals being sorted by density when the magma solidified, as well as from rhythmical deposition within the magmatic chamber (Sørensen 2001). The Kangerluarssuk zone covers a surface of 2 × 3 km, 400 m above sea level. Kakortokite has been studied for decades, especially for the following elements: Zr, Y, Ta, Nb and REEs. The Kringlerne deposit is being explored by the Tanbreez company.

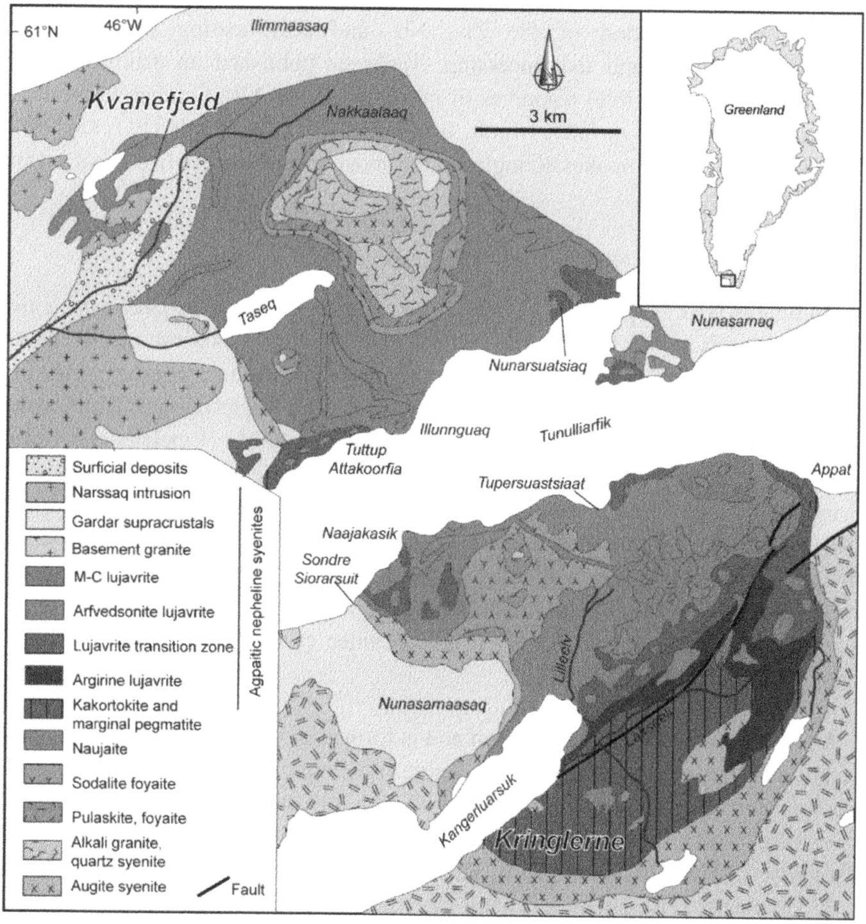

Figure 1.38. *Ilímaussaq complex including the Kvanefjeld and Kringlerne deposits*

Eudialyte is the main mineral of economic interest and carries Zr, Nb and REE (Ferguson 1964; Andersen et al. 1988; Johnsen et al. 2001; Sørensen and Kalvig 2011; GEUS 2014). The three types of cumulate layers have been analyzed for Zr content by Bohse et al. (1971): black kakortokite (1.1%), red kakortokite (3.8%) and white kakortokite (1.3%). The eudialyte contains 12 wt.% ZrO_2, 2 wt.% RE_2O_3 and 1% wt.% Nb_2O_3, on average (Borst et al. 2016). Moreover, the eudialyte shows low contents of Th and U (similar to contents in the surrounding rock). The eudialyte can be extracted easily from the surrounding materials using magnetic separation, and extraction of the REEs requires acid treatment (Tanbreez 2014). Indeed, recent studies have shown that eudialyte is often replaced by complex assemblages of secondary minerals resulting from the late magmatic interaction of fluids rich in Na and F. This alteration fractionates the major elements within these secondary minerals with separation of the Zr-, Nb- and REE-bearing phases, which complexifies the ore and its processing. Tanbreez published an estimate of the inferred (JORC compatible) resources of more than 4,300 Mt of eudialyte ore, with an average of 1.8% ZrO_2, 0.2% Nb_2O_5, 0.5% LREOs and 0.15% HREOs (which is 27.95 Mt REO), which makes Kringlerne the most substantial REE deposit on the planet (Tanbreez 2014).

1.6.4.3. *Motzfeldt Sø Nb-Ta-U-Zr-REE deposit*

Within the alkaline province at Gardar, alkaline complexes have been established inside and outside the Gardar rift. In addition to the Nunarssuit and Paatusoq complexes, the Igaliko complex is the most substantial in the province, formed of four intrusions (Motzfeldt, Qôroq Nord, Qôroq Sud, Idglerfigssalik) and some smaller intrusive bodies, dated between 1,280 and 1,140 Ma (Emeleus and Upton 1976; Upton and Emeleus 1987; Upton et al. 2003; Bartels 2013). The Motzfeldt intrusion (Figure 1.39) is situated to the northeast of the Igaliko complex and covers a surface of 20 × 15 km (McCreath et al. 2012). It is a pluton of multiple syenite intrusions intersecting the batholite of the Julianehåb, which dates back to the Proterozoic age, and the Gardar metasediments, which date back to between 1,280 and 1,220 Ma (Emeleus and Harry 1970; Tukiainen et al. 1984; Upton et al. 2003; Schoenenberger and Markl 2008).

Motzfeldt Sø is the oldest intrusion and is formed of whole and altered nepheline syenite and a border of arfvedsonite syenite. A complex network of hyperalkaline microsyenite sills and pegmatites associated with a hydrothermal alteration allowed the emplacement of Th-U-Nb-Ta-Zr-REE mineralization at the top of the intrusion (Thomassen 1988; Tukiainen 1988). The main ore is formed of pyrochlore (Nb, Ta, U, REE), thorite (Th), zircon (Zr) and bastnäsite (REE, Th) (Tukiainen 1986). The pyrochlore contains 1.3–8.3% Ta. The first estimate of resources by GEUS dates back to the 1980s and gives a mineralized volume of more than 500 Mt @

0.14% Nb, 120 ppm Ta, 60 ppm U and 90 ppm Th (Thomassen 1988; Tukiainen 1988; GEUS 2014). The estimated resources are less than 30 Mt, with a cutoff grade of 250 ppm Ta. Where Nb resources are concerned, the figures give 130 Mt @ 0.4–1.0% Nb_2O_5. In 2010, an exploration permit was obtained by Ram Resources to study the prospects of Aries, Merino, Rams Head and Romney. The Aries site has inferred resources of 340 Mt @ 120 ppm Ta_2O_5, 1,850 ppm Nb_2O_5, 4,600 ppm ZrO_2 and 2,600 ppm REO (Ram Resources Ltd. 2012); in other words, 1.56 Mt ZrO_2, 0.884 Mt REO, 0.629 Mt Nb_2O_5 and 41,000 t Ta_2O_5.

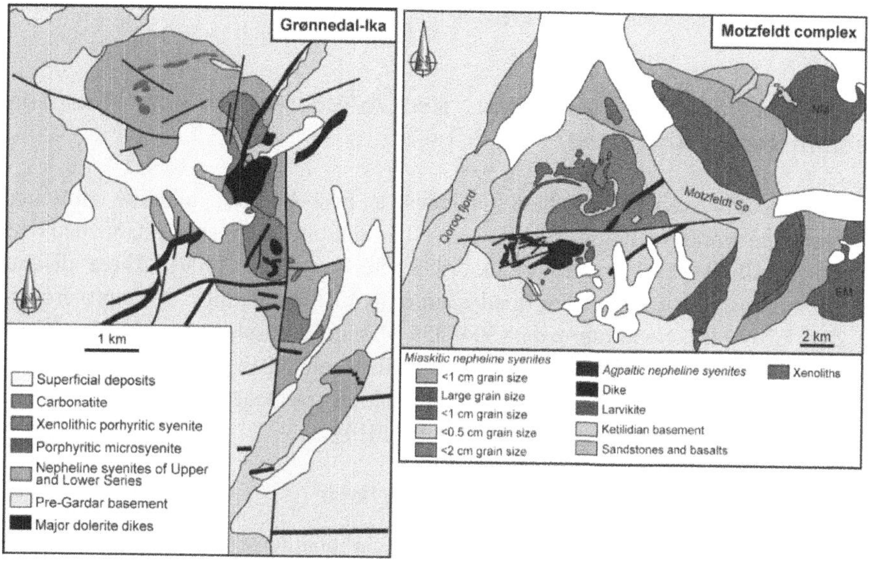

Figure 1.39. *Alkaline complexes at Motzfeldt and Grønnedal-Ika*

1.6.4.4. *Grønnedal-Ika alkaline complex*

In southern Greenland, the Grønnedal-Ika complex (8 × 3 km) is a nepheline layered syenite intrusion intersecting enclaves of porphyritic syenite and a carbonatite intrusion (Callisen 1943; Emeleus 1964; Pearce et al. 1997; Halama et al. 2005). This complex (Figure 1.39) has been dated back to 1,299 ± 17 Ma using the Rb–Sr method (Blaxland et al. 1978). The core is formed of a calcite-siderite-magnetite carbonatite. The Grønnedal-Ika complex seems favorable to REE potential (Sørensen and Kalvig 2011).

1.6.4.5. *Qassiarsuk alkaline complex*

In southern Greenland, the alkaline complex at Qassiarsuk (1.2 Ga) is situated to the north of the Precambrian Gardar rift, oriented E–W (Stewart 1970; Andersen 1997). This complex, intercalated in the sandstone, includes a series of alkaline silicate tufs and a carbonatite (discontinuous horizons of tuffs and sills). The massif extends from Qassiarsuk to the Eriksfjord fjord to the east, and to Tasiusaq Creek to the west. The carbonatite varies from calcitic carbonatite to ankerite-dolomite ferrocarbonatite. This carbonatite is one of the best targets for REE potential in the Gardar province (Sørensen and Kalvig 2011).

1.6.5. *Carbonatites from the Neoproterozoic to the Mesozoic (southwestern Greenland)*

The southwest of Greenland includes a post-Archean alkaline province, including networks of dykes and sills, as well as alkaline complexes established over nearly a billion years (Larsen et al. 1983; Secher et al. 2009). Three distinct magmatic episodes can be described according to their datings: Mesoproterozoic (1,284–1,209 Ma), Neoproterozoic (604–555 Ma) and Jurassic (165–155 Ma). Three carbonatite massifs are studied in particular for their REE and Nb: the Sarfartoq carbonatite (Ediacarian), the Qaqarssuk carbonatites (or Qeqertaasaq) and Tikiusaaq from the upper Jurassic (GEUS 2014; Stendal 2014).

1.6.5.1. *Sarfartoq Ediacaran Nb-REE carbonatite*

In southwestern Greenland, the Sarfartoq carbonatite (66.482°N–51.229°W) is situated approximately 60 km southwest of Kangerlussuaq airport (Figure 1.40). This complex was established at the boundary between the Archean craton to the south and the Paleoproterozoic Nagssugtoqidian orogenic belt to the north (Secher 1986). Carbonatite dates back to 564 Ma (age Rb/Sr on phlogopite) (Secher et al. 2009) and was discovered by GEUS in 1976 using airborne geophysical surveys and field work (Secher 1976, 1986; Secher and Larsen 1980). The massif covers 90 km^2 and the carbonatite occupies 10 km^2. The surrounding rock is represented by Archean orthogneisses reworked during the Paleoproterozoic. The Sarfartoq complex is a steeply dipping conical body, whose concentrically layered core is formed of ferrocarbonatite rich in apatite and magnetite (with ankerite rauhaugite dominant). The carbonatite can be divided into a central zone (>50% carbonatite, ~1 km^2) and an external zone (<50% carbonatite, 1–3 km, ~9 km^2). The central part is surrounded by concentric bands and carbonatite dykes (beforsite, ferrodolomite dominant), with intercalations of fenitized panels of surrounding rock (light gray aegirine fenite). Hydrothermal processes have allowed the formation of Nb minerals (pyrochlore) and REEs in the dykes and deformed zones 50–200 m in width, surrounding the core of the massif, with the latter sometimes marked by Th and U anomalies (Bedini 2009). The main minerals in the complex are dolomite, ankerite, apatite, phlogopite, richterite, arfvedsonite and magnetite, and some additional phases include pyrochlore, zircon and Nb-rutile (Secher and Larsen 1980).

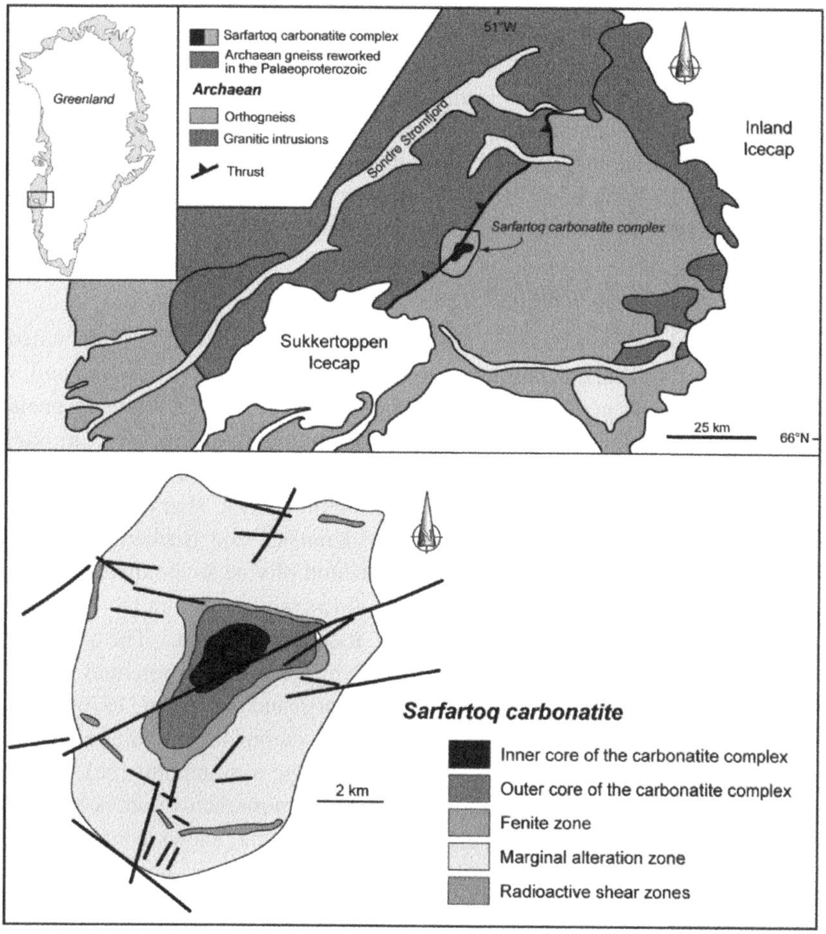

Figure 1.40. *Sarfartoq carbonatite (Secher et al. 2009)*

The carbonatite complex at Sarfartoq has been prospected for diamonds, as well as the following elements: P, Nb, Ta and REE (Sørensen and Kalvig 2010; Druecker and Simpson 2012; GEUS 2014). Since 2003, Hudson Resources has obtained two permits on 783 km^2 to explore Nb (2008) and REEs (2009). The estimate of (JORC compatible) resources gives measured resources of 23,478 t @ 5.95% Nb

(8.51% Nb_2O_5) with a cutoff grade of 3% Nb, and inferred resources of 64,301 t @ 3.89% Nb (5.56% Nb_2O_5) with a cutoff grade of 1% Nb (New Millenium Resources in 2002). For the REEs, Hudson Resources mentions that they are associated with ferrocarbonatite, fenite, hematite alterations and iron carbonates that coincide with the Th anomalies (Druecker and Simpson, 2012). The main REE-bearing minerals are bastnäsite-(Ce), synchysite-(Ce), synchysite-(Nd) and monazite-(Ce). In 2012, Hudson Resources published an estimate of (NI43-101 compatible) resources for the Sarfartoq project with indicated resources of 5.88 Mt @ 1.77% REO and inferred resources of 2.46 Mt @ 159% REO, based on a cutoff grade of 1% and underground exploitation (Druecker and Simpson 2012). Nd and Pr represent 25–54% of all REEs. A recovery rate of 6% has been announced, but it has not yet been subjected to metallurgical tests.

1.6.5.2. *Qaqarssuk/Qeqertaasaq Upper Jurassic Nb-REE-P*

The Qaqarssuk or Qeqertaasaq carbonatite (65.383°N–51.667°W) is situated in the southwest of Greenland, approximately 135 km northeast of Nuuk, and was discovered in 1969 (Vuotovesi 1974). The carbonatite intrudes Archean gneisses and dates back to 165.0 ± 0.9 Ma (Rb–Sr age on phlogopite) (Secher et al. 2009). The massif covers 15 km^2 and is formed of several concentric carbonatite dykes surrounded by a fenitized zone in the surrounding rock (Larsen et al. 1983; Kunzendrof and Secher 1987; Knudsen 1989; Knudsen and Buchardt 1991). The first stage is represented by dolomite carbonatite and olivine søvite dykes, while the late stage is represented by dykes of fine-grained dolomite carbonatite, ferrocarbonatite, søvite, silico-søvite and REE carbonatite. The Qaqarssuk carbonatite has been widely explored for its Nb, REE and P potential since the 1970s by Kryolit Selskabet Oresund (Gothenborg and Pederson 1976; Nielsen 1976). The Nb mineralizations are pyrochlore cumulates in the søvite and pyrochlore precipitations associated with fenite in the surrounding rock, near the carbonatite dykes. The pyrochlore shows contents of more than 15 wt.% Nb_2O_5 and up to 20 wt.% Ta_2O_5. Apatite, present in the silico-søvite, shows content between 3.5 and 6.0 wt.% P_2O_5.

Knudsen (1991) shows that carbonatite dykes are rich in REEs. Since 2010, Nuna Minerals has explored the massif by taking surface and trench samples, mapping the facies, and undertaking a drilling campaign totaling 1,590 m (Størensen and Kalvig 2011; Stendal 2014). The REE mineralizations are restricted to the center of the complex in the dykes (1–4.5 m in width and some hundreds of meters in length). The average content, according to surface samples taken over a zone of 1.5 km^2, is 2.4% REO. The best intersection by drilling amounts to 4.5% REO over nearly 4.7 m. Ancylite is the main REE-bearing in the dykes in the form of well-developed crystals (0.1-0.3 mm) associated with other phases, such as burbankite

($(Na,Ca)_3(Sr,Ba,REE)_3(CO_3)_5$), huanghoite ($BaREE(CO_3)_2F$) and qaqarssukite ($BaREE(CO_3)_2F$) (Grice et al. 2006). LREEs dominate, and the concentrations of some elements are 50% Ce, 27% La, 16% Nd, 5% Pr and 2% other REEs.

1.6.5.3. *Tikiusaaq Upper Jurassic P-REE carbonatite*

In southwestern Greenland, the Tikiusaaq carbonatite (64°N–46.75°W) is situated 100 km south of Nuuk. This complex was discovered by GEUS in 2005 thanks to an integrated framework/stream sediment geochemistry/aeromagnetism (Steenfelt et al. 2006). The massif seems to cover a surface of 100 km^2 and intrudes the strongly deformed and metamorphosed Archean basement, which is formed of granite, orthogneiss, amphibolite and anorthosite. The core of the massif (3.3 × 2 km) is formed of vertical and closely spaced blades and carbonatite (up to a few meters in thickness) of two types: early white calcitic carbonatite (calcite and planes with green mica, magnetite and apatite) and late gray dolomite carbonatite (ferrous dolomite and ankerite) (Nuna Minerals 2010). A fenitized zone is visible around the carbonatite core. Within the surrounding rock, numerous carbonatite veins (from 1 mm to 20 cm) are described. The Tikiusaaq carbonatite dates back to 158 ± 2 Ma (age U/Pb on zircon; Frei et al. 2006; Steenfelt et al. 2006), which is equivalent to the Qaqarssuk carbonatite (165 Ma).

The Tikiussaq complex has been explored by Nuna Minerals for REEs and diamonds (Nuna Minerals 2010). The exploration campaign focused on the core of the massif, with 78 carbonatite samples that have confirmed the REE concentration within the late ankerite carbonatite, while phosphates are present in the calcitic carbonatite. The contents amount to 9.6% REO (an average of 0.6% REO, dominated by LREEs) and up to 8.5% P_2O_5 (an average of 1.6% P_2O_5). The REE minerals are ancylite, churchite and parasite. The REE composition is broken down as follows: Ce (47%), La (33%), Nd (12%), Pr (4%) and other REEs (4%). The level of U is low. Nuna Minerals estimates that the mineralized body is 750 m in length and 100 m in width and extends to at least 500 m in depth. The important anomaly in Th indicates a substantial REE deposit, the potential resource of which is estimated to be more than 200,000 t @ 1–5% REO (Stendal 2014), even if further exploratory studies are needed.

1.6.6. *Milne Land Jurassic–Cretaceous REE-Ti-Zr paleoplacer*

In eastern Greenland, the island of Milne Land includes a Jurassic–Cretaceous paleoplacer with heavy mineral sands (70.685°N–25.920°W) (Sørensen and Kalvig 2011; Stembera 2012; GEUS 2014; Stendal 2014). These zircon and monazite sands have been detected through Th anomalies identified by airborne radiometry and a sampling campaign carried out by Nordisk Mineselskab and Research Establishment RISØ in 1968. Marine sandstones have been dated from the middle Jurassic to the lower Cretaceous and form part of the Charcot Bugt Formation (Birkelund et al. 1978, 1984; Birkelund and Callomon 1985; Larsen et al. 2003). The paleoplacer is situated at the base of this sedimentary formation, and the main anomaly is located at Bays Fjelde ("Hill 800" site), measuring 500 m in diameter and 40–50 m in width. The heavy mineral sands present as irregular lenses of 10–40 cm in thickness within an arkosic sandstone unit and breccia of approximately 20 m in width (Harpøth et al. 1986). The heavy minerals are as follows: monazite, xenotime, rutile, zircon, garnet and ilmenite with potential in REEs, Ti, Zr and Th. The detrital level of 20 m contains 5 Mt @ 1.0–3.8% Zr and 0.5–1.9% REO (Harpøth et al. 1986).

In 1990, metallurgical tests carried out by Coffs Harbour Rutile concluded that the economic extraction of monazite, zircon and garnet was possible. An estimate of resources for the "Hill 800" site gives 3.7 Mt @ 1.1% zircon, 0.5% monazite, 2.6% anatase, 3.1% garnet and 0.03% xenotime (Sørensen and Kalvig 2011). Monazite is the mineral containing REEs (90% LREE and 10% HREE). Since 2010, the CGRG (Czech Geological Research Group) has explored the Milne Land paleoplacer. In 2011, the CGRG collected 105 samples and made radiometric measurements, demonstrating three Zr-REE-Ti anomalies with levels of 2.0 wt.% ZrO_2, 2.5 wt.% REO and 35 wt.% TiO_2 (Stembera 2012).

1.6.7. *Alkaline complexes from the Paleogene (central and eastern Greenland)*

At the start of the Paleogene, central-eastern Greenland was marked by the opening of the North Atlantic, leading to the separation of Greenland from Europe. Magmatic rocks from this period outcrop between the latitudes 68°N and 75°N (Brooks and Nielsen 1982; Nielsen 1987; Saunders et al. 1997; Nielsen 2002). This magmatic event is characterized by vast basalt plateaus and alkaline intrusions established between 62 and 25 Ma (Tegner et al. 1998, 2008; Brooks et al. 2004). REE occurrences have been described at Kap Simpson, Kap Parral and others associated with the Gardiner complex (Stembera 2012; GEUS 2014; Stensgaard et al. 2016; EURARE 2017).

1.6.7.1. *Kap Simpson (Bjørnedal-Traill) Nb-REE alkaline complex*

Kap Simpson is situated on the central-eastern coast of Greenland and is one of the most important alkaline intrusions from the Paleogene (Nielsen 1987). This complex includes an alkaline syenite to the northwest and a system of concentric syenite dykes within a volcano-sedimentary unit to the northeast (Harpøth et al.

1986). The southeastern part is a caldeira called the "Dreibuchten Zone" (Schaub 1942; Harpøth et al. 1986). The Kap Simpson complex has been explored by Nordisk Mineselskab and AMAX (Martens and Seemann 1973; Schassberger and Spieth 1978; Geyti 1979; Schassberger and Newall 1980; Damtoft and Grahl-Madsen 1982). Intense hydrothermal alteration has been described across the whole massif, especially in the Dreibuchten zone, with silicification, pyritization and clay enrichment. Harpøth et al. (1986) mention a potential in Mo, fluorine and base metals. Moreover, the samples collected in 1969 by Nordmine have revealed high REE levels.

Further work has allowed the discovery of quartz veins and Nb-REE, 15 cm broad and 30 m long, within Jurassic sediments close to the contact with the Bjørnedal-Traill syenite (72.341°N–22.766°W) (Martens and Seemann 1973; Harpøth et al. 1986). The samples contained up to 3.2% Nb, 3.0% REE and 0.3% Y. The main REE-bearing minerals are euxenite, monazite, bastnäsite, samarskite and fergusonite. More recently, the CGRG (Czech Geological Research Group) explored the Bjørnedal-Traill site for Zr, Nb, Ta, REE and Mo potential (Stembera 2012). In 2011, samples were collected and showed levels up to 2.5 wt.% Zr, 0.3 wt.% Nb and 1.5 wt.% REE.

1.6.7.2. *Kap Parral alkaline complex*

Kap Karral is situated just to the north of Kap Simpson, on the central-eastern coast of Greenland. This cap is formed of an alkaline syenite intrusion similar to that at Kap Simpson (Nielsen 2002). It remains unexplored, although it may form Nb and REE potential.

1.6.7.3. *Gardiner ultramafic alkaline complex*

At the core of the Kangerlussuaq fjord in the center-east of Greenland, the Gardiner massif (68.618°N–33.151°W) is an ultrabasic alkaline complex that dates black to 50.3 ± 1.4 Ma, forming a REE occurrence (Woodley 1987; Campbell et al. 1997; Orris and Grauch 2002). The intrusion is formed of an ultrabasic suite and a system of late concentric melilitolite, syenite and carbonatite dykes (Frisch and Keusen 1977; Nielsen 1979, 1980, 1994; Johnsen et al. 1985). The REE-bearing minerals are perovskite (and loparite in the syenite) and apatite associated with other additional phases, such as titanite and magnetite. Perovskite concentrates more REEs than apatite, even though these two minerals are rich in LREEs (Campbell et al. 1997).

Perovskite is present in dunite, melilitolite and the associated rocks, and may form a modal proportion of nearly 20% in some cases where it is mentioned. Frisch and Keusen (1977) carried out a chemical analysis using a microprobe along a perovskite grain profile, showing that the perovskite is a calcium titanate with minor substitutions of Na, Nb and LREE, and contents of approximately 7 wt.% RE_2O_3. Analyses carried out by Campbell et al. (1997) on some grains of perovskite (melilitolite dike) and loparite (syenite) have given contents of 4.11% REO and 35.5% REO, respectively.

1.7. The origin of the rare earth deposits in Europe and Greenland

The conclusion of this summary is unarguable: Europe includes many occurrences of REEs associated with varied geological environments. The geological processes at the origin of these occurrences can thus be observed at all stages of the geological cycle, from the high temperature and high pressure conditions encountered in the mantle (up to 1,200°C and 1.5 GPa) to more surficial environments that are sensitive to the climate.

These very different processes may be of magmatic, metamorphic, hydrothermal or diagenetic origin, or may quite simply have origins associated with the regolith. Thus, the primary deposits are essentially associated with igneous alkaline rocks and carbonatites, whereas the secondary deposits, especially placers or bauxite, generally have a source in the fertile terrain of the European subsoil. One point is essentially to understand the main mechanisms at the origin of REE enrichment in endogenous environments as in exogenous ones, as well as the geological cycle of REEs in the continental crust (Figure 1.41).

1.7.1. *Partial melting and crystal fractionation*

The hyperalkaline carbonatites and magmas (including kimberlites) are enigmatic witnesses of the deep carbon cycle linking diamonds in the Earth's crust to the volcanic emission of CO_2 and other gases (Green and Wallace 1988; Nelson et al. 1988; Hammouda and Keshav 2015; Foley and Fischer 2017). These two types of magma are often linked and are generally encountered in hotspot- or rift-type environments, as observed in extended intracontinental domains (Woolley and Kjarsgaard 2008; Goodenough et al. 2016). They are uncommon (approximately 5% of volcanism in volume), and their compositions and physical-chemical properties are part of the spectrum of conventional silicate magmas. They are particularly rich in volatile constituents (CO_2, H_2O, halogens and sulfides in order of abundance) and poor in silica (Le Bas 1981; Le Maitre 2005).

Another peculiarity of these magmas is the wide range of differentiation temperatures, ranging from more than 1,300°C to less than 500°C (Weidendorfer et al. 2017). The cold temperatures correspond to the only carbonatites currently emitted by the Doinyo Lengaï volcano in Tanzania (Keller and Krafft 1990). It was in the course of this particular magmatic differentiation that gigantic geochemical anomalies developed to the point of forming deposits of REEs and other rare metals. The formation of these deposits remains an exception, however. In fact, out of nearly 500 carbonatites listed globally, only five are exploited today for their exceptional tonnage and REE content (Verplanck et al. 2016). The reason behind such evolutions is debated. A deeper investigation on the formation modes and differentiation modes of these highly specific magmas is key.

Carbonatites are mostly formed by magmatic immiscibility: two liquids, one a silicate and the other a carbonate, coexist in equilibrium without mixing chemically, similar to oil and water (Kjarsgaard and Hamilton 1989; Kjarsgaard et al. 1995; Weidendorfer et al. 2017). Over the course of this immiscibility, the REEs are shared very variably between the two liquids, a process that remains poorly understood. Using experiments at high temperature and high pressure, recent works (Nabyl et al. 2020) modeled the behavior of REEs during the formation of immiscible carbonatites. Nabyl et al. (2020) demonstrate that REEs follow calcium and can be overconcentrated in carbonatites coexisting with fairly viscous phonolite-type liquids. On the other hand, although carbonatites do not form at the stage known as phonolitic, they focus on no or few REEs. The latter therefore remain in the silicate liquid, and given their incompatible behavior, the fractionated crystallization processes will be concentrated in highly differentiated magmas, such as those in hyperalkaline complexes.

Magmatic mechanisms are thus able to concentrate REEs highly in the carbonatites at levels comparable to the large, world-class deposits. It is very possible, however, that the final stages of the magmatism of carbonatites, in other words, degassing at the transition between magmatic and hydrothermal processes, are the root of the still poorly understood remobilization of REEs. Indeed, enrichments in REEs in carbonatites are often described as resulting from late magmatic and hydrothermal processes (Bühn et al. 2002; Nadeau et al. 2016). At this stage, hot aqueous fluids rich in halogens, phosphorus and sulfides are surely able to redistribute REEs from carbonatites. Today, decoding all these magmatic and hydrothermal processes remains the key to understanding the formation of these geochemical anomalies and suggesting “proxies”, that is, easy-to-use mineralogical indicators, which are nevertheless able to record and rebuild the magmatic and hydrothermal stages leading to the formation of a rare metal deposit.

1.7.2. *Chlorinity of hydrothermal systems*

The behavior of REEs in solution can be considered a scientific paradox. Since the 19th century and the first attempts at solvent extraction from rare ores such as *yttria* (i.e. gadolinite), cerite or even samarskite, it has been known that REEs are soluble in aqueous solutions (Mosander 1843; Marignac 1880). Until the second half of the 20th century, these REEs have been considered immobile by geologists (Menzies et al. 1979; Michard and Albarède 1986), even though the latter, contained in monazite or bastnäsite, were processed and separated by solvation in industrial applications.

Moreover, it seems that REEs can be mobile in hydrothermal fluids (Juteau et al. 1979; Wood 1990a, 1990b; Lottermoser 1992) and that the latter can mobilize REEs during the leaching of rocks and minerals to then concentrate them in many instances. These processes are at the origin of economic deposits in many contexts. They may involve the following:

– exsolved fluid phases of magmatic liquids such as Strange Lake in Canada (Gysi et al. 2016) or Lofdal in Namibia (Bodeving et al. 2017);

– but also, although rarer, fluids resulting from non-magmatic processes such as in the Browns Range dome in northwestern Australia (Nazari-Dehkordi et al. 2018) and the Maw Zone in Athabasca in Canada (Rabiei et al. 2017).

The growing role of REEs in energy transition involves the development of innovative technologies to extract these elements from so-called unconventional minerals (e.g. eudialyte, steenstrupine, gadolinite and allanite). This also reminds us of the understanding that is needed about the chemical mechanisms responsible for the transport of REEs in solution, their fractionation and concentration in the natural processes that lead to the genesis of unconventional economic concentrations of hydrothermal affiliation.

At the atomic scale, lanthanides are characterized by the gradual filling of the electron sublayer 4f, deeply buried and protected by the external layers 5 $s^2 5p^6$. This peculiarity makes them relatively insensitive to their chemical environment, hence their supposedly immobile behavior. Mainly present in trivalent state, Pearson's rule (1963) predicts their behavior in acid solutions, which allows them preferential interaction with hard inorganic ligands. The most stable are $F^- > OH^- > Cl^-$ when monovalent complexing agents are considered and $CO_3^{2-} > SO_4^{2-} > P_2O_5^{2-}$ for divalent ones. Recent work (Migdisov 2009, 2004, 2019) has made it possible to evaluate the main factors governing the transport and deposit of REEs in hydrothermal fluids. In particular, it has revealed the major role of Cl^-, as well as F^- in their complexation in solution and their deposition. Experiments including thermodynamic models suggest that the main ligands responsible for the transport of REEs in hydrothermal conditions are chloride and sulfate. On the other hand, the presence of fluoride considerably reduces the solubility of REEs in these hydrothermal solutions, contrary to all expectations. Indeed, the presence of F^- could facilitate the trapping of REEs and contribute to the effective precipitation of REE minerals.

The most effective precipitation mechanism would therefore involve neutralizing an acidic fluorinated fluid rich in REEs using carbonated formations of sedimentary or magmatic origin, which would, for example, trigger precipitation by saturation of REE fluorocarbonates such as bastnäsite. The addition of phosphate complexes would, however, contribute to the precipitation of monazite and/or xenotime. REEs transported in the form of sulfate complexes would be destabilized by the addition of barium and the precipitation of insoluble barite. These results nonetheless remain preliminary insofar as too little research has focused on the speciation in solution of carbonated complexes such as REE-hydrolated complexes.

1.7.3. *Exogenous rare earth cycle*

Exogenous REE deposits form in conditions called surface conditions, and result from interactions between rocks, biosphere and atmosphere. The rocks, like their constituent minerals, undergo chemical and physical transformations in contact with surface agents such as air, wind, water, ice and the living world. It makes sense to distinguish two types of exogenous deposits:

- those associated with regolith;
- more distal ones located in sedimentary basins.

In the environment of the regolith, reactions of hydrolysis, oxidation, hydration, or decarbonation trigger a chemical alteration of the rocks and minerals, in addition to physical phenomena (Freyssinet et al. 2005). It must be specified that the solubility of REEs in solution is moreover strongly dependent on the temperature (Wood 1990a, 1990b); they can be considered very poorly soluble, or indeed immobile, in ambient conditions. The time factor will also be preponderant in the genesis of exogenous REE deposits in aqueous conditions. Thus, enrichment in insoluble elements (Fe, Al, REE) is residual and strongly dependent – partially or wholly – on the leaching of soluble elements in very aggressive climates (Mg, Ca). The hydrodynamism of underground waters will allow the recombination of insoluble elements in neoformed minerals, mainly clays, hydrophosphates or carbonates.

In the ionic clay-type deposits often associated with the weathering of granites, REEs are adsorbed to the surface of halloysite/kaolinite-type neoformed clays

(Roaldset 1973; Chi and Tian 2008; Borst et al. 2020). The adsorption of REEs to the surface of clays is favored under pH neutral conditions (Coppin et al. 2002), such as those encountered in deposits at the interface between saprolite and laterite (Li et al. 2020). The character of ionic clay-type deposits, rich in LREEs or HREEs, can be explained by:

– effects linked to relief (i.e. topographic gradients) (Li et al. 2020);

– the morphology and stability of crystalline surfaces (Coppin et al. 2002);

– variations in the solubility of REEs in surface waters when these elements are complexified by inorganic anions (e.g. CO_3^{2-}, F^-), simple organic anions, or colloidal organo-mineral fractions (Tuduri et al. 2011; Rillard et al. 2019);

– adsorption/resorption phenomena at the surface of minerals (Pourret and Davranche 2013; Liu et al. 2017; Li et al. 2020). These numerous, very small deposits (several tens of thousands of tonnes) are widely exploited in China despite their (very) low contents between 0.03 and 0.35% REO (Chi and Tian 2008; Li et al. 2020).

Thus, in a laterite setting, the weathering of rocks initially rich in REEs (e.g. carbonatite, granite) will be able to produce secondary REE deposits that are even more fertile (Borst et al. 2020). In bauxitization settings, REEs are thus alternatively adsorbed and then desorbed in line with the progression of weathering toward the bottom. The neutralization of surface waters by carbonated formations can lead to the neoformation of hydroxylated bastnäsite in addition to REEs adsorbed to the surface of the clays (Mongelly 1997).

Finally, let us specify that recent research has shown that the processing of karst bauxites using the Bayer process makes it possible to concentrate REEs in waste produced from the manufacture of alumina: red mud (Deady et al. 2016).

Not all the minerals that form rocks, including those carrying REEs, present the same stability in relation to surface agents. Over the course of physical weathering, rocks fragment, disintegrate and the most resistant, insoluble minerals are taken up and leached by the wind and water. Placer-type deposits thus formed in sedimentary environments. These are accumulations of heavy minerals in sands and gravels separated by gravitational processes during their transportation by water and/or wind. With time and subsidence, diagenetic processes consolidated these placers, transforming them into paleoplacers. In fact, the placers are generally recent, of Cenozoic or Quaternary age, although some older ones date back to the Precambrian. The main REE minerals encountered are monazite, xenotime, fergusonite, euxenite and allanite (Donnot et al. 1973; Morteani 1991).

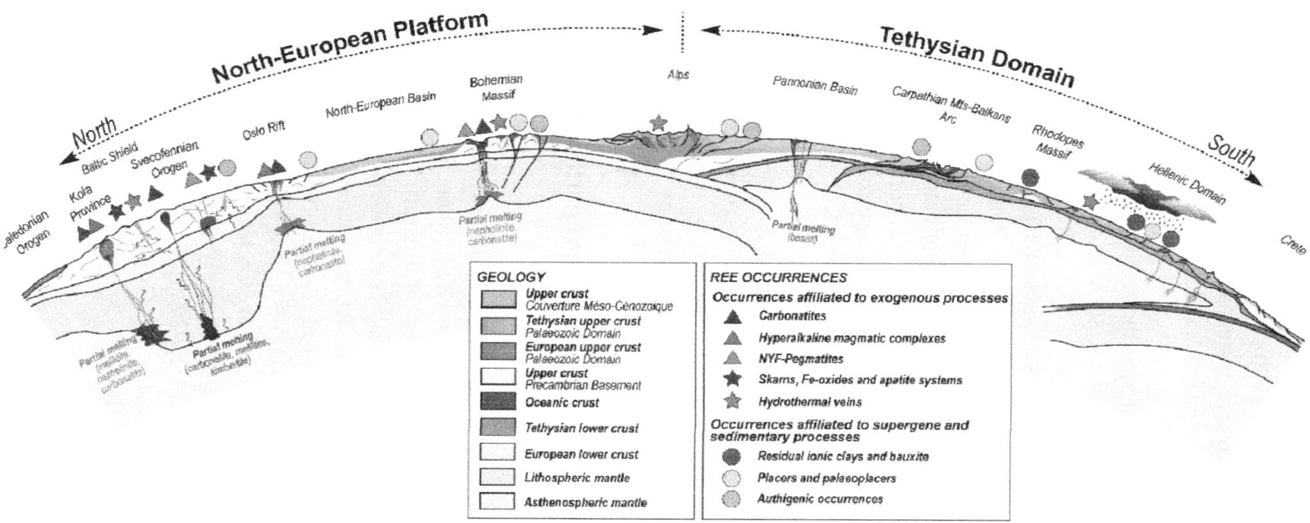

Figure 1.41. *Synthetic geodynamic section of Europe with the main types of REE deposits*

In a more distal fashion, the oceans are often considered to be in a stable chemical state, thus displaying an elementary balance maintained by input and output rates (Elderfield et al. 1990). However, the total loss of REEs – the quantity of REEs washed onto coastal plateaus and into the oceans – is underestimated (Rasmussen et al. 1998; Lacan and Jeandel 2005).

Recent work has demonstrated the major roles of the coarse colloidal organic fraction on the transport of REEs, such as that of the salinity gradient on their speciation in estuaries (Freslon et al. 2014; Rousseau et al. 2015). REEs are thus gradually eliminated through coagulation of this colloidal fraction, which therefore sediments the bottom of continental plateaus. Pourret and Tuduri (2017) suggest that the diagenesis of some silico-clastic rocks enriched in both organic matter and REEs can lead, under rising temperatures, to the formation of authigenic deposits of gray monazites characterized by a bell spectrum (i.e. enriched in average REEs). Another accumulation process, this time of microorganisms and algae, but in a marine environment, can also produce concentrations of phosphates called phosphorites under the action of diagenesis. These rocks, which are particularly rich in apatite, can in some settings include REEs, as well as uranium. High REE concentrations have also been indicated in manganese nodules, iron and manganese crusts (Kato et al. 2011; Bau et al. 2014; Josso et al. 2017), which are being studied for their economic potential.

1.8. Strengths and weaknesses of rare earths deposits in Europe and Greenland

Although Europe was the first region in the world where REEs were exploited (at Bastnäs-Bergslagen, Sweden), no REE mine is currently active on the Old Continent (except that of Lovozero in Russia). The political desire for better knowledge of the potential of mineral resources in Europe and Greenland strengthened toward the end of the 2000s. Awareness of China's hegemony in the mining of many commodities resulted in the European initiative on raw materials in 2008. The European Union (EU) now wished to secure its supplies in accordance with a list of strategic commodities (i.e. Critical Raw Materials, CRM) published in 2011, in which REEs feature prominently (HREEs and LREEs) in the last list published in 2020 (Blengini et al. 2020). Moreover, the EU imports 95% of the REEs it uses from China (Gislev and Grohol 2018). Some member states are also taking their own initiatives:

– France with COMES (Committee for Strategic Metals);

– Denmark with MiMa (Center for Minerals and Materials);

– the Netherlands with a government program on the circular economy.

The EU is also co-financing networks and expert platforms on mineral resources with:

– ERA-MIN (Network on the Industrial Handling of Raw Materials for European Industries);

– ERECON (European Rare Earths Competency Network);

– EIP (European Innovation Partnership);

– ETP (European Technology Platform on Sustainable Mineral Resources);

– AdWG (Ad hoc Working Group);

– SCRREEN (Solutions for Critical Raw Materials – A European Experts Network);

– research projects and databases with EURARE (Development of a sustainable exploitation scheme for Europe's Rare Earth ore deposits), Mine-rals4EU or Promine.

All these initiatives have made it possible in the past few years to improve knowledge of the REE potential of Europe and Greenland (Ahonen et al. 2015; Goodenough et al. 2016; Machacek and Kalvig 2016; Balomenos et al. 2017; Gislev and Grohol 2018; Lauri et al. 2018).

Europe and Greenland have undeniable REE potential, and the few deposits already studied and enjoying a NI43-101 or JORC compatible resource estimate (Norra Kärr, Kvanefjeld, Kringlerne, Motzfeldt, Olserum, Sarfartoq, Fen and Delitzsch-Storkwitz) demonstrate that the Old Continent has no reason to envy other parts of the globe and could indeed be self-sufficient (Figure 1.42).

Indeed, it is important to note that in the current state of knowledge of global REE reserves, Europe represents merely less than 1% of these estimated reserves (Figure 1.43). These estimates rely on data from deposits and use, and on worldwide feasibility studies for different types of deposits. In 2015, depending on higher or lower estimates, global reserves formed between 137 and 375 Mt REO, and China had an important place here. Where Greenland is concerned, this territory, affiliated to Denmark, includes between 8 and 11% of global REE reserves estimated in 2015 (Figure 1.43).

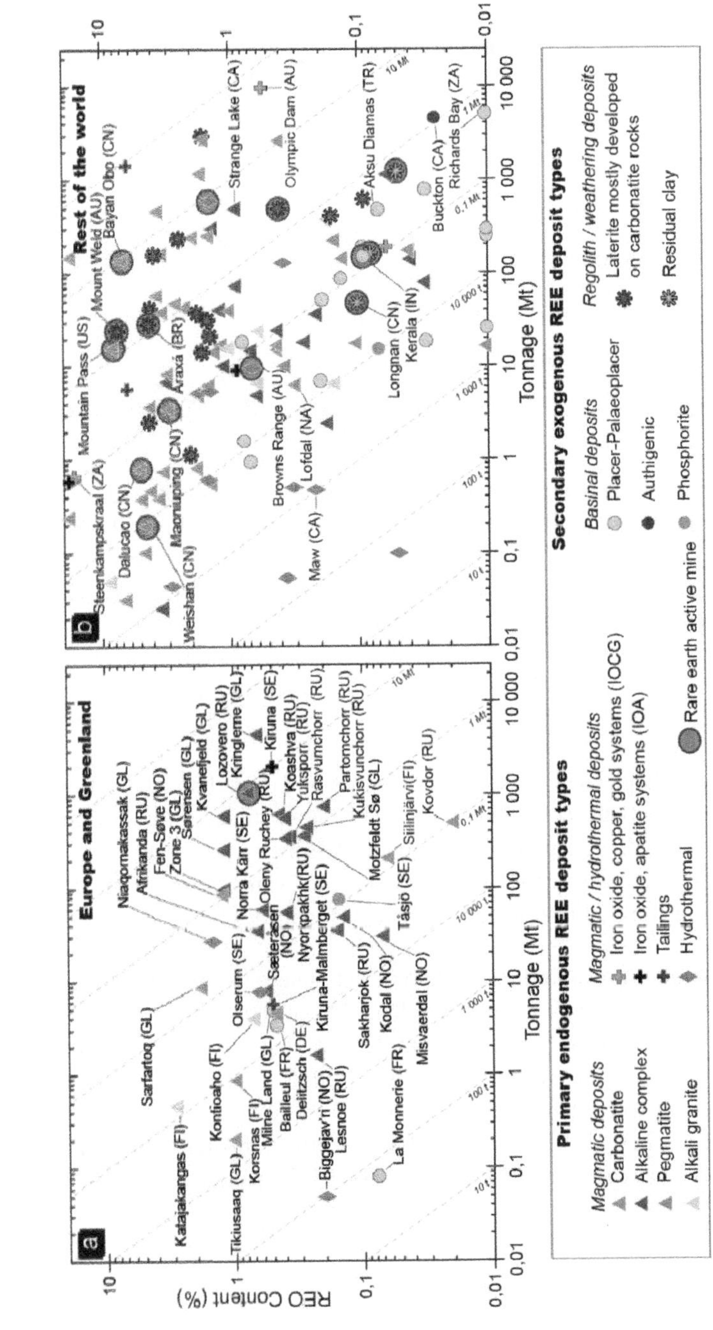

Figure 1.42. *Grade/tonnage relationship for the different types of localized rare earth deposits*

COMMENT ON FIGURE 1.42.– *(a) In Europe and Greenland and (b) in the rest of the world. Abbreviations: AU (Australia), BR (Brazil), CA (Canada), CN (China), DE (Germany), FI (Finland), FR (France), GL (Greenland), GR (Greece), IN (India), KE (Kenya), NA (Namibia), NO (Norway), RU (Russia), SE (Sweden), TR (Turkey), US (United States), ZA (South Africa). Data for Europe, Greenland (for this study, see the table of deposits in Appendix 1) and the rest of the world are presented (Weng et al. 2015; Zhou et al. 2017 and references therein).*

The geodynamic settings in which REE deposits were emplaced are numerous, but only some settings make it possible to concentrate REEs in economically beneficial quantities (Figures 1.41 and 1.42). Thus, the main REE resources are associated with alkaline complexes and with carbonatites that were formed in the setting of intracontinental rifting (Chakhmouradian 2012b; Wall 2014; Goodenough et al. 2016). This rule also applies to Europe and Greenland (Gardar in Greenland, the Svecofennian in Scandinavia).

Other occurrences of REEs have also been highlighted, such as heavy mineral placers (Greece, France, Germany, Spain) or the laterite bauxite deposits of the Mediterranean belt (Charles et al. 2013; Goodenough et al. 2016; EURARE 2017; Mondillo et al. 2019).

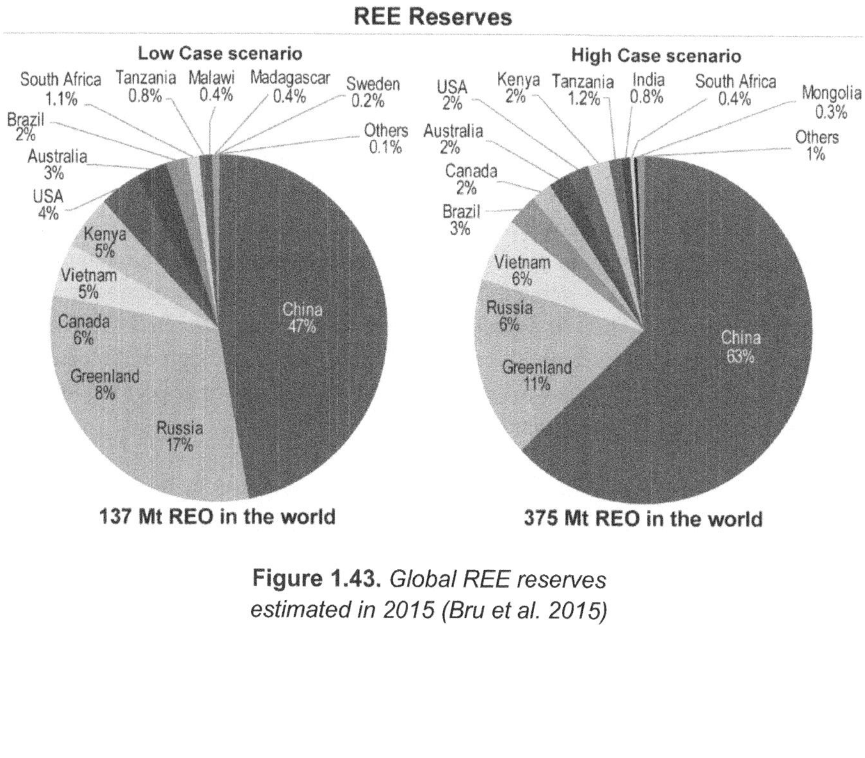

Figure 1.43. *Global REE reserves estimated in 2015 (Bru et al. 2015)*

Moreover, the many other REE occurrences show that further exploration could increase Europe's and Greenland's known potential. Beyond primary deposits, REEs can also be the subject of by-products resulting from the re-processing of mining waste. Although REEs are less abundant than in end-of-life electronic products, the volumes of mining waste are much greater. The best example is that of the Kiruna-Malmberget iron mines, the tailings of which contain 5.6 Mt of apatite with 0.5% REO.

The main question hinges, of course, on the economic viability of REE exploitation in Europe and Greenland. Indeed, there are too few economic studies available on the most promising deposits and on the possibility of comparing them with other deposits in the world via the use of international standards (e.g. NI43-101, JORC). On the other hand, there are few companies that wish to take the risk of investing in exploration, given the fairly unpredictable setting of REEs (global demand, supply, price (Bru et al. 2015)). In addition, the very substantial investment needed to develop innovative techniques with a view to processing different ores depending on the geology of the deposits is a major restriction. The key point is to be able to separate the different REEs, while still respecting a balance between the economic cost and the environmental impact. The environment is indeed very important to European citizens, those who are increasingly consuming REEs in the form of new digital technologies, but also in the form of the "ecological" and energy transition (Machacek and Kalvig 2016; Sebastiaan et al. 2017).

The quality of the deposit remains essential and is guided by its geology. A clever balance has to be found between a tonnage and a grade, between the proportion of HREEs and that of LREEs, between the REE-bearing minerals from which useful commodities can be extracted without investing colossal sums in R&D, or between the other related commodities (e.g. Nb, U, Zr, Ta, P). The geology is also independent of communication routes; although a deposit may present a tonnage and an interesting grade, if it is geographically difficult to access, the economic stakes may drop markedly. This is especially the case for the immense deposits in Greenland, where infrastructure is very often absent: roads, to begin with, electrical installations, housing and bases for miners to live in, a deep-water port for logistics and, if necessary, a processing plant.

A challenge also lies in finding local, qualified manpower without bringing in too many external staff. In Greenland, the hypothesis of the massive influx of Chinese workers is a sensitive subject for this country, which has 56,000 inhabitants and is half the size of the EU. In fact, the development of Kvanefjeld reached an important milestone in 2018 under the impetus of Chinese support and the growing interest of the Middle Empire in Greenland (Andersson et al. 2018a). Shenghe

Resources Holding Co. Ltd. is the main shareholder in the Australian junior, Greenland Minerals, which is developing the project. In January 2019, the formation of a joint venture between Shenghe and China National Nuclear Corp was announced. It was entrusted with the import, export and trading of REOs and U_3O_8. An agreement was made to take 32 kt/year of concentrate, as well as the chance for Shenghe to become the main shareholder (60% of shares). The last obstacle involved requests for mining permits.

The inhabitants of Greenland and Europe also increasingly consider it important to respect the environment. Like any industrial activity, the mining industry has an impact on the environment. Nevertheless, the application of good practices and the existence of a strict regulatory framework in these regions of the world may allow mines to be exploited while respecting the environment and populations, and in particular, acquiring the SLO Social Licence to Operate and at least the approbation of all stakeholders. Thus, although it appears promising, the Norra Kärr project (Sweden) is still awaiting environmental authorization from the Swedish government. The exploitation of an REE deposit can lead to a substantial impact on the environment and on individuals if certain conditions are not respected: gaseous effluent linked to the use of acids when processing ores, management of tailings, which can impact the hydrosphere and engenders risks of dam failure, management of radioactive elements that can be associated with REEs, the impact of exploitation on the countryside, etc. Although techniques and good practices are known and the legislative framework is already very restrictive in Europe and Greenland, public opinion is becoming increasingly hostile to mining activity. This situation is broadly contradictory, between, on the one hand, the need for a whole continent to ensure its supply of a raw material on which it is very much dependent and, on the other hand, a population that wishes to benefit from a living environment associated with the latest technologies without accepting some downsides. For now, the increasing power of technologies known as green technologies and the policy on the European scale to move toward a carbon-free economy through a Green Pact (EC 2019a, 2019b) do not seem to resolve the situation, which also depends largely on global-scale challenges.

Another sizable challenge is the recycling of REEs on a European scale. Recent figures demonstrate that REEs remain very little recycled: 0% for Dy, Er (and Sc); 1% for La, Ce, Nd, Sm, Gd, Ho, Tm, Yb and Lu; 10% for Pr; 22% for Tb; 38% for Eu and 31% for Y (Gislev and Grohol 2018). Application sectors seem to be a priority for recycling REEs: permanent magnets (Nd, Pr, Dy, Tb, Sm), phosphors (Eu, Tb, Y, Ce, Gd, La), batteries (La, Ce, Nd, Pr), polishing components (Ce) and catalysts (La, Ce, Pr, Nd, Y; Ahonen et al. 2015).

1.9. Conclusion

The "vitamins of modern industry" have come a long way since 1794. REEs are already used in numerous domains, both current, cutting edge ones, and those to come, since their substitutability is often synonymous with a loss of performance and increased investment in the face of recycling, which is still limited and costly, while eco-design products are still not widespread. The European Union has defined REEs as critical and strategic metals for its economy. With a view to the independence of its supply, the Old Continent has benefitted from a panorama of its geological potential in REEs through diverse projects, initiatives and expert networks over recent years. The geological potential of the REEs of Europe and Greenland is situated in intracratonic areas associated with old rifting zones, which have given rise to the intrusion of alkaline complexes (Scandinavia, Gardar) (Figures 1.10, 1.35, 1.41 and 1.42). The rare deposits benefitting from an estimate of resources compatible with international standards demonstrate their global size and their capacity to palliate the continent's need for at least some decades.

Nevertheless, if Europe wishes to increase its independence, a voluntary and assumed investment in R&D remains indispensable to develop ore processing techniques adapted to the REE-bearing minerals specific to the European and Greenland deposits. It is by developing this competency in the extraction and separation of these metals that Europe will be able to secure its supplies and develop a real value chain for REEs. Beyond this technical-economic engagement, Europe and Greenland must also gain the confidence and approbation of stakeholders in the extraction industry and in civil society. The societal, social and environmental challenges are in fact as important, and will be even more so in the future, as the technical-economic challenges. Although the geology is favorable to REEs in Europe and Greenland, this is still not entirely the case with regard to procuring a social license to operate from European citizens.

1.10. Acknowledgments

This summary was the result of the ANR-11-ECOT-002 ASTER "*Analyse Systémique des rare earths–Flux et Stocks*" (Systematic Analysis of Rare Earths and Elements – Flows and Stocks) project and a collaboration with the team from the EURARE project since 2014. The authors are keen to thank the BRGM for its support, as well as Dr. Decrée, the editor of this work, for her help in reviewing this chapter.

1.11. References

Abdelnour, S.A., El-Hack, M.E.A., Khafaga, A.F., Noreldin, A.E., Arif, M., Chaudhry, M.T., Losacco, C., Abdeen, A., Abdel-Daim, M.M. (2019). Impacts of rare earth elements on animal health and production: Highlights of cerium and lanthanum. *Sci. Total Environ.*, 672, 1021–1032.

Adamson, O.J. (1944). The petrology of the Norra Kärr district: An occurrence of alkaline rocks in southern Sweden. *GFF*, 437, 113–255.

Agard, J. (1975). Évaluation des ressources minerales françaises, les ressources en bauxite de la France. Report, BRGM, Orléans.

Ager, D.V. (1980). *The Geology of Europe*. McGraw-Hill, Maidenhead.

Ahonen, H., Arvanitidis, N., Auer, A., Baillet, E., Bellato, N., Binnemans, K., Blengini, G.A., Bonato, D., Brouwer, E., Brower, S. et al. (2015). Strengthening the European rare earths supply-chain. Challenges and policy options. Report, The European Rare Earths Competency Network.

Äikäs, O. (1990). Torium-ja niobi-lantanidimalmiaiheet Vuolijoen Otanmäessä. Report, Geological Survey of Finland, Espoo.

Al-Ani, T. and Sarapää, O. (2009). Rare earth elements and their mineral phases in Jammi carbonatite veins and fenites on the south side of Sokli carbonatite complex, NE Finland. Report, Geological Survey of Finland, Espoo.

Al-Ani, T. and Sarapää, O. (2013). Geochemistry and mineral phases of REE in Jammi carbonatite veins and fenites, southern end of the Sokli complex, NE Finland. *Geochemistry: Exploration, Environment, Analysis*, 13, 217–224.

Al-Ani, T., Konnunaho, K., Sarapää, O. (2011). Microprobe analysis on REE-minerals of Enontekiö (Palkiskuru and Palovaara) and Kuusamo (Hoinkilehto) hydrothermal mineralizations and Kortejärvi-Laivajoki carbonatites, Northern Finland. Report, Geological Survey of Finland, Espoo.

Alkane Resources Ltd. (2020). Export finance Australia confirms interest in Dubbo project financing [Online]. Available at: https://investors.alkane.com.au/site/PDF/c905578e-e9e1-4e05-9daf-0e82adeabbad/EFAConfirmsInterestinDubboProjectFinancing.

Andersen, O. (1926). *Feltspat I. Norges Geologiske Undersøkelse*. Aschehoug, Oslo.

Andersen, T. (1986). Magmatic fluids in the Fen carbonatite complex, SE Norway: Evidence of mid-crustal fractionation from solid and fluid inclusions in apatite. *Contributions to Mineralogy and Petrology*, 93, 491–503.

Andersen, T. (1988). Evolution of peralkaline calcite carbonatite magma in the Fen complex, southeast Norway. *Lithos*, 22, 99–112.

Andersen, T. (1996). Sr, Nd and Pb isotopic data of the Alnö carbonatite complex. Abstract volume. *22nd Nordic Geological Winter Meeting*, Reykjavik.

Andersen, T. (1997). Age and petrogenesis of the Qassiarsuk carbonatite-alkaline silicate volcanic complex in the Gardar rift, South Greenland. *Mineralogical Magazine*, 61, 499–513.

Andersen, T. and Seiersten, M. (1994). Deep cumulates in a shallow intrusion: Origin and crystallization history of a pyroxenite (jacupirangite s.l.) body in the Larvik Pluton, Oslo Region, South Norway. *Neues Jb. Miner. Monat.*, 6, 255–274.

Andersen, S., Bohse, H., Steenfelt, A. (1988). The southern part of the Ilímaussaq complex South Greenland 1:20,000. Report, Geological Survey of Greenland, Copenhagen.

Andersen, T., Andresen, A., Sylvester, A.G. (2002). Timing of late- to post-tectonic Sveconorwegian granitic magmatism in South Norway. *Norwegian Geological Survey Bulletin*, 440, 5–18.

Andersson, U.B. and Öhlander, B. (2004). The late Svecofennian magmatism. The Transscandinavian Igneous Belt (TIB) in Sweden. *Geological Survey of Finland, Special Papers*, 37, 102–104.

Andersson, U.B., Holtstam, D., Broman, C. (2013). Additional data on the age and origin of the Bastnäs-type REE deposits, Sweden. Mineral deposit research for a high-tech world. *Proceedings of the 12th Biennial SGA Meeting 4*, Upssala.

Andersson, P., Zeuthen, J.W., Kalvig, P. (2018a). Chinese mining in Greenland: Arctic access or access to minerals? In *Arctic Yearbook 2018*, Heininen, L. and Exner-Pirot, H. (eds). Arctic Portal [Online]. Available at: https://arcticyearbook.com/arctic-yearbook/2018.

Andersson, S.S., Wagner, T., Jonsson, E., Fusswinkel, T., Leijd, M., Berg, J.T. (2018b). Origin of the high-temperature Olserum-Djupedal REE-phosphate mineralization, SE Sweden: A unique contact metamorphic-hydrothermal system. *Ore Geology Reviews*, 101, 740–764.

Armands, G. (1964). Geologiska undersökningar i Tåsjö-området under 1963 och 1964. Report, AB Atomenergi KOP-102.

Armands, G. (1970). A uranium-bearing layer from the Lower Ordovician, Sweden. *Geologiska Föreningens i Stockholm Förhandlingar*, 92, 481–490.

Arzamastsev, A.A., Bea, F., Arzamastseva, L.V., Montero, P. (2005). Proterozoic Gremyakha-Vyrmes polyphase massif, Kola Peninsula: An example of mixing basic and alkaline mantle melts. *Petrology*, 14, 361–389.

Arzamastsev, A.A., Yakovenchuk, V., Pakhomovsky, Y., Ivanyuk, G. (2008). The Khibina and Lovozero alkaline massifs: Geology and unique mineralization. Field Guide Book, Geological Institute of the Russian Academy of Science, Apatity.

Asch, K. (2003). The 1:5 million international geological map of Europe and adjacent areas: Development and implementation of a GIS-enabled concept. Geologisches Jahrbuch, Hannover.

Atwood, D.A. (2012). *The Rare Earth Elements: Fundamentals and Applications. Encyclopedia of Inorganic and Bioinorganic Chemistry*. John Wiley & Sons, New York.

Auboin, J. (1980). Geology of Europe: A synthesis. *Episodes*, 1, 3–8.

Avannaa Resources Ltd. (2010). Karrat Rare Earth Project, West Greenland [Online]. Available at: www.avannaa.com.

Baadsgaard, H., Chaplin, C., Griffin, W.L. (1984). Geochronology of the Gloserheia pegmatite, Froland, southern Norway. *Norsk Geol. Tidsskr*, 2, 111–119.

Balaram, V. (2019). Geoscience frontiers rare earth elements: A review of applications, occurrence, exploration, analysis, recycling, and environmental impact. *Geoscience Frontiers*, 10, 1285–1303.

Balomenos, E., Davris, P., Deady, E., Yang, J., Panias, D., Friedrich, B., Binnemans, K., Seisenbaeva, G., Dittrich, C., Kalvig, P. et al. (2017). The EURARE project: Development of a sustainable exploitation scheme for Europe's rare earth ore deposits. *Johnson Matthey Technol. Rev.*, 61, 142–153.

Bao, Z. and Zhao, Z. (2008). Geochemistry of mineralization with exchangeable REY in the weathering crusts of granitic rocks in South China. *Ore Geol. Rev.*, 33, 519–535.

Bárdossy, G. (1982). *Karst Bauxites. Bauxite Deposits on Carbonate Rocks*. Elsevier, Amsterdam.

Bárdossy, G. and Dercourt, J. (1990). Les gisements de bauxites téthysiens (Méditerranée, Proche et Moyen-Orient) ; cadre paléogeographique et contrôles génétiques. *Bulletin de la Société Géologique de France*, 7(6), 869–888.

Bárdossy, G. and Pantó, G. (1973). Trace mineral and element investigation on bauxites by electron probe. *3e Congrès ICSOBA*. Comité International pour l'Étude des Bauxites, de l'Alumine et de l'Aluminium, Nice.

Bárdossy, G., Pantó, G., Várhegyi, G. (1976). Rare metals of Hungarian bauxites and conditions of their utilization. *Travaux ICSOBA*. Comité International pour l'Étude des Bauxites, de l'Alumine et de l'Aluminium.

Bartels, A. (2013). The Gardar province. In *Geological Excursion to South Greenland. Excursion Book for the 12th Biennal SGA Meeting*, Kolb, J. (ed.). SGU, Uppsala.

Barth, T.W.W. and Ramberg, I.B. (1966). The Fen circular complex. In *Carbonatites*, Tuttle, O.F. and Gittins, J. (eds). John Wiley and Sons, New York.

Barton, M.D. and Johnson, D.A. (1996). Evaporitic-source model for igneous-related Fe oxide-(REE-Cu-Au-U) mineralization. *Geology*, 24, 259–262.

Batieva, I.D. and Bel'kov, I.V. (1984). *The Sakharjok Alkaline Massif, Its Rocks and Minerals*. Kola Branch of the Academy of Sciences of the USSR, Apatity.

Bau, M., Schmidt, K., Koschinsky, A., Hein, J., Kuhn, T., Usui, A. (2014). Discriminating between different genetic types of marine ferro-manganese crusts and nodules based on rare earth elements and yttrium. *Chemical Geology*, 381, 1–9.

Bayanova, T.B., Kirnarskii, Y.M., Levkovich, N.V. (1997). U-Pb dating of baddeleyite from rocks of the Kovdor massif. *Trans. Russ. Acad. Sci., Earth Sci. Sec.*, 356, 1094.

Bedini, E. (2009). Mapping lithology of the Sarfartoq carbonatite complex, southern West Greenland, using HyMap imaging spectrometer data. *Remote Sensing of the Environment*, 113, 1209–1219.

Belolipetskyi, A.P., Britvin, S.N., Voloshin, A.V., Gordienko, V.V., Zozulya, D.R., Kozyreva, L.V., Kulakov, A.N., Mets, O.F., Petrov, S.I., Subbotin, V.V. et al. (1992). *Rare Metals Mineralization of the Kola Metallogenic Province*. Kola Branch of the Academy of Sciences of the USSR, Apatity.

Berger, A., Kokfeldt, T.F., Kolb, J. (2014). Exhumation rates in the Archean from pressure-time paths: Example from the Skjoldungen Orogen (SE Greenland). *Precambrian Research*, 255(3), 774–490.

Bergstøl, S. (1972). The jacupirangite at Kodal, Vestfold, Norway. *Mineralium Deposita*, 7, 233–246.

Berl, R. (1996). Lazulith in Österreich, Teil 1. Bemerkungen zu einigen Lazulithfundgebieten von Niederösterreich und Steiermark. *Der Steirische Mineralog*, 10, 5–14.

Bernhard, F., Walter, F., Ettinger, K., Taucher, J., Mereiter, K. (1998). Pretulite, $ScPO_4$, a new scandium mineral from the Styrian and Lower Austrian lazulite occurrences, Austria. *American Mineralogist*, 83, 625–630.

Bingen, B., Nordgulen, Ø., Viola, G. (2008). A four phase model for the Sveconorwegian orogeny, SW Scandinavia. *Norwegian Journal of Geology*, 88, 43–72.

Birkelund, T. and Callomon, J.H. (1985). The Kimmeridgian ammonite faunas of Milne Land, central East Greenland. *Bulletin Grønlands Geologiske Undersøgelse*, 153, 56.

Birkelund, T., Callomon, J.H., Fürsich, F.T. (1978). The Jurassic of Milne Land, central East Greenland. *Grønlands Geologiske Undersøgelse*, 90, 99–106.

Birkelund, T., Callomon, J.H., Fürsich, F.T. (1984). The stratigraphy of the Upper Jurassic and Lower Cretaceous sediments of Milne Land, central East Greenland. *Bulletin Grønlands Geologiske Undersøgelse*, 147, 56.

Bizzarro, M., Simonetti, A., Stevenson, R.K., David, J. (2002). Hf isotope evidence for a hidden mantle reservoir. *Geology*, 30(9), 771–774.

Bjørlykke, H. (1937). The granite pegmatites of southern Norway. *American Mineralogist*, 22, 241–255.

Blaxland, A.B. (1977). Agpaitic magmatism at Norra Kärr, Rb-Sr isotopic evidence. *Lithos*, 10, 1–8.

Blaxland, A.B., van Breemen, O., Emeleus, C.H., Anderson, J.G. (1978). Age and origin of the major syenite centers in the Gardar Province of South Greenland: Rb-Sr studies. *Geological Society of America Bulletin*, 89, 231–244.

Blengini, G.A., Latunussa, C.E.L., Eynard, U., Torres de Matos, C., Wittmer, D., Georgitzikis, K., Pavel, C., Carrara, S., Mancini, L., Unguru, M. et al. (2020). Study on the EU's list of critical raw materials. Report, European Commission. doi: 10.2873/11619.

Blichert-Toft, J., Rosing, M.T., Lesher, C.E., Chauvel, C. (1995). Geochemical constraints on the origin of the Late Archean Skjoldungen alkaline igneous province, SE Greenland. *Journal of Petrology*, 36, 515–561.

Bodeving, S., Williams-Jones, A.E., SwIndian, S. (2017). Carbonate–silicate melt immiscibility, REE mineralising fluids, and the evolution of the Lofdal Intrusive Suite, Namibia. *Lithos*, 268–271, 383–398.

Bohse, H. and Andersen, S. (1981). Review of the stratigraphic divisions of the kakortokite and lujavrite in southern Ilímaussaq. *Grønlands Geologiske Undersøgelse*, 103, 53–62.

Bohse, H., Brooks, C.K., Kunzendorf, H. (1971). Field observations on the kakortokites of the Ilimaussaq intrusion, South Greenland, including mapping and analyses by portable X-ray fluorescence equipment for zirconium and niobium. Report, GEUS.

Bollingberg, H., Hopgood, A.M., Kalsbeek, F. (1976). Some minor and trace elements in Archaean marbles and metamorphosed silico-carbonatites from the Fiskenæsset region. *Grønlands Geologiske Undersøgelse*, 73, 86–90.

Bonazzi, P., Bindi, L., Parodi, G. (2003). Gatelite-(Ce), a new REE-bearing mineral from Trimouns, French Pyrenees: Crystal structure and polysomatic relationships with epidote and törnebo-hmite-(Ce). *American Mineralogist*, 88, 223–228.

Borst, A.M., Waight, T., Smit, M., Friis, H., Nielsen, T. (2014). Alteration of eudialyte and implications for the REE, Zr and Nb resources of the layered kakortokites in the Ilimaussaq intrusion, South West Greenland. *1st European Rare Earth Resources Conference*, Milos.

Borst, A.M., Friis, H., Andersen, T., Nielsen, T., Waight, T., Smit, A. (2016). Zirconosilicates in the kakortokites of the Ilímaussaq complex, South Greenland: Implications for fluid evolution and high-field-strength and rare-earth element mineralization in agpaitic systems. *Mineralogical Magazine*, 80, 5–30.

Borst, A.M., Smith, M.P., Finch, A.A., Estrade, G., Villanova-de-Benavent, C., Nason, P., Marquis, E., Horsburgh, N.J., Goodenough, K.M., Xu, C. et al. (2020). Adsorption of rare earth elements in regolith-hosted clay deposits. *Nature Communications*, 11(1), 4386.

Bottke, H. (1981). *Lagerstättenkunde des Eisens – Geochemie, Genese, Typengliederung, wirtschaftl.* Verlag Glückauf, Essen.

Bouch, J.E., Naden, J., Shepherd, T.J., Young, B., Benham, A.J., Mckervey, J.A., Sloane, H.J. (2008). Stratabound Pb-Zn-Ba-F mineralization in the Alston Block of the North Pennine Ore-field (England) – Origins and emplacement. Report, British Geological Survey.

Brandenberger, E. (1931). Die Kristallstruktur von Koppit, Zeitsch. *Krist.*, 76, 322–334.

Brooks, C.K. and Nielsen, T.F.D. (1982). The Phanerozoic development of the Kangerdlugssuaq area, East Greenland. *Meddelelser om Grønland Geoscience*, 9, 31.

Brown, G.M. (1963). The Skye granites. *Mineralogical Magazine*, 33, 533–562.

Browne, A. (2008). Report on current resource estimates for Kläppicäcken and Duobblon uranium properties, and review of Tåsjö uranium project, northern Sweden. Report, Mawson Resources.

Bru, K., Christmann, P., Labbé, J.F., Lefebvre, G. (2015). Panorama mondial 2014 du marché des terres rares. Report, BRGM RP-64330-FR.

Brueckner, H.K. and Rex, D.C. (1980). K-Ar and Rb-Sr geochronology and Sr isotopic study of the Alnö alkaline complex, northeastern Sweden. *Lithos*, 13, 111–119.

Bühn, B., Rankin, A.H., Schneider, J., Dulski, P. (2002). The nature of orthomagmatic, carbonatitic fluids precipitating REE, Sr-rich fluorite: Fluid-inclusion evidence from the Okorusu fluorite deposit, Namibia. *Chemical Geology*, 186(1), 75–98.

Burnotte, E., Pirard, E., Michel, G. (1989). Genesis of grey monazites: Evidence from the Palaeozoic of Belgium. *Economic Geology*, 84, 1417–1429.

Calderón, S. (1910). *Los Minerales de España*. Eduardo Anas, Madrid.

Callisen, K. (1943). Igneous rocks of the Ivigtut region, Greenland. Part 1: The nepheline syenites of the Grønnedal-Ika area. *Meddelelser om Grønland*, 131(8), 74.

Campbell, L.S., Henderson, P., Wall, F., Nielsen, T.F. (1997). Rare earth chemistry of perovskite group minerals from the Gardiner Complex, East Greenland. *Mineralogical Magazine*, 61, 197–212.

Cann, J.R. and Banks, D.A. (2001). Constraints on the genesis of the mineralization of the Alston Block, Northern Pennine Orefield, northern England. *Proceedings of the Yorkshire Geological Society*, 53, 187–196.

Carlborg, H. (1923). Ekonomisk-teknisk beskrivning. In *Riddarhytte malmfält i Skinnskattebergs socken i Västmanlands län. Kungl. Kommerskollegium and Sveriges Geologiska Undersökning*. Victor Pettersson, Stockholm.

Castor, S.B. and Hedrick, J.B. (2006). Rare earth elements. In *Industrial Minerals and Rocks*, Kogel, J.E., Trivedi, N.C., Barker, J.M., Krukowski, S.T. (eds). Society for Mining, Metallurgy and Exploration, Englewood.

Černý, P. and Ercit, T.S. (2005). The classification of granitic pegmatites revisited. *Canadian Mineralogist*, 43, 2005–2026.

Chakhmouradian, A.R. and Wall, F. (2012a). Rare earth elements: Minerals, mines, magnets (and more). *Elements*, 8, 333–340.

Chakhmouradian, A.R. and Zaitsev, A.N. (2012b). Rare earth mineralization in igneous rocks: Sources and processes. *Elements*, 8, 347–353.

Charles, N., Tuduri, J., Guyonnet, D., Melleton, J., Pourret, O. (2013). Rare earth elements in Europe and Greenland: A geological potential? An overview. *Proceedings of the 12th Biennial SGA Meeting*, Uppsala.

Chi, R. and Tian, T. (2008). *Weathered Crust Elution-deposited Rare Earth Ores*. Nova Science Publishers Inc., New York.

Choubert, G. and Faure-Muret, A. (1976). Sheet 9 Europe. Scale 1:10,000,000. With explanatory notes by G. Luetting and F. Delany. Report, Geological Atlas of the World, UNESCO and Commission for the Geological Map of the World.

Christensson, U. (2013). Characterising the alteration of the contact to the Norra Kärr alkaline complex, southern Sweden. BA Thesis, University of Gothenburg.

Clausen, S. and Alvaro, J.J. (2007). Lower Cambrian shelled phosphorites from the northern Montagne Noire, France. *Geological Society of London*, 275, 17–28.

Cliff, R.A., Rickard, D., Blake, K. (1990). Isotope systematics of the Kiruna magnetite ores, Sweden. Part 1: Age of the ore. *Economic Geology*, 85, 1770–1776.

Cloethingh, S., van Wees, J.D., Ziegler, P.A., Lenkey, L., Beekman, F., Tesauro, M., Förster, A., Norden, B., Kaban, M., Hardebol, N. et al. (2010). Lithosphere tectonics and thermo-mechanical properties: An integrated modeling approach for enhanced geothermal systems exploration in Europe. *Earth-Science Reviews*, 102, 159–206.

Cockburn, A.M. (1935). The geology of St Kilda. *Trans. Roy. Soc. Edinburgh*, 58, 511–547.

Cooper, D.C., Bashan, I.R., Smith, T.K. (1983). On the occurrence of an unusual form of monazite in panned stream sediments in Wales. *Geological Journal*, 18, 121–127.

Cooper, D.C., Rollin, K., Howells, M.F., Morgan, D.J. (1984). Regional geochemical and geophysical surveys in the Berwyn dome and adjacent areas, North Wales. Mineral Reconnaissance Programme Report, British Geological Survey.

Coppin, F., Berger, G., Bauer, A., Castet, S., Loubet, M. (2002). Sorption of lanthanides on smectite and kaolinite. *Chemical Geology*, 182(1), 57–68.

Cornelius, H.P. (1931). Neue Lazulithfunde im Mürztal. *Verhandlungen der Geologischen Bundesanstalt*, 4, 93–94.

Cornet, A. and Sapinart, J. (1967). Monazite à europium armoricaine. Travaux préliminaires au test d'exploitation envisagé à La Monnerie, sur le ruisseau d'Aron (Ille-et-Vilaine). Report, BRGM, DRMM-67, A-8.

Dahlgren, S. (2005). Miljøgeologisk undersøkelse av lavradioaktivt slagg fra ferroniobproduksjon ved Norsk bergverk på Søve 1956-1965. Report, Regiongeologen.

Dahlgren, S., Corfu, F., Heaman, L.M. (1996). U-Pb isotopic time constraints, and Hf and Pb source characteristics of the Larvik plutonic complex, Oslo palaeorift. Geodynamic and geochemical implications for the rift evolution. *Journal of Conference Abstracts*, 1, 120.

Damtoft, J.S. and Grahl-Madsen, L. (1982). Geological and VLF geophysical mapping of the Bredhorn Ba-Pb prospect and preliminary mapping on SE Traill Ø for porphyry molybdenum mineralisation. Internal NM report 2/81.

Deady, E.A., Mouchos, E., Goodenough, K., Williamson, B.J., Wall, F. (2016). A review of the potential for rare-earth element resources from European red muds: Examples from Seydişehir, Turkey and Parnassus-Giona, Greece. *Mineralogical Magazine*, 80, 43–61.

Debrand-Passard, S., Courbouleix, S., Lienhardt, M.J. (1984). Synthèse géologique du sud-est de la France. Report, BRGM.

Decrée, S., Ihlen, P.M., Schiellerup, H., Hallberg, A., Demetriades, A., Raha, M., Soesoo, A. (2017). Potential of phosphate deposits in Europe. *SGA News*, 41, 1–20.

Decrée, S., Savolainen, M., Mercadier, J., Debaille, V., Höhne, S., Frimmel, H., Baele, J.M. (2020). Geochemical and spectroscopic investigation of apatite in the Siilinjärvi carbonatite complex: Keys to understanding apatite forming processes and assessing potential for rare earth elements. *Applied Geochemistry*, 123, 104778.

Dickin, A.P., Mootbath, S., Welke, H.J. (1981). Isotope, trace element and major element geochemistry of tertiary igneous rocks. *Royal Society of Edinburgh Transactions*, 72, 159–170.

Dietzel, C.A.F., Kristandt, T., Dahlgren, S., Giebel, R.J., Marks, M.A.W., Wenzel, T., Markl, G. (2019). Hydrothermal processes in the Fen alkaline-carbonatite complex, southern Norway. *Ore Geology Reviews*, 111, 102969.

Donnot, M., Guiges, J., Lulzac, Y., Magnien, A., Parfenoff, A., Picot, P. (1973). Un nouveau type de gisement d'Europium : la monazite grise à europium en nodules dans les schistes paléozoïques de Bretagne. *Mineralium Deposita*, 8, 7–18.

Druecker, M. and Simpson, R.G. (2012). Technical report on the Sarfartoq project, West Greenland. Hudson Resources Inc.

Dushyantha, N., Batapola, N., Ilankoon, I., Rohitha, S., Premasiri, R., Abeysinghe, B., Ratnayake, N., Dissanayake, K. (2020). The story of rare earth elements (REEs): Occurrences, global distribution, genesis, geology, mineralogy and global production. *Ore Geology Reviews*, 122, 103521.

Eilu, P., Hallberg, A., Bergman, T., Bjerkgård, T., Feoktistov, V., Korsakova, M., Krasotkin, S., Lampio, E., Litvinenko, V., Philippov, N. et al. (2013). Fennoscandian ore deposit database. Annual update [Online]. Available at: http://en.gtk.fi/informationservices/databases/fodd/Indiax.html.

Elderfield, H., Upstill-Goddard, R., Sholkovitz, E.R. (1990). The rare earth elements in rivers, estuaries, and coastal seas and their significance to the composition of ocean waters. *Geochimica et Cosmochimica Acta*, 54(4), 971–991.

Eliseev, N.A. and Fedorov, E.E. (1953). Lovozero pluton and related ore deposits. *Transactions of Laboratory of Precambrian Rocks*, 1, 308.

Emeleus, C.H. (1964). The Grønnedal-Ika alkaline complex, South Greenland. The structure and geological history of the complex. *Meddelelser om Grønland*, 172, 1–75.

Emeleus, C.H. and Gyopari, M.C. (1992). *British Tertiary Volcanic Province*. Chapman and Hall, London.

Emeleus, C.H. and Harry, W.T. (1970). The Igaliko nepheline syenite complex. General description. *Grønlands Geologiske Undersøgelse*, 85, 116.

Emeleus, C.H. and Upton, B.G.J. (1976). The Gardar period in southern Greenland. In *Geology of Greenland*, Escher, A. and Watt, W.S. (eds). Grønlands Geologiske Undersøgelse, Copenhagen.

Emsbo, P., McLaughlin, P.I., Breit, G.N., du Bray, E.A., Koenig, A.E. (2015). Rare earth elements in sedimentary phosphate deposits: Solution to the global REE crisis? *Gondwana Research*, 27, 776–785.

EURARE (2017). Development of a sustainable exploitation scheme for Europe's rare earth ore deposits [Online]. Available at: http://eurare.brgm-rec.fr/

European Commission (2011). Tackling the challenges in commodity and markets and on raw materials. Communication from the Commission to the European Parliament, the European Council, The Council, The European Economic and Social Committee of the Regions, Brussels.

European Commission (2017). The 2017 list of critical raw materials for the EU. Communication from the Commission to the European Parliament, the European Council, The Council, The European Economic and Social Committee of the Regions, Brussels.

European Commission (2019a). The European Green Deal. Communication from the Commission to the European Parliament, the European Council, The Council, The European Economic and Social Committee of the Regions, Brussels.

European Commission (2019b). Masterplan for a competitive transformation of EU energy-intensive industries enabling a climate-neutral, circular economy by 2050. Report, High-Level Group on Energy-Intensive Industries. doi: 10.2873/854920.

Evans, A.M. (2000). *Ore Geology and Industrial Minerals – An Introduction*. Blackwell Science, Oxford.

Evans, J. and Zalasiewicz, J. (1996). U/Pb, Pb/Pb and Sm/Nd dating of authigenic monazite: Implications for the diagenetic evolution of the Welsh Basin. *Earth and Planetary Science Letters*, 144(3), 421–433.

Exley, R.A. (1980). Microprobe studies of REE-rich accessory minerals: Implications for Skye granite petrogenesis and REE mobility in hydrothermal systems. *Earth and Planetary Science Letters*, 48, 97–110.

Farrow, C. (1974). The geology of the Skjerstad area, Nordland, North Norway. PhD Thesis, University of Bristol.

Faul, H., Elmore, P.L.D., Brannock, W.W. (1959). Age of the Fen carbonatite (Norway) and its relation to the intrusives of the Oslo region. *Geochim. Cosmochim. Acta*, 17, 153–156.

Favreau, G. (1994). Trimouns, Ariège, Frankreich: Seltenerden-mineralien aus dem Dolomit. *Lapis*, 12(94), 18–39.

Fedele, L., Plant, A.J., De Vivo, B., Lima, A. (2008). The REE distribution over Europe: Geogenic and anthropogenic sources. *Geochemistry: Exploration, Environment, Analysis*, 8, 3–18.

Ferguson, J. (1964). Geology of the Ilímaussaq alkaline intrusion, South Greenland: Description of map and structure. *Meddelelser om Grønland*, 172, 1–82.

Ferrero, A., Ruíz, J., Vidal, A. (1989). Investigation about rare earths in north-west peninsular. Galicia (España). *Cuaderno Lab. Xeolóxico de Laxe*, 14, 255–270.

Finger, F., Roberts, M.P., Haunschmid, B., Schermaier, A., Steyrer, H.P. (1997). Variscan granitoids of central Europe: Their typology, potential sources and tectonothermal relations. *Mineral. Petrol.*, 61, 67–96.

Fitton, J.G. and Upton, B.G. (1987). Alkaline igneous rocks. *Geological Society of London*, 30.

Floor, P., Corretgé, G., Montero, P. (2007). The Ordovician Galiñeiro peralkaline gneiss complex, Vigo, NW Spain. *Goldschmidt Conference Abstracts*, A286.

Foley, S.F. and Fischer, T.P. (2017). An essential role for continental rifts and lithosphere in the deep carbon cycle. *Nature Geoscience*, 10(12), 897–902.

Foley, S.F., Venturelli, G., Green, D.H., Toscani, L. (1987). The ultrapotassic rocks: Charateristics, classification, and constraints for petrogenetic models. *Earth Science Review*, 24, 81–134.

Fortuné, J.P. (1971). Contribution à l'étude minéralogique et génétique des talcs pyrénéens. PhD Thesis, Université Paul Sabatier, Toulouse.

Fortuné, J.P., Gavoille, B., Thiébaud, J. (1980). Le gisement de talc de Trimouns près de Luzenac (Ariège). *26th International Geological Conference*, Paris.

Frei, D., Hollis, J.A., Gerdes, A., Karlsson, C., Vasquez, P., Franz, G., Harlov, D., Johansson, L., Knudsen, C. (2006). Advanced in situ trace element and geochronological microanalysis of geomaterials by laser ablation techniques. *Geological Survey of Denmark and Greenland Bulletin*, 10, 25–28.

Freslon, N., Bayon, G., Toucanne, S., Bermell, S., Bollinger, C., Chéron, S., Etoubleau, J., Germain, Y., Khripounoff, A., Ponzevera, E. et al. (2014). Rare earth elements and neodymium isotopes in sedimentary organic matter. *Geochimica et Cosmochimica Acta*, 140, 177–198.

Freyssinet, P., Butt, C.R.M., Morris, R.C., Piantone, P. (2005). Ore-forming processes related to lateritic weathering. In *Economic Geology 100th Anniversary Volume*, Hedenquist, J.W., Thompson, J.F.H., Goldfarb, R.J., Richards, J.P. (eds). Society of Economic Geologists Inc., Littleton.

Frietsch, R. (1977). The iron ore deposits in Sweden. In *The Iron Ore Deposits of Europe and Adjacent Areas. Explanatory Note to the International Map of Iron Ore Deposits of Europe*, Zitzrnau, A. (ed.). BGR, Hanover.

Frietsch, R. and Perdahl, J.A. (1995). Rare earth elements in apatite and magnetite in Kiruna-type iron ores and some other iron ore types. *Ore Geology Reviews*, 9, 489–510.

Frietsch, R., Tuisku, P., Martinsson, O., Perdahl, J.A. (1997). Early Proterozoic Cu-(Au) and Fe ore deposits associated with regional Na-Cl metasomatism in northern Fennoscandia. *Ore Geology Reviews*, 12, 1–34.

Frisch, W. and Keusen, H. (1977). The Gardiner intrusion, an ultramafic complex at Kangerdlugssuaq, East Greenland. *Grønlands Geologiske Undersøgelse Bulletin*, 122, 62.

Gamaletsos, P.N., Godelitsas, A., Kasama, T., Church, N.S., Douvalis, A.P., Göttlicher, J., Steininger, R., Boubnov, A., Pontikes, Y., Tzamos, E. et al. (2017). Nano-mineralogy and geochemistry of high-grade diasporic karst-type bauxite from Parnassos-Ghiona mines, Greece. *Ore Geology Reviews*, 84, 228–244.

Gamaletsos, P.N., Godelitsas, A., Filippidis, A., Pontikes, Y. (2019). The rare earth elements potential of Greek bauxite active mines in the light of a sustainable REE demand. *Journal of Sustainable Metallurgy*, 5(1), 20–47.

Garde, A.A. (1978). *The Lower Proterozoic Mârmorilik Formation, East of Mârmorilik, West Greenland. Meddelelser om Grønland 200*. Busck, Koøbenhavn.

Gates, P.E., Horlacher, C.F., Reed, G. (2012). Preliminary economic assessment NI43-101 technical report for the Norra Kärr (REE -Y-Zr) deposit, Gränna, Sweden. Report, Runge Pincock Minarco, prepared for Tasman Metals Ltd., Gränna.

Gates, P.E., Horlacher, C.F., Reed, G. (2013). Amended and restated preliminary economic assessment NI43-101 technical report for the Norra Kärr (REE-Y-Zr) deposit, Gränna, Sweden. Report, Runge Pincock Minarco, prepared for Tasman Metals Ltd., Gränna.

Gee, D.G., Fossen, H., Henriksen, N., Higgins, A.K. (2008). From the early Paleozoic platforms of Baltica and Laurentia to the Caledonide orogen of Scandinavia and Greenland. *Episodes*, 31, 44–51.

Geijer, P. (1931). The iron ores of the Kiruna type. *Sveriges Geol. Unders.*, 367, 1–39.

Geijer, P. (1961). The geological significance of the cerium mineral occurrences of the Bastnäs-type in central Sweden. *Arkiv Mineral. Geol.*, 3, 99–105.

Geijer, P. and Ödman, O.H. (1974). The emplacement of Kiruna iron ores and related deposits. *Sver. Geol. Unders.*, 700, 1–48.

Gerasimovsky, V.I., Volkov, V.P., Kogarko, L.N., Polyakov, A.I., Balashov, U. (1966). *Geochemistry of Lovozero Alkaline Massif. Part I: Geology and Petrology. Part 2: Geochemistry*. Australian National University Press, Canberra.

GEUS, Bureau of Minerals and Petroleum (2010). Map of Greenland geology and selected mineral occurrences [Online]. Available at: http://geus.dk/geuspage-uk.htm.

GEUS, Bureau of Minerals and Petroleum (2014). Greenland mineral resources portal [Online]. Available at: http://geus.dk/geuspage-uk.htm.

Geyti, A. (1979). Drilling and mapping in the EGMO concession area. Report, Nordisk Minselskab Prospecting.

Gislev, M. and Grohol, M. (2018). Report on critical raw materials and the circular economy. Report, Publications Office of the European Union. doi: 10.2873/331561.

GME (2012). Kvanefjeld concentrator pilot plant campaign complete; results exceed expectations. Report, Company Announcement.

GME (2014). EURARE project update: Kvanefjeld bulk ore sample selected for demonstration plant operations. Report, Company Announcement.

Goliáš, V. (2001). Thorium occurrences in the Czech Republic and their mineralogy. In *Uranium Deposits*, Kribek, B. and Zeman, J. (eds). Czech Geological Survey, Brno.

Goodenough, K.M., Schilling, J., Jonsson, E., Kalvig, P., Charles, N., Tuduri, J., Deady, E.A., Sadeghi, M., Schiellerup, H., Müller, A. et al. (2016). Europe's rare earth element resource potential: An overview of REE metallogenetic provinces and their geodynamic setting. *Ore Geol. Rev.*, 72, 838–856.

Gothenborg, J. and Pedersen, J.L. (1976). Exploration of the Qaqarssuk carbonatite complex. 1975. Part II. Internal report, Kryolitselskabet Øresund A/S, 1 plate. Geological Survey of Denmark and Greenland.

Green, D. and McCallum, D. (2005). Kainosite-(Y) from the Strontian Mines, Highland region, Scotland. *UK Journal of Mines and Minerals*, 26, 23–26.

Green, D. and Wallace, M.E. (1988). Mantle metasomatism by ephemeral carbonatite melts. *Nature*, 336(6198), 459–462.

Green, D., Bell, R., Moreton, S. (2005). Drusy cavity minerals including the first Irish danalite from Lindsay's Leap, Mourne Mountains. *UK Journal of Mines and Minerals*, 25, 25–30.

Grew, E.S., Essene, E.J., Peacor, D.R., Su, S.C., Asami, M. (1991). Dissakisite-(Ce), a new member of the epidote group and the Mg analogue of allanite-(Ce), from Antarctica. *American Mineralogist*, 76, 1990–1997.

Grice, J.D., Gault, R.A., Rowe, R., Johnsen, O. (2006). Qaqarssukite-(Ce), a new barium-ceriuml fluorcarbonate mineral species from Qaqarssuk, Greenland. *Canadian Mineralogist*, 44, 1137–1146.

Gupta, C.K. and Krishnamurthy, N. (2005). *Extractive Metallurgy of Rare Earths*. CRC Press, Boca Raton.

Gustafsson, B. (1979). Uranuppslag inom Norrbotten och Västerbotten. Report, Sveriges Geologiska Undersökning.

Guyonnet, D., Planchon, M., Rollat, A., Escalon, V., Tuduri, J., Charles, N., Vaxelaire, S., Dubois, D., Fargier, H. (2015). Material flow analysis applied to rare earth elements in Europe. *J. Clean. Prod*, 107, 215–228.

Gysi, A.P., Williams-Jones, A.E., Collins, P. (2016). Lithogeochemical vectors for hydrothermal processes in the Strange Lake peralkaline granitic REE-Zr-Nb deposit. *Economic Geology*, 111(5), 1241–1276.

Halama, R., Venneman, T., Siebel, W., Markl, G. (2005). The Grønnedal-Ika carbonatite-syenite complex, South Greenland: Carbonatite formation by liquid immiscibility. *Journal of Petrology*, 46, 191–217.

Hallberg, A. (2003). Styles of hydrothermal alteration and accompanying chemical changes in the Sången formation, Bergslagen, Sweden, and adjacent areas. Economic Geology Research and Documentation 2, 2001–2002. *Sver. Geol. Undersök.*, 113, 4–35.

Halliday, A.N., Aftalion, M., Parsons, I., Dickin, A.P., Johnson, M.R.W. (1987). Syn-orogenic alkaline magmatism and its relation to the Moine thrust zone and the thermal state of the lithosphere in NW Scotland. *Journal of the Geological Society of London*, 144, 611–618.

Hamilton, M.A., Pearson, D.G., Thompson, R.N., Kelley, S.P., Emeleus, C.H. (1999). Rapid eruption of Skye lavas inferred from precise U/Pb and Ar/Ar dating of the Rum and Cuillin plutonic complexes. *Nature*, 394, 260–263.

Hammouda, T. and Keshav, S. (2015). Melting in the mantle in the presence of carbon: Review of experiments and discussion on the origin of carbonatites. *Chem. Geol.*, 418, 171–188.

Harding, R.R. (1966). The major ultrabasic and basic intrusions of St Kilda, Outer Hebrides. *Transactions of the Royal Society of Edinburgh*, 66, 419–444.

Harding, R.R., Merriman, R.J., Nancarrow, P.H.A. (1982). A note on the occurrence of chevkinite, allanite, and zirkelite on St. Kilda, Scotland. *Mineralogical Magazine*, 46, 445–448.

Harding, R.R., Merriman, R.J., Nancarrow, P.H.A. (1984). St. Kilda: An illustrated account of the geology. Report, British Geological Survey.

Harker, A. (1904). *The Tertiary Igneous Rocks of Skye*. Arkose Press, Hawthorne.

Harlov, D.E., Andersson, U.B., Foerster, H., Nystrom, J.O., Dulski, P., Broman, C. (2002). Apatite-monazite relations in the Kiirunavaara magnetite-apatite ore, northern Sweden. *Chemical Geology*, 191, 47–72.

Harpøth, O., Pedersen, J.L., Schønwandt, H.K., Thomassen, B. (1986). The mineral occurrences of central East Greenland. *Meddelelser om Grønland Geoscience*, 17, 138.

Hedrick, J.B. (1985). Rare earth minerals and metals. Minerals yearbook. *Geological Survey*, 1, 791–803.

Hedrick, J.B. and Templeton, D.A. (1991). Rare-earth minerals and metals. Minerals yearbook. *Geological Survey*, 825–844.

Hedrick, J.B., Sinha, S.P., Kosynkin, V.D. (1997). Loparite, a rare earth ore (Ce, Na, Sr, Ca) (Ti, Nb, Ta, Fe^{3+})O_3. *Journal of Alloys and Compounds*, 250, 467–470.

Henderson, P. (1984). *Rare Earth Element Geochemistry. Developments in Geochemistry 2.* Elsevier, Amsterdam.

Henderson, I.H.C. and Ihlen, P.M. (2004). Emplacement of polygeneration pegmatites in relation to Sveco-Norwegian contractional tectonics: Examples from southern Norway. *Precambrian Research*, 133, 207–222.

Henderson, G. and Pulvertaft, T.C.R. (1987). Descriptive text to the geological map of Greenland, 1:100,000, Mârmorilik 71 V.2 Syd, Nûgâtsiaq 71 V.2 Nord, Pangnertôq 72 V.2 Syd. GEUS, Copenhagen.

Henriksen, N., Higgins, A.K., Kalsbeek, F. (2009). Greenland from Archaean to Quaternary. Descriptive text to the 1995 geological map of Greenland, scale 1:2,500,000. *Geological Survey of Denmark and Greenland Bulletin*, 18.

Himmi, R. (1975). Outokumpu Oy:n Korsnäsin ja Petolahden kaivosten vaiheita. *Vuoriteollisuus-Bergshanteringen*, 33, 35–38.

Hitzman, M.W., Oreskes, N., Einaudi, M.T. (1992). Geological characteristics and tectonic setting of Paleoproterozoic iron oxide (Cu-U-Au-REE) deposits. *Precambrian Research*, 58, 241–287.

Hode Vuorinen, J. and Hålenius, U. (2005). Nb-, Zr- and LREE-rich titanite from the Alnö alkaline complex: Crystal chemistry and its importance as a petrogenetic indicator. *Lithos*, 83, 128–142.

Högdahl, K., Andersson, U.B., Eklund, O. (2004). The Transscandinavian Igneous Belt (TIB) in Sweden: A review of its character and evolution. *Geological Survey of Finland*, 37, 123.

Holtstam, D. (2004). The Bastnäs-type REE deposits. In *The Bastnäs-type REE-mineralizations in North-western Bergslagen, Sweden – A Summary with Geological Background and Excursion Guide*, Andersson, U.B. (ed.). Sveriges geologiska undersökning, Uppsala.

Holtstam, D. and Andersson, U.B. (2007). The REE minerals of the Bastnäs-type deposits, south-central Sweden. *The Canadian Mineralogist*, 45, 1073–1114.

Hornig-Kjarsgaard, I. (1998). Rare earth elements in sövitic carbonatites and their mineral phases. *Journal of Petrology*, 39, 2105–2121.

Hugg, R. (1985). Katajakangas. Geologinen malmiarvio. Report, Rautaruukki Oy Malminetsintä.

Hugg, R. and Heiskanen, V. (1986). Otanmäen alueen niobi-lantaniditutkimukset, tilanne 31.12.1985 (Lisäys 15.5.1986 L. Pekkarinen). Report, Rautaruukki Oy Malminetsintä.

Hyslop, E.K., Gillanders, R.J., Hill, P.G., Fakes, R.D. (1999). Rare-earth bearing minerals fergusonite and gadolinite from the Arran granite. *Scottish Journal of Geology*, 35, 65–69.

IGE Nordic AB (2007). The rare earth deposit of Olserum. Report, Sweden Internal.

IGM (1998). Mineral potential of Portugal [Online]. Available at: http://www.lneg.pt/CienciaParaTodos/edicoes_online/diversos/potential.

Ihlen, P., Schiellerup, H., Gautneb, H., Skar, O. (2013). Characterization of apatite resources in Norway and their REE potential – A review. *Ore Geology Reviews*. doi: 10.1016/j.oregeorev.2013.11.003.

Inverno, C., Oliveira, D., Viegas, L., Lencastre, J., Salgueiro, R. (1998). REE-enriched Ordovician quartzites in Vale de Cavalos, Portalegre, Portugal. *Proceedings Volume of GeoCongress '98, Pretoria*. Geological Society of South Africa, Johannesburg.

Inverno, C., Oliveira, D., Rodrigues, L. (2007). Inventariação e prospecção de terras raras nas regiões fronteiriças da Beira Baixa e do Norte Alentejo. Report, INETI, Alfragide.

Isokangas, P. (1975). The mineral deposits in Finland. PhD Thesis, University of Helsinki.

Ixer, R.A., Young, B., Stanley, C.J. (1996). Bismuth-bearing assemblages from the Northern Pennine Orefield. *Min. Mag.*, 60, 317–324.

Jacquemin, D. (2020). The Phosphatic chalk of the Mons Basin, Belgium. Petrography and geo-chemistry of the Ciply Phosphatic Chalk and implications on its genesis. *Memoirs of the Geological Survey of Belgium*, 64, 46.

Jia, Y.-H. and Liu, Y. (2020). REE enrichment during magmatic–hydrothermal processes in carbonatite-related REE deposits: A case study of the Weishan REE deposit, China. *Minerals*, 10(1), 25.

Johnsen, O. and Grice, J.D. (1999). The crystal chemistry of the eudialyte group. *Canadian Mineralogist*, 37, 865–891.

Johnsen, O., Petersen, O.V., Medenbach, O. (1985). The Gardiner complex, a new locality in Greenland. *Mineral. Record*, 16, 485–494.

Johnsen, O., Grice, J.D., Gault, R.A. (2001). The eudialyte group: A review. In *The Ilímaussaq Complex, South Greenland: Status of Mineralogical Research with New Results*, Sørensen, H. (ed.). Geology of Greenland Survey Bulletin, Copenhagen.

Jones, A.P., Wall, F., Williams, C.T. (1996). *Rare Earth Minerals: Chemistry, Origin and Ore Deposits*. Chapman & Hall, London.

Josso, P., Pelleter, E., Pourret, O., Fouquet, Y., Etoubleau, J., Cheron, S., Bollinger, C. (2017). A new discrimination scheme for oceanic ferromanganese deposits using high field strength and rare earth elements. *Ore Geology Reviews*, 87, 3–15.

Judge, W.D., Xiao, Z.W., Kipouros, G.J. (2017). Application of rare earths for higher efficiencies in energy conversion. In *Rare Metal Technology 2017*, Kim, H., Alam, S., Neelameggham, N.R., Oosterhof, H., Ouchi, T., Guan, X. (eds). Springer, Berlin.

Juteau, T., Noack, Y., Whitechurch, H., Courtois, C. (1979). Mineralogy and geochemistry of alteration products in Holes 417A and 417D basement samples (Deep Sea Drilling Project Leg 51). In *Initial Reports of the Deep Sea Drilling Project*, Donnelly, T., Francheteau, J., Bryan, W., Robinson, P., Flower, M., Salisbury, M. (eds). U.S. Government Printing Office, Washington.

Kalashnikov, A.O., Konopleva, N.G., Pakhomovsky, Y.A., Ivanyuk, G.Y. (2016). Rare earth deposits of the Murmansk Region, Russia – A review. *Economic Geology*, 111(7), 1529–1559.

Kalatha, S., Perraki, M., Economou-Eliopoulos, M., Mitsis, I. (2017). On the origin of Bastnaesite-(La,Nd,Y) in the Nissi (Patitira) Bauxite Laterite deposit, Lokris, Greece. *Minerals*, 7(3), 19.

Kalsbeek, F., Pulvertaft, T.C.R., Nutman, A.P. (1998). Geochemistry, age and origin of metagrey-wackes from the Palaeoproterozoic Karrat Group, Rinkian belt, West Greenland. *Precambrian Research*, 91, 383–399.

Kärenlampi, K., Kontinen, A., Hanski, E., Huhma, H., Lahaye, Y., Krause, J., Heinig, T. (2020). Age and origin of the Nb-Zr-REE mineralization in the Paleoproterozoic A1-type granitoids at Otanmäki, central Finland. *Bulletin of the Geological Society of Finland*, 92(1), 39–71.

Kärkkäinen, N. and Huhta, P. (1993). Korsnäsin lyijypitoisten kalkkilohkareiden selvitys v. 1989-1990. *Geological Survey of Finland*, Report no. 3577, 23.

Kato, Y., Fujinaga, K., Nakamura, K., Takaya, Y., Kitamura, K., Ohta, J., Toda, R., Nakashima, T., Iwamori, H. (2011). Deep-sea mud in the Pacific Ocean as a potential resource for rare-earth elements. *Nature Geoscience*, 4(8), 535–539.

Keller, J. (1981). Carbonatitic volcanism in the Kaiserstuhl alkaline complex: Evidence for highly fluid carbonatitic melts at the earth's surface. *Journal of Volcanology and Geothermal Research*, 9, 423–431.

Keller, J. and Krafft, M. (1990). Effusive natrocarbonatite activity of Oldoinyo Lengai, June 1988. *Bulletin of Volcanology*, 52(8), 629–645.

Kimbell, G.S., Young, B., Millward, D., Crowley, Q.G. (2010). The North Pennine batholith (Wear-dale Granite) of northern England: New data on its age and form. *Proceedings of the Yorkshire Geological Society*, 58, 107–128.

Kirchheimer, F. (1957). *Bericht über das Vorkommen von Uran in Baden Württemberg*. Abh. des Geol. Land, Baden-Württemberg.

Kjarsgaard, B.A. and Hamilton, D.L. (1989). The genesis of carbonatites by immiscibility. In *Genesis and Evolution Carbonatites*, Bell, K. (ed.). Unwin-Hyman, London.

Kjarsgaard, B.A., Hamilton, D.L., Peterson, T.D. (1995). Peralkaline nephelinite/carbonatite liquid immiscibility: Comparison of phase compositions in experiments and natural lavas from Ol-doinyo Lengai. In *Carbonatite Volcanism: Oldoinyo Lengai and the Petrogenesis of Natrocarbonatites*, Bell, K. and Keller, J. (eds). Springer, Berlin.

Kleinhanns, I.C., Fischer-Gödde, M., Hansen, B.T. (2012). Sr-Nd isotope and geochemical characterisation of the paleoproterozoic Västervik formation (Baltic Shield, SE Sweden): A southerly exposure of Svecofennian meta-siliciclastic sediments. *International Journal of Earth Sciences*, 101, 39–55.

Knudsen, C. (1989). Pyrochlore group minerals from the Qaqarssuk carbonatite complex. In *Lanthanides, Tantalum and Niobium*, Saupé, F. (ed.). Springer-Verlag, Berlin.

Knudsen, C. (1991). Petrology, geochemistry and economic geology of the Qaqarssuk carbonatite complex, southern West Greenland. *Monograph Series on Mineral Deposits*, 29, 110.

Knudsen, C. and Buchardt, B. (1991). Carbon and oxygen isotope composition of carbonates from the Qaqarssuk carbonatite complex, southern West Greenland. *Chemical Geology*, 86, 263–274.

Kogarko, L.N. (1987). Alkaline rocks of the eastern part of the Baltic Shield (Kola Peninsula). In *Alkaline Igneous Rocks*, Fitton, J.G. and Upton, B.G.J (eds). Geological Society Publications, London.

Kogarko, L.N., Kononova, V.A., Orlova, M.P., Woolley, A.R. (1995). *Alkaline Rocks and Carbonatites of the World. Part 2: Former USSR*. Chapman and Hall, London.

Kolb, J., Thrane, K., Bagas, L. (2013). Field relationship of high-grade Neo- to Mesoarchaean rocks of South East Greenland: Tectonometamorphic and magmatic evolution. *Gondwana Research*, 23, 471–492.

Kontio, M. and Pankka, H. (2006). Tutkimustyöselostus Savukosken kunnassa valtausalueella Jammi 1, kaiv. rek. nro 7819/1 tehdyistä malmitutkimuksista 2004-2005. Report, Geological Survey of Finland, M06/4723/2006/1/10.

Korneliussen, A., McEnroe, S.A., Nilsson, L.P., Schiellerup, H., Gautneb, H., Meyer, G.B., Størseth, L.R. (2000). An overview of titanium deposits in Norway. *Norges geologiske undersøkelse Bulletin*, 436, 27–38.

Korovkin, V.A., Turyleva, L.V., Rudenko, D.G., Juravlev, A.V., Kluchnikova, G.N. (2003). Mineral resources of the northwest Russian Federation. *VSEGEI Cartographic Factory*, Saint-Petersburg, 519 [in Russian].

Korsakova, M., Krasotkin, S., Stromov, V., Iljina, M., Lauri, L., Nilsson, N.P. (2012). Metallogenic areas in Russian part of the Fennoscandian shield. In *Mineral Deposits and Metallogeny of Fennoscandia*, Eilu, P. (ed.). Geological Survey of Finland, Espoo.

Kraemer, D., Viehmann, S., Banks, D., Sumoondur, A.D., Koeberl, C., Bau, M. (2019). Regional variations in fluid formation and metal sources in MVT mineralization in the Pennine Orefield, UK: Implications from rare earth elements and yttrium distribution, Sr-Nd isotopes and fluid inclusion compositions of hydrothermal vein fluorites. *Ore Geology Reviews*, 107, 960–972.

Kramm, A. and Kogarko, L.N. (1994). Nd and Sr isotope signatures of the Khibina and Lovozero agpaitic centres, Kola alkaline province, Russia. *Lithos*, 32, 225–242.

Kramm, U., Kogarko, L.N., Kononova, V.A., Vartiainen, H. (1993). The Kola alkaline province of the CIS and Finland: Precise Rb-Sr ages define 380-360 Ma age range for all magmatism. *Lithos*, 30, 33–44.

Kresten, P. and Morogan, V. (1986). Fenitization at the Fen complex, southern Norway. *Lithos*, 19, 27–42.

Krištiak, J. and Záliš, Z. (1994). *Database of Radioactive Sites of the Czech Uranium Industry MS*. Geofond Praha, Holesovice.

Krűger, J.C., Romer, R.L., Kämpf, H. (2013). Late cretaceous ultramafic lamprophyres and carbonatites from the Delitzsch Complex, Germany. *Chemical Geology*, 353, 140–150.

Krumrei, T.V., Villa, I.M., Marks, M.A., Markl, G. (2006). A $^{40}Ar/^{39}Ar$ and U/Pb isotopic study of the Ilímaussaq complex, South Greenland: Implications for the ^{40}K decay constant and for the duration of magmatic activity in a peralkaline complex. *Chemical Geology*, 227, 258–273.

Kunzendorf, H. and Secher, K. (1987). Dispersion of niobium and phosphorus in soil overlying the Qaqarssuk carbonatite complex, Southern West Greenland. *Journal of Geochemical Exploration*, 28, 285–296.

Kusiak, M.A., Dunkley, D.J., Suzuki, K., Kachlik, V., Kedzior, A., Lekki, J., Oplustil, S. (2010). Chemical (non-isotopic) and isotopic dating of Phanerozoic zircon. A case study of durbachite from the Třebíč Pluton, Bohemian Massif. *Gondwana Research*, 17, 153–161.

Lacan, F. and Jeandel, C. (2005). Neodymium isotopes as a new tool for quantifying exchange fluxes at the continent–ocean interface. *Earth and Planetary Science Letters*, 232(3), 245–257.

Lacomme, A. and Fontan, F. (1971). Sur la présence de monazite dans les Pyrénées. *Comptes-Rendus de l'Académie des Sciences*, 272, 1193–1194.

Lacomme, A., Hottin, A.M., Laval, M. (1993). La monazite grise du massif de l'Arize (Pyrénées françaises). Report, BRGM, R37041 GEO-SGN.

Larsen, R.B. (2002). The distribution of rare-earth elements in K-feldspar as an indicator of petrogenetic processes in granitic pegmatites: Examples from two pegmatite fields in southern Norway. *Canadian Mineralogist*, 40, 137–151.

Larsen, A.O. (ed.) (2010). *The Langesundsfjord: History, Geology, Pegmatites and Minerals*. Verlag, Berlin.

Larsen, L.M. and Sørensen, H. (1987). The Ilímaussaq intrusion-progressive crystallization and formation of layering in an agpaitic magma. In *Alkaline Igneous Rocks*, Fitton, J.G. and Upton, B.G.J. (eds). Geological Society Publications, London.

Larsen, L.T., Rex, D., Secher, K. (1983). The age of carbonatites, kimberlites and lamprophyres from southern West Greenland: Recurrent alkaline magmatism during 2500 million years. *Lithos*, 16, 215–221.

Larsen, M., Piasecki, S., Surlyk, F. (2003). Stratigraphy and sedimentology of a basement-onlapping shallow marine sandstone succession, the Charcot Bugt Formation, Middle-Upper Jurassic, East Greenland. *Geological Survey of Denmark and Greenland Bulletin*, 1, 893–930.

Larsen, B.T., Olaussen, S., Sundvoll, B., Heeremans, M. (2008). The Permo-Carboniferous Oslo Rift through six stages and 65 million years. *Episodes*, 31, 52–58.

Lauri, L.S., Eilu, P., Brown, T., Gunn, G., Kalvig, P., Sievers, H. (2018). Identification and quantification of primary CRM resources in Europe. Report, SCRREEN Deliverable 3.1.

Laval, M. and Dromart, G. (1990). Les terres rares dans les shales noirs mésozoïques. Report, BRGM, DAM/DEX 1756.

Laval, M., Prian, J.P., Hottin, A.M., Orgeval, J.J. (1990). Les rare earths dans les formations cambriennes des Cévennes et de la Montagne Noire (Massif central français). Report, BRGM, DAM/DEX R31672.

Le Bas, M.J. (1981). Carbonatite magmas. *Mineralogical Magazine*, 44(334), 133–140.

Le Maitre, R.W., Streckeisen, A., Zanettin, B., Le Bas, M., Bonin, B., Bateman, P. (2005). *Igneous Rocks: A Classification and Glossary of Terms*. Cambridge University Press.

Lefebvre, G. (2017). Sursaut sur le marché des terres rares en 2017 [Online]. Available at: http://www.mineralinfo.fr/ecomine/sursaut-marche-terres-rares-en-2017.

Lehijärvi, M. (1960). The alkaline district of Iivaara, Kuusamo, Finland. *Bulletin of the Geological Commission of Finland*, 185, 1–62.

Lehtonen, M.I., Kujala, H., Kärkkäinen, N., Lehtonen, A., Mäkitie, H., Mänttäri, I., Virransalo, P., Vuokko, J. (2005). Etelä-Pohjanmaan liuskealueen kallioperä. Summary: Pre-quaternary rocks of the south Ostrobothnian schist belt. Report, Geological Survey of Finland.

Lemarchand, R. (1970). Résultats des travaux de reconnaissance et premier pronostic sur les grès à zircon-rutile-monazite de Bailleul (Orne). Report, BRGM, 70-RME-011-RMM.

Lemoine, M. (1984). *Les marges continentales actuelles et fossiles autour de la France*. Masson, Paris.

Li, M.Y.H., Zhou, M.-F., Williams-Jones, A.E. (2020). Controls on the dynamics of rare earth elements during subtropical hillslope processes and formation of regolith-hosted deposits. *Economic Geology*, 115(5), 1097–1118.

Lie, A. and Østergaard, C. (2011). The Fen carbonatite complex, Ulefoss, South Norway. Summary of historic work and data. *21st North in Commission for REE Minerals*, Svendborg.

Lie, A. and Østergaard, C. (2014). The Fen rare earth element deposit, Ulefoss, South Norway. Summary of historic work and data. *21st North in Commission for REE Minerals*, Svendborg.

Limbourg, Y. (1986). Découverte de monazite grise en nodules dans l'Eodévonien du Synclinorium de Neufchateau (Belgium). *Bulletin de la Société belge de géologie*, 95, 281–285.

Lindberg, P.A. (1985). Fe-Ti-P mineralizations in the Larvikite-lardalite complex, Oslo Rift. In *Titaniferous Magnetite, Ilmenite and Rutile Deposits in Norway*, Korneliussen, A. and Robins, B. (eds). Norges geologiske undersøkelse Bulletin, Trondheim.

Lindh, A. and Persson, P.O. (1990). Proterozoic granitoids rocks of the Baltic shield – Trends of development. In *Mid-Proterozoic Laurentia-Baltica*, Gower, C.F., Rivers, T., Ryan, A.B. (eds). Special Papers, Geological Association of Canada, St. John's, NL.

Lipin, B.R. and McKay, G.A. (1989). Geochemistry and mineralogy of rare earth elements. Review in mineralogy. *Mineralogical Society of America*, 21, 348.

Liu, H., Pourret, O., Guo, H., Bonhoure, J. (2017). Rare earth elements sorption to iron oxyhydroxide: Model development and application to groundwater. *Applied Geochemistry*, 87, 158–166.

Lottermoser, B.G. (1990). Rare-earth element mineralization within the Mt. Weld carbonatite late-rite, Western Australia. *Lithos*, 24, 151–167.

Lottermoser, B.G. (1992). Rare earth elements and hydrothermal ore formation processes. *Ore Geology Reviews*, 7(1), 25–41.

Luckscheiter, B. and Morteani, G. (1980). Microthermometrical and chemical studies of fluid inclusions in minerals from Alpine veins from the penninic rocks of the central and western Tauern Window (Austria/Italy). *Lithos*, 13, 61–77.

Luossavaara-Kiirunavaara, A.B. (2018). By Per Juntti, Mine waste will now be utilised. Press release, LKAB.

Machacek, E. and Kalvig, P. (2016). Assessing advanced rare earth element-bearing deposits for industrial demand in the EU. *Resources Policy*, 49, 186–203.

Maksimović, Z.J. (1976). Genesis of some Mediterranean karstic bauxite deposits. Travaux ICSOBA. Report, Comité International pour l'Etude des Bauxites, de l'Alumine et de l'Aluminium.

Maksimović, Z.J. (1987). Trace elements in the Jurassic bauxites of Montenegro and their genetic significance. Special volume devoted to acad. *Acad. Montenegro Sci. Arts*, 47–60.

Maksimović, Z.J. and Pantó, G. (1996). Authigenic rare earth minerals in karst-bauxites and karstic nickel deposits. In *Rare Earth Minerals: Chemistry, Origin and Ore Deposits*, Jones, A.P., Wall, F., Williams, C.T. (eds). Springer, Berlin.

Maksimović, Z.J. and Roaldset, E. (1976). Lanthanide elements in some Mediterranean karstic bauxite deposits. Travaux ICSOBA. Report, Comité International pour l'Étude des Bauxites, de l'Alumine et de l'Aluminium.

Maòour, J., Procházka, J., Vavøín, I. (1992). Zircon in sediments of the upper Lužnice River and its relationship to aeroradiometric anomalies. *Geol. Prùzk.*, 34, 129–131.

Marcos, A., Perez-Estaun, A., Pulgar, J.A., Bastida, F., Vargas, I. (1980). Mapa geológica de España. Scale 1:50,000 (Becerreá). *Instituto Geológico y Minero de España*, Sheet n°2.

Marignac, C. (1880). Sur les terres de la samarskite. *Annales de Chimie et de Physique*, 5(20), 535–557.

Marmo, V., Hoffren, V., Hytönen, K., Kallio, P., Lindholm, O., Siivola, J. (1966). On the granites of Honkamäki and Otanmäki, Finland, with special reference to the mineralogy of accessories. *Bulletin of the Geological Commission of Finland*, 221, 34.

Marmolejo-Rodríguez, A.J., Caetano, M., de Pablo, H., Vale, C., Prego, R. (2008). Yttrium in the Vigo Ria (NW Iberian Peninsula): Sources, distribution, and background levels. *Ciencias Marinas*, 34, 399–406.

Martens, P.N. and Seemann, U. (1973). Prospektionsbericht 1972. Traill Ø. Report, Nordisk Mineselskab A/S, Geological Survey of Denmark and Greenland archives, GEUS.

Martinsson, O. (1997). Tectonic setting and metallogeny of the Kiruna greenstones. PhD Thesis, Luleå University of Technology.

Martinsson, O. (2007). Apatite-iron deposits. Metallogeny and tectonic evolution of the northern Fennoscnadian shield: Field trip guidebook. *Geological Survey of Finland*, 54, 113.

McCreath, J.A., Finch, A.A., Simonsen, S.L., Donaldson, C.H., Armour-Brown, A. (2012). India-pendent ages of magmatic and hydrothermal activity in alkaline igneous rocks: The Motzfeldt Centre, Gardar Province, South Greenland. *Contribution to Mineralogy and Petrology*, 163, 967–982.

McLennan, S.M. and Taylor, S.R. (2012). Geology, geochemistry, and natural abundances of the rare earth elements. In *The Rare Earth Elements: Fundamentals and Applications*, Atwood, D.A. (ed.). John Wiley & Sons, New York.

Meert, J.G., Torsvik, T.H., Eide, E.A., Dahlgren, S. (1998). Tectonic significance of the Fen Province, S. Norway: Constraints from geochronology and palaeomagnetism. *Journal of Geology*, 106, 553–564.

Meert, J.G., Walderhaug, H.J., Torsvik, T.H., Hendriks, B.W.H. (2007). Age and paleomagnetic signature of the Alnø carbonatite complex (NE Sweden): Additional controversy for the Neo-proterozoic paleoposition of Baltica. *Precambrian Research*, 154(3), 159–174.

Melfos, V. and Voudouris, P.C. (2012). Geological, mineralogical and geochemical aspects for critical and rare metals in Greece. *Minerals*, 2, 300–317.

Menzies, M., Seyfried, W., Blanchard, D. (1979). Experimental evidence of rare earth element immobility in greenstones. *Nature*, 282(5737), 398–399.

Michard, A. and Albarède, F. (1986). The REE content of some hydrothermal fluids. *Chemical Geology*, 55(1), 51–60.

Migdisov, A.A. and Williams-Jones, A.E. (2014). Hydrothermal transport and deposition of the rare earth elements by fluorine-bearing aqueous liquids. *Mineralium Deposita*, 49(8), 987–997.

Migdisov, A.A., Williams-Jones, A.E., Wagner, T. (2009). An experimental study of the solubility and speciation of the rare earth elements (III) in fluoride- and chloride-bearing aqueous solutions at temperatures up to 300°C. *Geochimica et Cosmochimica Acta*, 73(23), 7087–7109.

Migdisov, A.A., Guo, X., Nisbet, H., Xu, H., Williams-Jones, A.E. (2019). Fractionation of REE, U, and Th in natural ore-forming hydrothermal systems: Thermodynamic modeling. *The Journal of Chemical Thermodynamics*, 128, 305–319.

Milodowski, A.E. and Zalasiewicz, J.A. (1991). Redistribution of rare earth elements during diagenesis of turbidite/hemipelagite mudrock sequences of Llandovery age from central Wales. *Geological Society Special Publication*, 57, 101–124.

Mitchell, R.H. and Brunfelt, A.O. (1975). Rare earth element chemistry of the Fen alkaline complex, Norway. *Contributions to Mineralogical Petrology*, 52, 247–259.

Mitrofanov, F.P., Zozulya, D.R., Bayanova, T.B., Levkovich, N.V. (2000). The world's oldest anorogenic alkali granitic magmatism in the Keivy structure on the Baltic Shield, Dokl. *Earth Sci.*, 374, 1145–1148.

Moine, B., Fortuné, J.P., Moreau, P., Viguier, F. (1989). Comparative mineralogy, geochemistry, conditions of formation of two metasomatic talc and chlorite deposits: Trimouns (Pyrenees, France) and Rabenwald (eastern Alps, Austria). *Economic Geology*, 84, 1398–1416.

Mondillo, N., Balassone, G., Boni, M., Chelle-Michou, C., Cretella, S., Mormone, A., Putzolu, F., Santoro, L., Scognamiglio, G., Tarallo, M. (2019). Rare earth elements (REE) in Al- and Fe-(oxy)-hydroxides in bauxites of Provence and Languedoc (Southern France): Implications for the potential recovery of REEs as by-products of bauxite mining. *Minerals*, 9, 504. doi: 10.3390/min9090504.

Mongelli, G. (1997). Ce-anomalies in the textural components of Upper Cretaceous karst bauxites from the Apulian carbonate platform (southern Italy). *Chemical Geology*, 140(1), 69–79.

Montero, P., Floor, P., Corretgé, G. (1998). The accumulation of rare-earth and high-field-strength elements in peralkaline granitic rocks: The Galiñeiro orthogneissic complex, northwestern Spain. *Canadian Mineralogist*, 36, 638–700.

Montero, P., Bea, F., Corretgé, L.G., Floor, P., Whitehouse, M.J. (2009). U-Pb ion microprobe dating and Sr and Nd isotope geology of the Galiñeiro Igneous Complex: A model for the pera-luminous/peralkaline duality of the Cambro-Ordovician magmatism of Iberia. *Lithos*, 107(3), 227–238.

Morogan, V. (1988). Fenitization, hallmark of the ijolite-carbonatite magmatic association. PhD Thesis, Stockholm University.

Morogan, V. and Woolley, A.R. (1988). Fenitization at the Alnö carbonatite complex, Sweden: Distribution, mineralogy and genesis. *Contributions to Mineralogy and Petrology*, 100, 169–182.

Morteani, G. (1991). The rare earths; their minerals, production and technical use. *European Journal of Mineralogy*, 3(4), 641–650.

Morteani, G. and Preinfalk, C. (1996). REE distribution and REE carriers in laterites formed on the alkaline complexes of Araxa and Catalao (Brazil). In *Rare Earth Minerals: Chemistry, Origin and Ore Deposits*, Jones, A.P., Wall, F., Williams, C.T. (eds). Springer, Berlin.

Mosander, C.G. (1843). On the new metals, lanthanium and didymium, which are associated with cerium; and on erbium and terbium, new metals associated with yttria. *Philosophical Magazine*, 23(152), 241–254.

Mott, A.V., Bird, D.K., Grove, M., Rose, N., Bernstein, S., Mackay, H., Krebs, J. (2013). Karrat Isfjord: A newly discovered Paleoproterozoic carbonatite-sourced REE deposit, central West Greenland. *Economic Geology*, 108(6), 1471–1488.

Müller, A., Kearsley, A., Spratt, J., Seltmann, R. (2012). Petrogenetic implications of magmatic garnet in granitic pegmatites from southern Norway. *Canadian Mineralogist*, 50, 1095–1115.

Müller, A., Spratt, J., Thomas, R., Williamson, B.J., Seltmann, R. (2018). Alkali-F-rich albite zones in evolved NYF pegmatites: The product of melt–melt immiscibility. *Canadian Mineralogist*, 56(4), 657–687.

Mullis, J. (1996). P-T-t path of quartz formation in extensional veins of the Central Alps. *Schweiz. Mineral. Petrogr. Mitt.*, 76, 159–164.

Mussett, A.E., Hodgson, B.N., Skehorn, R.R. (1987). Palaeomagnetism and age of the quartz-porphyry intrusions, Isle of Arran, Scotland. *Scottish Journal of Geology*, 23, 9–22.

Mutanen, T. (2011). Alkalikiviä ja appiniitteja. Raportti hankken "Magmatismi ja malminmuodos-tus II" toiminnasta 2002-2005. Report, Geological Survey of Finland.

Nabyl, Z., Massuyeau, M., Gaillard, F., Tuduri, J., Iacono-Marziano, G., Rogerie, G., Le Trong, E., Di Carlo, I., Melleton, J., Bailly, L. (2020). A window in the course of alkaline magma differentiation conducive to immiscible REE-rich carbonatites. *Geochimica et Cosmochimica Acta*, 282, 297–323.

Nadeau, O., Stevenson, R., Jébrak, M. (2016). Evolution of Montviel alkaline–carbonatite complex by coupled fractional crystallization, fluid mixing and metasomatism – Part I: Petrography and geochemistry of metasomatic aegirine–augite and biotite: Implications for REE–Nb mineralization. *Ore Geology Reviews*, 72, 1143–1162.

Nazari-Dehkordi, T., Spandler, C., Oliver, N.H.S., Wilson, R. (2018). Unconformity-related rare earth element deposits: A regional-scale hydrothermal mineralization type of Northern Australia. *Economic Geology*, 113(6), 1297–1305.

Nelson, D.R., Chivas, A.R., Chappell, B.W., McCulloch, M.T. (1988). Geochemical and isotopic systematics in carbonatites and implications for the evolution of ocean-island sources. *Geochimica et Cosmochimica Acta*, 52(1), 1–17.

NGU (2013). Metallogenic map of Norway [Online]. Available at: http://geo.ngu.no/kart/mineralressurser.

Nielsen, B.L. (1976). Economic minerals. In *Geology of Greenland*, Escher, A. and Watt, W.S. (eds). Geological Survey of Greenland, Copenhagen.

Nielsen, T.F.D. (1979). The occurrence and formation of Ti-aegirines in peralkaline syenites. An example from the Tertiary ultramafic alkaline Gardiner complex, East Greenland. *Contrib. Mineral. Petrol.*, 69, 235–244.

Nielsen, T.F.D. (1980). The petrology of a melilitolite, melteigite, carbonatite and syenite ring dike system, in the Gardiner complex, East Greenland. *Lithos*, 13, 181–197.

Nielsen, T.F.D. (1987). Tertiary alkaline magmatism in East Greenland: A review. In *Alkaline Igneous Rocks*, Fitton, J.G. and Upton, B.G.J. (eds). Geological Society Special Publication, London.

Nielsen, T.F.D. (1994). Alkaline dyke swarms of the Gardiner complex and the origin of ultramafic alkaline complexes. *Geochemistry International*, 31, 37–56.

Nielsen, T.F.D. (2002). Palaeogene intrusions and magmatic complexes in East Greenland, 66 to 75°N. Report, GEUS.

Nielsen, T.F.D. and Rosing, M.T. (1990). The Archaean Skjoldungen alkaline province, South-East Greenland. Report, Grønlands Geologiske Undersøgelse 148.

Nielsen, B.K. and Steenfelt, A. (1979). Intrusive events at Kvanefjeld in the Ilímaussaq igneous complex. *Bulletin of the Geological Society of Denmark*, 27, 143–155.

Nonnon, M. (1984). Découverte de monazite grise en nodules et d'or alluvionnaire dans le massif de la Croix-Scaille. *Bulletin de la Société belge de géologie*, 93, 307–314.

Nonnon, M. (1989). Une nouvelle occurrence de monazite grise en Belgium. *Bulletin de la Société belge de géologie*, 98, 73–76.

Notholt, A.J.G., Sheldon, R.P., Davidson, D.F. (eds) (1989). *Phosphate Deposits of the World. Volume 2: Phosphate Rocks Resources*. Cambridge University Press.

Nuna Minerals (2010). Commercially interesting rare earth element (REE) grades at two separate projects [Online]. Available at: http://nunaminerals.com.

Nykänen, J., Laajoki, K., Karhu, J. (1997). Geology and geochemistry of the early Proterozoic Kortejärvi and Laivajoki carbonatites, Central Fennoscandian Shield, Finland. *Bull. Soc. Finland*, 69(1–2), 5–30.

Olerud, S. (1985). Metallogeny of Finnmark, North Norway. *Nor. Geol. Unders. Bull.*, 403, 183–196.

Olerud, S. (1988). Davidite-loveringite in early Proterozoic albite felsite in Finnmark, north Norway. *Mineralogical Magazine*, 52, 400–402.

Orris, G.J. and Grauch, R.I. (2002). Rare earth element mines, deposits, and occurrences. Report, USGS.

Özlü, N. (1983). Trace-element content of "karst bauxites" and their parent rocks in the Mediterranean belt. *Mineralium Deposita*, 18, 469–476.

Paarma, H. (1970). A new find of carbonatite in north Finland, the Sokli plug in Savukoski. *Lithos*, 3, 129–133.

Papunen, H. (1986). Suomen metalliset malmiesiintymät. Suomen malmigeologia: Metalliset malmiesiintymät. *Geological Society of Finland*, 86, 49.

Papunen, H. and Lindsjö, O. (1972). Apatite, monazite and allanite: Three rare earth minerals from Korsnäs, Finland. *Bulletin of the Geological Society of Finland*, 44, 123–129.

Parák, T. (1973). Rare earths in the apatite iron ores of Lapland and some data about the Sr, Th and U content of these ores. *Economic Geology*, 68, 210–221.

Parák, T. (1975). The origin of the Kiruna iron ores. *Sver. Geol. Unders.*, 8, 709.

Parák, T. (1985). Phosphorus in different types of ore, sulfide in the iron deposit and the type and origin of ores at Kiruna. *Economic Geology*, 80, 646–665.

Pardillo, F. and Soriano, V. (1929). Hallazgo de la monacita en las arenas de la Ría de Vigo. *Asoc. Esp. Prog. Cienc., Con. Barc.*, 4, 141.

Parga Pondal, I. (1935). Arena monacítica de la Ría de Arosa (Galicia). *An. Soco Esp. Fís. Quím.*, 33, 466–469.

Parseval, P. (1992). Minéralogie et géochimie du gisement de talc et chlorite de Trimouns (Pyrénées, France). PhD Thesis, Université Paul Sabatier, Toulouse.

Parseval, P., Moine, B., Fortuné, J.P., Ferret, J. (1993). Fluid-mineral interactions at the origin of the Trimouns talc and chlorite deposit (Pyrénées, France). In *Current Research in Geology Applied to Ore Deposits*, Hach-Ali, F., Torres-Ruiz, J., Gervilla, F. (eds). Springer, Berlin.

Parseval, P., Fontan, F., Aigouy, T. (1997). Composition chimique des minéraux de terres rares de Trimouns (France). *Comptes-Rendus de l'Académie des Sciences*, 324, 625–630.

Pavel, C.C., Lacal-Arántegui, R., Marmier, A., Schüler, D., Tzimas, E., Buchert, M., Jenseit, W., Blagoeva, D. (2017). Substitution strategies for reducing the use of rare earths in wind turbines. *Resources Policy*, 52, 349–357.

Pearce, N.J.G., Leng, M.J., Emeleus, C.H., Bedford, C.M. (1997). The origins of carbonatites and related rocks from the Grønnedal-Ika Nepheline Syenite complex, South Greenland: C-O-Sr isotope evidence. *Mineralogical Magazine*, 61, 515–529.

Pearson, R.G. (1963). Hard and soft acids and bases. *Journal of the American Chemical Society*, 85(22), 3533–3539.

Pergamalis, F., Karageorgiou, D.E., Koukoulis, A. (2001b). The location of Tl, REE, Th, U, Au deposits in the seafront zones of Nea Peramos-Loutra Eleftheron area, Kavala (N. Greece) using γ radiation. *Bulletin of the Geological Society of Greece*, 34, 1023–1029.

Pergamalis, F., Karageorgiou, D.E., Koukoulis, A., Katsikis, I. (2001a). Mineralogical and chemical composition of sand ore deposits in the seashore zone Nea Peramos-Loutra Eleftheron Kavala (N. Greece). *Bulletin of the Geological Society of Greece*, 34, 845–850.

Perissoratis, C., Moorby, S.A., Angelopoulos, I., Cronan, D.S., Papavasiliou, C., Konispoliatis, N., Sakellariadou, F., Mitropoulos, D. (1988). Mineral concentrations in the recent sediments of Eastern Macedonia, Northern Greece: Geological and geochemical considerations. In *Mineral Deposits within the European Community*, Boissonnas, J. and Omenetto, P. (eds). Springer, Berlin.

Piilonen, P.C., McDonald, A.M., Poirier, G., Rowe, R., Larsen, A.O. (2018). The mineralogy and crystal chemistry of alkaline pegmatites in the Larvik plutonic complex, Oslo rift valley, Norway. Part 1. Magmatic and secondary zircon: Implications for petrogenesis from trace-element geochemistry. *Mineralogical Magazine*, 76(3), 649–672.

Piret, P., Deliens, M., Pinet, M. (1990). La trimounsite-(Y), nouveau silicotitanate de terres rares de Trimouns, Ariège, France : $(TR)_2Ti_2SiO_9$. *European Journal of Mineralogy*, 2, 725–729.

Plant, J.A. and Jones, D.G. (1999). Development of regional exploration criteria for buried carbonate-hosted mineral deposits: A multidisciplinary approach in northern England. Report, BGS WP/91/1.

Plant, J.A., Whittaker, A., Demetriades, A., De Vivo, B., Lexa, J. (2005). The geological and tectonic framework of Europe. In *Geochemical Atlas of Europe. Part 1: Background Information, Methodology and Maps*, Salminen, R. (ed.). FOREGS, GTK, Nottingham.

Pourret, O. and Davranche, M. (2013). Rare earth element sorption onto hydrous manganese oxide: A modeling study. *Journal of Colloid and Interface Science*, 395, 18–23.

Pourret, O. and Tuduri, J. (2017). Continental shelves as potential resource of rare earth elements. *Scientific Reports*, 7(1), 5857.

Pozhilenko, V.I., Gavrilenko, B.V., Zhirov, D.V., Zhabin, S.V. (2002). *Geology of Ore Areas of the Murmansk Oblast*. KSC RAS, Apatity.

Prego, R., Caetano, M., Vale, C., Marmolejo-Rodríguez, J. (2009). Rare earth elements in sediments of the Vigo Ria, NW Iberian Peninsula. *Continental Shelf Research*, 29(7), 896–902.

Prego, R., Caetano, M., Bernárdez, P., Brito, P., Ospina-Alvarez, N., Vale, C. (2012). Rare earth elements in coastal sediments of the northern Galicia shelf: Influence of geological features. *Continental Shelf Research*, 35, 75–85.

Pusstinen, K. (1971). Geology of the Siilinjärvi carbonatite complex, Eastern Finland. *Bulletin of the Geological Commission of Finland*, 249, 1–43.

Rabiei, M., Chi, G., Normand, C., Davis, W.J., Fayek, M., Blamey, N.J.F. (2017). Hydrothermal rare earth element (xenotime) mineralization at Maw Zone, Athabasca Basin, Canada, and its relationship to unconformity-related uranium deposits. *Economic Geology*, 112(6), 1483–1507.

Ram Resources Ltd. (2012). Ram Resources to build rare earths resource inventory at Motzfeldt with a $3m capital raising [Online]. Available at: www.ramresources.com.au.

Rasmussen, B., Buick, R., Taylor, W.R. (1998). Removal of oceanic REE by authigenic precipitation of phosphatic minerals. *Earth and Planetary Science Letters*, 164(1), 135–149.

Rauchenstein-Martinek, K., Wagner, T., Wälle, M., Heinrich, C. (2014). Gold concentrations in metamorphic fluids: A LA-ICPMS study of fluid inclusions from the Alpine orogenic belt. *Chemical Geology*, 385, 70–83.

Read, D., Cooper, D.C., McArthur, J.M. (1987). The composition and distribution of nodular monazite in the Lower Palaeozoic rocks of Great Britain. *Mineralogical Magazine*, 51, 271–280.

Reed, G.C. (2013). Amended and restated technical report for Olserum REE deposit, southern Sweden. Report, Prepared for Tasman Metals Ltd. by Reed Leyton Consulting.

Renberg, R. (2014). Recovery of Rare Earth elements from magnetic waste in the WEEE recycling industry and tailings from the iron ore industry. Project "REEcovery", Luleå University of Technology [Online]. Available at: www.ltu.se/research/subjects/Malmgeologi/Forskningsprojekt/REEcover?l=en.

Rillard, J., Pourret, O., Censi, P., Inguaggiato, C., Zuddas, P., Toulhoat, P., Gombert, P., Brusca, L. (2019). Behavior of rare earth elements in an aquifer perturbed by CO_2 injection: Environmental implications. *Science of the Total Environment*, 687, 978–990.

Roaldset, E. (1973). Rare earth elements in Quaternary clays of the Numedal area, southern Norway. *Lithos*, 6(4), 349–372.

Rollat, A., Guyonnet, D., Planchon, M., Tuduri, J. (2016). Prospective analysis of the flows of certain rare earths in Europe at the 2020 horizon. *Waste Management*, 49, 427–436.

Romer, R.L. and Smeds, S.A. (1994). Implications of U/Pb ages of columbite-tantalites from granitic pegmatites for the Palaeoproterozoic accretion of 1.90-1.85 Ga magmatic arcs to the Baltic Shield. *Precambrian Research*, 67(1), 141–158.

Rosa, D., Salgueiro, R., Inverno, C., de Oliveira, D., Guimarães, F. (2010). Occurrence and origin of alluvial xenotime from Central Eastern Portugal (Central Iberian Zone/Ossa-Morena Zone). *Comunicações Geológicas*, 97, 63–72.

Rosenblum, S. and Mosier, E.L. (1983). Mineralogy and occurrence of Europium-rich dark monazite. Report, US Geological Survey Professional Paper 1181.

Rosing, M.T., Lesher, C.E., Thomsen, H.S., Blichert-Toft, J. (1992). Alkaline magmatism in the Archaean Skjoldungen province, East Greenland (abstract). *Eos Transactions, American Geophysical Union*, 73, 331.

Roure, F., Brun, J.P., Colletta, B., van den Driessche, J. (1992). Geometry and kinematics of extensional structures in the Alpine Foreland Basin of southeastern France. *Journal of Structural Geology*, 14, 503–519.

Rousseau, T.C.C., Sonke, J.E., Chmeleff, J., van Beek, P., Souhaut, M., Boaventura, G., Seyler, P., Jeandel, C. (2015). Rapid neodymium release to marine waters from lithogenic sediments in the Amazon estuary. *Nature Communications*, 6(1), 7592.

Ryghaug, P. (1983). Geochemical exploration methods for niobium and rare earth elements at Sæteråsen, Vestfold volcanic area, Oslo Graben. *Norsk geologisk tidsskrift*, 63, 1–13.

Sadeghi, M., Morris, G.A., Carranza, E.J.M., Ladenberger, A., Andersson, M. (2013). Rare earth elements distribution and mineralization in Sweden: An application of principal component analysis to FOREGS soil geochemistry. *Journal of Geochemical Exploration*, 133, 160–175.

Sæther, E. (1957). *The Alkaline Rock Province of the Fen Area in Southern Norway*. Bruns Bokhandel i kommisjon, Trondheim.

Sahlström, F., Jonsson, E., Högdahl, K., Troll, V.R., Harris, C., Jolis, E.M., Weis, F. (2019). Interaction between high-temperature magmatic fluids and limestone explains "Bastnäs-type" REE deposits in central Sweden. *Scientific Reports*, 9(1), 15203.

Salgueiro, R., Inverno, C., de Oliveira, D.P.S., Guimarães, F., Lencastre, J., Rosa, D. (2020). Alluvial nodular monazite in Monfortinho (Idanha-a-Nova, Portugal): Regional distribution and genesis. *Journal of Geochemical Exploration*, 210, 106444.

Sarapää, O. (1996). *Proterozoic Primary Kaolin Deposits at Virtasalmi, South-eastern Finland*. Geological Survey of Finland, Espoo.

Sarapää, O. and Sarala, P. (2011). Mäkäräselkä REE-Au exploration target, Sodankylä, northern Finland. *25th International Applied Geochemistry Symposium 2011, 22-26 August 2011, Rovaniemi, Finland. Geochemical and Indicator Mineral Exploration Methods and Ongoing Projects in the Glaciated Terrains in Northern Finland: Excursion Guide, 26-30 August 2011: Vuorimiesyhdistys-Finnish Association of Mining and Metallurgical Engineers*, Rovaniemi.

Sarapää, O. and Sarala, P. (2013). Rare earth element and gold exploration in glaciated terrain – Example from the Mäkärä area, Northern Finland. *Geochemistry: Exploration, Environment, Analysis*, 13, 131–143.

Sarapää, O., Al Ani, T., Lahti, S.I., Lauri, L.S., Sarala, P., Torppa, A., Kontinen, A. (2013). Rare earth exploration potential in Finland. *Journal of Geochemical Exploration*, 133, 25–41.

Saunders, A.D., Fitton, J.G., Kerr, A.C., Norry, M.J., Kent, R.W. (1997). The North Atlantic Igneous Province. In *Large Igneous Provinces. Geophysical Monograph*, Mahoney, J.J. and Coffin, M.F. (eds). American Geophysical Union, Washington.

Schärer, U., de Parseval, P., Polvé, M., de Saint-Blanquat, M. (1999). Formation of the Trimouns talc-chlorite deposit (Pyrenees) from persistent hydrothermal activity between 112 and 97 Ma. *Terra Nova*, 11, 30–37.

Schassberger, H.T. and Newall, G.C. (1980). The east Greenland molybdenum project 1979. Internal NM-report, AMAX.

Schassberger, H.T. and Spieth, V. (1978). Molybdenum exploration in NE Greenland 1977. Internal NM-report, AMAX.

Schaub, H.P. (1942). Zur Geologie der Traill Insel (Nordost-Grønland). *Eclogae Geol. Helv.*, 35, 1–54.

Scherer, E., Münker, C., Mezger, K. (2001). Calibration of the lutetium-hafnium clock. *Science*, 293, 683–687.

Scheyder, E. (2019). U.S. Army will fund rare earths plant for weapons development. *Reuters* [Online]. Available at: https://www.reuters.com/article/us-usa-rareearths-army-exclusive-idUSKBN1YF0HU.

Scheyder, E. (2020). Pentagon resumes rare earths funding program after review. *Reuters* [Online]. Available at: https://uk.reuters.com/article/us-usa-rareearths/pentagon-resumes-rare-earths-funding-program-after-review-idUSKCN24M2Z4.

Schilling, J., Wu, F.Y., McCammon, C., Wenzel, T., Marks, M.A.W., Pfaff, K., Jacob, D.E., Markl, G. (2011). The compositional variability of eudialyte-group minerals. *Miner. Magazine*, 75, 87–115.

Schleicher, H., Keller, J., Kramm, U. (1990). Isotope studies on alkaline volcanics and carbonatites from the Kaiserstuhl, Federal Republic of Germany. *Lithos*, 26, 21–35.

Schoenenberg, J. and Markl, G. (2008). The magmatic and fluid evolution of the Motzfeldt intrusion in South Greenland: Insights into the formation of agpaitic and miaskitic rocks. *Journal of Petrology*, 49, 1549–1577.

Sebastiaan, D., Mancheri, N., Tukker, A., Brown, T., Petavratzi, E., Espinoza, L. (2017). Report on the current use of critical raw materials. Report, Solutions for Critical Raw Materials – A European Expert Network (SCRREEN).

Secher, K. (1976). Airborne radiometric survey between 66° and 69°N, southern and central West Greenland. *Grønlands Geologiske Undersøgelse*, 80, 65–67.

Secher, K. (1980). Distribution of radioactive mineralization in central West Greenland. *Grønlands Geologiske Undersøgelse*, 100, 61–65.

Secher, K. (1986). Exploration for the Sarfartoq carbonatite complex, southern West Greenland. *Grønlands geol. Unders*, 128, 89–101.

Secher, K. and Larsen, L.M. (1980). Geology and mineralogy of the Sarfartoq carbonatite complex, southern West Greenland. *Lithos*, 13, 199–212.

Secher, K., Heaman, L.M., Nielsen, T.F.D., Jensen, S.M., Schjøth, F., Creaser, R.A. (2009). Timing of kimberlite, carbonatite, and ultramafic lamprophyre emplacement in the alkaline province located 64°-67° N in southern West Greenland. *Lithos*, 112S, 400–406.

Seifert,W., Kämpf, H., Wasternack, J. (2000). Compositional variation in apatite, phlogopite and other accessory minerals of the ultramafic Delitzsch complex, Germany: Implication for cooling history of carbonatites. *Lithos*, 53, 81–100.

Shaw, M.H. and Gunn, A.G. (1993). Rare earth elements in alkaline intrusions, North-West Scotland. Report, British Geological Survey.

Shepherd, T.J., Darbyshire, D.P.F., Moore, G.R., Greenwood, D.A. (1982). Rare earth element and isotopic geochemistry of the North Pennine ore deposits. In *Symposium Jules Agard ; gîtes filoniens Pb-Zn-F-Ba de basse température du domaine varisque d'Europe et d'Afrique du Nord*, Anonymous (ed.). BRGM, Orléans.

Sinisi, R. (2018). Mineralogical and geochemical features of Cretaceous bauxite from San Giovanni Rotondo (Apulia, Southern Italy): A provenance tool. *Minerals*, 8(12), 19.

Sjöqvist, A.S.L., Cornell, D.H., Andersen, T., Erambert, M., Ek, M., Leijd, M. (2013). Three compositional varieties of rare-earth element ore: Eudialyte-group minerals from the Norra Kärr alkaline complex, southern Sweden. *Minerals*, 3, 94–120.

Škoda, R. and Novák, M. (2007). Y, REE, Nb, Ta, Ti-oxide (AB_2O_6) minerals from REL-REE euxenite-subtype pegmatites of the Třebíč Pluton, Czech Republic: Substitutions and fractionation trends. *Lithos*, 95, 43–57.

Škoda, R., Novák, M., Houzar, S. (2006). Granitic NYF pegmatites of the Třebíč Pluton. *Acta Mus. Moraviae, Sci. Geol.*, 91, 129–176.

Smeds, S.A. (1990). Regional trends in mineral assemblages of Swedish Proterozoic granitic pegmatites and their geological significance. *Geologiska Föreningens I Stockholm Förhandlingar*, 112, 227–242.

Smirnov, V.I. (1977). *Ore Deposits of the USSR 3*. Pitman, San Francisco.

Smith, R.T., Cooper, D.C., Bland, D.J. (1994). The occurrence and economic potential of nodular monazite in south-central Wales. Mineral Reconnaissance Programme Report 130. British Geological Survey Technical Report WF/94/1.

Sørensen, H. (2001). The Ilímaussaq complex, South Greenland: Status of mineralogical research with new results. *Geology of Greenland Survey Bulletin*, 190, 55–64.

Sørensen, L.L. and Kalvig, P. (2011). The rare earth element potential in Greenland. *Geology and Ore*, 20(2011).

Sørensen, H., Hansen, J., Bondam, J. (1969). Preliminary account of the geology of the Kvanefjeld area of the Ilímaussaq intrusion, South Greenland. *GEUS*, 18, 1–40.

Sørensen, H., Bailey, J.C., Rose-Hansen, J. (2011). The emplacement and crystallization of the U-Th-REE rich agpaitic and hyperagpaitic lujavrites at Kvanefjeld, Ilímaussaq alkaline complex, South Greenland. *Bulletin of the Geological Society of Denmark*, 59, 69–92.

Steenfelt, A. (2012). Rare earth elements in Greenland: Known and new targets identified and characterised by regional stream sediment data. *Geochemistry: Exploration, Environment, Analysis*, 12, 313–326.

Steenfelt, A., Hollis, J.A., Secher, K. (2006). The Tikiusaaq carbonatite: A new Mesozoic intrusive complex in southern West Greenland. *GEUS Bulletin*, 10, 41–44.

Steensgaard, B.M., Kolb, J., Nielsen, T.F.D., Olsen, S.D., Pilbeam, L., Lieber, D., Clausen, A. (2010). The mineral resource assessment project, South-East Greenland: Year one. *Geological Survey of Denmark and Greenland Bulletin*, 20, 59–62.

Steensgaard, B.M., Stendal, H., Kalvig, P., Hanghøj, K. (2016). Review of potential resources for critical minerals in Greenland. MiMa Report 2016/3.

Stembera, J. (2012). Exploration projects in East Greenland. Oral presentation, Greenland Day Congress, CRGR (Czech Geological Research Group), Perth.

Stendal, H. (2014). Critical minerals in Greenland. Oral presentation, Mines and Money Congress [Online]. Available at: www.minesandmoney.com/hongkong.

Stewart, J.W. (1970). Precambrian alkaline-ultramafic/carbonatite volcanism at Qagssiarssuk, South Greenland. *Bulletin Grønlands Geologiske Undersøgelse*, 84, 70.

Sturesson, U. (1995). Llanvirnian (Ord.) iron ooids in Baltoscandia: Element mobility, REE distribution patterns, and origin of the REE. *Chemical Geology*, 125, 45–60.

Sulovský, P. (2000). Comparison of chemistry of the Třebíč durbachite and other durbachites. *Geol. Výzk. Morave Slez. v r.*, 1999, 135–140.

Tanbreez Mining Greenland A/S (2014). Tanbreez [Online]. Available at: http://tanbreez.com.

Tasman Metals Ltd. (2010a). Tasman expands Otanmäki REE Project in Finland [Online]. Available at: www.tasmanmetals.com.

Tasman Metals Ltd. (2010b). Tasman expands rare earth element portfolio with the Bastnäs Project, Sweden [Online]. Available at: www.tasmanmetals.com.

Tasman Metals Ltd. (2010c). Tasman acquires the past producing Korsnäs REE-Pb Mine, Finland [Online]. Available at: www.tasmanmetals.com.

Tasman Metals Ltd. (2012). Tasman expands rare earth element project portfolio in Finland [Online]. Available at: www.tasmanmetals.com.

Tasman Metals Ltd. (2013). Tasman submits mining lease application over Olserum heavy rare earth element project, Sweden [Online]. Available at: www.tasmanmetals.com.

Tegner, C., Duncan, R.A., Bernstein, S., Brooks, C.K., Bird, D.K., Storey, M. (1998). ^{40}Ar-^{39}Ar geochronology of tertiary mafic intrusions along the East Greenland rifted margin: Relation to flood basalts and the Iceland hotspot track. *Earth and Planetary Science Letters*, 156, 75–88.

Tegner, C., Brooks, C.K., Duncan, R.A., Heister, L.E., Bernstein, S. (2008). ^{40}Ar-^{39}Ar ages of intrusions in East Greenland: Rift-to drift transition over the Iceland hotspot. *Lithos*, 101, 480–500.

Teisseyre, W. (1903). Der Paleozoische Horst yon Podolien und die ihn umgehenden Senkugsfel-der. Beiträige zur Paleontologic und Geologie Osterreichs-Ungarn und d. *Orients*, 15, 29.

Thirlwall, M.F. (1988). Geochronology of Late Caledonian magmatism in northern Britain. *Journal of the Geological Society of London*, 145, 951–967.

Thomassen, B. (1988). The Motzfeldt 87 project. Report, Grønlands Geologiske Undersøgelse, 88/1.

Thompson, P., Mussett, A.E., Dagley, P. (1987). Revised ^{40}Ar-^{39}Ar age for granites of the Mourne Mountains, Ireland. *Scottish Journal of Geology*, 23, 215–220.

Thorpe, R.S., Potts, P.J., Sarre, M.B. (1977). Rare earth evidence concerning the origin of granites of the Isle of Skye, northwest Scotland. *Earth and Planetary Science Letters*, 36, 111–120.

Thybo, H., Pharaoh, T., Guterch, A. (eds) (1999). Geophysical investigations of the trans-European suture zone. *Tectonophysics*, 314(1), 1–5.

Thybo, H., Pharaoh, T., Guterch, A. (eds) (2002). The trans European suture zone II. *Tectonophysics*, 360, 314.

Tichomirowa, M., Whitehouse, M.J., Gerdes, A., Götze, J., Schulz, B., Belyatsky, B.V. (2013). Different zircon recrystallization types in carbonatites caused by magma mixing: Evidence from U-Pb dating, trace element and isotope composition (Hf and O) of zircons from two Precambrian carbonatites from Fennoscandia. *Chemical Geology*, 353, 173–198.

Tornquist, A.J.H. (1908). Die Feststellung des Siidwestrandes des Baltisch Russischen Schildes. *Schriften der Physikalisch-Okonomische Gesellschaft zu Konigsberg*, 49, 1–12.

Trägårdh, J. (1991). Metamorphism of magnesium-altered felsic volcanic rocks from Bergslagen, central Sweden. A transition from Mg-chlorite to cordierite-rich rocks. *Ore Geology Reviews*, 6, 485–497.

Tuduri, J., Pourret, O., Chauvet, A., Barbanson, L., Gaouzi, A., Ennaciri, A. (2011). Rare earth elements as proxies of supergene alteration processes from the giant Imiter silver deposit (Morocco). In *Proceeding of the 11th Biennial SGA Meeting*, Barra, F., Reich, M., Campos, E., Tornos, F. (eds). Ediciones Universidad Católica del Norte, Antofagasta.

Tuduri, J., Chevillard, M., Colin, S., Gloaguen, E., Gouin, J., Potel, S., Pourret, O. (2013). Formation of monazite-(MREE) from Paleozoic shales: Role of host rock chemical composition and organic material. *Mineral. Mag.*, 77, 2362.

Tuduri, J., Charles, N., Guyonnet, G., Melleton, J., Pourret, P., Rollat, A. (2015). Project ANR ASTER. Rapport de Tâche 4. Potentialité de stocks géologiques de terres rares en Europe et au Greenland. Report, BRGM, RP-64910-FR.

Tuduri, J., Lefebvre, G., Charles, N., Nabyl, Z., Gaillard, F., Pourret, O. (2020). Lumière sur la géologie des terres rares, pourquoi tant d'attraits ? *Géologues*, 204, 48–54.

Tuisku, P. and Huhma, H. (2006). Evolution of migmatitic granulite complexes: Implications from Lapland Granulite Belt, Part II: Isotopic dating. *Bulletin of the Geological Society of Finland*, 78, 143–175.

Tukiainen, T. (1986). Pyrochlore in the Motzfeldt centre of the Igaliko nepheline syenite complex, South Greenland. *Grønlands Geologiske Undersøgelse*, 86, 3.

Tukiainen, T. (1988). Niobium-Tantalum mineralization in the Motzfeldt centre of the Igaliko nepheline syenite complex, South Greenland. In *Mineral Deposits within European Community*, Boissonnas, J. and Omenetto, P. (eds). Springer-Verlag, Berlin.

Tukiainen, T., Bradshaw, C., Emeleus, C.H. (1984). Geological and radiometric mapping of the Motzfeldt Centre of the Igaliko Complex, South Greenland. *Grønlands Geologiske Undersøgelse*, 120, 78–83.

Upton, B.G.J. (2013). Tectono-magmatic evolution of the younger Gardar southern rift, South Greenland. *Geological Survey of Denmark and Greenland Bulletin*, 29, 128.

Upton, B.G.J. and Emeleus, C.H. (1987). Mid-Proterozoic alkaline magmatism in southern Greenland: The Gardar province. In *Alkaline Igneous Rocks*, Fitton, J.G. and Upton, B.G.J. (eds). Geological Society of London.

Upton, B.G.J., Emeleus, C.H., Heaman, L.M., Goodenough, K.M., Finch, A.A. (2003). Magmatism of the mid-Proterozoic Gardar Province, South Greenland: Chronology, petrogenesis and geological setting. *Lithos*, 68, 43–65.

Van Wambeke, L., Brinck, J.W., Deutzmann, W., Gonfiantini, R., Hubaux, A., Métais, D., Omenetto, P., Tongiorgi, E., Verfaillie, G., Weber, K. et al. (1964). *Les roches alkalines et les carbonatites du Kaiserstuhl*. European Atomic Energy Community, Rome.

Vanhanen, E. (2001). Geology, mineralogy and geochemistry of the Fe-Co-Au-(U) deposits in the Paleoproterozoic Kuusamo Schist Belt, northeastern Finland. *Geological Survey of Finland Bulletin*, 399, 283.

Vaquero, C. (1979). Descubrimiento, por primera vez en España, de una monacita de facies aberrantes portado de Europia: Spain. *Boletín Geológico y Minero*, 90–4, 374–379.

Vartiainen, H. (1980). The petrography, mineralogy and petrochemistry of the Sokli carbonatite massif, northern Finland. *Geological Survey of Finland Bulletin*, 313, 126.

Vartiainen, H. (2001). Sokli carbonatite complex, northern Finland. Res Terrae. PhD Thesis, University of Oulu.

Vartiainen, H. and Vitikka, A. (1994). The late stage dykes of the Sokli massif and their tectonic control. *Geochemistry International*, 31, 103–106.

Vartiainen, H. and Woolley, A.R. (1974). The age of the Sokli carbonatite, Finland, and some relationships of the North Atlantic alkaline igneous province. *Bulletin of the Geological Society of Finland*, 46, 81–91.

Vartiainen, H. and Woolley, A.R. (1976). The petrography, mineralogy and chemistry of the fenites of the Sokli carbonatite intrusion, Finland. *Geological Survey of Finland Bulletin*, 280, 89.

Vazquez-Lopez, R. and Jézéquel, P. (2002). Expertise géologique des anomalies radioactives des plages de Camargue : plage de l'Espiquette et les Saintes Maries de la Mer Ouest. Report, BRGM, RP-51754-FR.

Verhulst, A., Balaganskaya, E., Kirnarsky, Y., Demaiffe, D. (2000). Petrological and geochemical (trace elements and Sr-Nd isotopes) characteristics of the Paleozoic Kovdor ultramafic, alkaline and carbonatite intrusion (Kola Peninsula, NW Russia). *Lithos*, 51, 1–25.

Verplanck, P.L., Mariano, A.N., Mariano, A.J. (2016). Rare earth element ore geology of carbonatites. In *Rare Earth and Critical Elements in Ore Deposits*, Verplanck, P.L. and Hitzman, M.W. (eds). Society of Economic Geologists, Littleton.

Vidal-Legaz, B., Mancini, L., Blengini, G.A., Pavel, C., Marmier, A., Blagoeva, D., Latunussa, C., Nuss, P., Dewulf, J., Nita, V. et al. (2016). Raw Materials Scoreboard: European Innovation Partnership on Raw Materials. Report, Office of the European Union, Luxembourg.

Vlasov, K.A., Kuz'menko, M.V., Yes'kova, Y.M. (1966). *The Lovozero Alkali Massif.* Hafner, New York.

Vuotovesi, T. (1974). A new find of carbonatite in West Greenland, the Qaqarssuk ring dike. PhD Thesis, University of Helsinki.

Wagner, T., Boyce, A.J., Erzinger, J. (2010). Fluid-rock interaction during formation of metamorphic quartz veins: A REE and stable isotope study from Rhenish massif, Germany. *American Journal of Science*, 310, 645–682.

Waight, T., Baker, J., Willigers, B.W. (2002). Rb isotope dilution analyses by MC-ICPMS using Zr to correct for mass fractionation: Towards improved Rb-Sr geochronology? *Chemical Geology*, 186, 99–116.

Walter, B.F., Parsapoor, A., Braunger, S., Marks, M.A.W., Wenzel, T., Martin, M., Markl, G. (2018). Pyrochlore as a monitor for magmatic and hydrothermal processes in carbonatites from the Kaiserstuhl volcanic complex (SW Germany). *Chemical Geology*, 498, 1–16.

Walters, A.S. (2011). A reconnaissance investigation of REE concentrations in the North Pennine Orefield. BGS Report IR/11/050.

Walters, A.S., Lusty, P., Hill, A. (2011). Rare earth elements. Report, BGS NERC.

Walters, A.S., Hughes, H.S.R., Goodenough, K.M., Gunn, A.G., Lacinska, A. (2012). Rare earth mineralization in the Cnoc nan Cuilean intrusion of the Loch Loyal syenite Complex, northern Scotland. *Geophysical Research Abstracts*, 14, EGU2012 4587.

Weidendorfer, D., Schmidt, M.W., Mattsson, H.B. (2017). A common origin of carbonatite magmas. *Geology*, 45, 507–510.

Weihed, P., Eilu, P., Kolb, J., Sandstad, J.S. (2013). Metallic minerals deposits in Fennoscandia, Greenland and NW Russia. *SGA News Bulletin*, 33, 9–14.

Weisenberger, T.B., Spürgin, S., Lahaye, Y. (2014). Hydrothermal alteration and zeolitization of the Fohberg phonolite, Kaiserstuhl Volcanic Complex, Germany. *International Journal of Earth Sciences*, 103, 2273–2300.

Welin, E. (1980). Tabulation of recalculated radiometric ages published 1960-1979 for rocks and minerals in Sweden. *GFF*, 101, 309–320.

Weng, Z.H., Jowitt, S.M., Mudd, G.M., Haque, N. (2013). Assessing rare earth element mineral deposit types and links to environmental impacts. *Applied Earth Science*, 122(2), 83–96.

Weng, Z.H., Jowitt, S.M., Mudd, G.M., Haque, N. (2015). A detailed assessment of global rare earth element resources: Opportunities and challenges. *Economic Geology*, 110(8), 1925–1952.

Wimmenauer, W. (1963). Beitrffge zur Petrographie des Kaiserstuhls. Teile VI-VII. *Neues Jahrb. Mineral., Abh.*, 99, 231–276.

Wimmenauer, W. (1974). The alkaline province of Central Europe and France. In *The Alkaline Rocks*, Søensen, H. (ed.). Wiley, London.

Windle, S. and Nesbitt, R. (1993). The genesis of grey monazites: Textural and geochemical evidence from Central Spain. In *Mineral Deposits Studies Group Annual Meeting 1993, Abstracts Volume*. The Natural History Museum, London.

Wood, S.A. (1990a). The aqueous geochemistry of the rare-earth elements and yttrium 1: Review of available low-temperature data for inorganic complexes and the inorganic REE speciation of natural waters. *Chemical Geology*, 82, 159–186.

Wood, S.A. (1990b). The aqueous geochemistry of the rare-earth elements and yttrium 2: Theoretical predictions of speciation in hydrothermal solutions to 350°C at saturation water vapor pressure. *Chemical Geology*, 88(1), 99–125.

Woolley, A.R. (1987). *Alkaline Rocks and Carbonatites of the World. Part 1: North and South America*. Cambridge University Press.

Woolley, A.R. and Kjarsgaard, B.A. (2008). Paragenetic types of carbonatites as indicated by the diversity and relative abundances of associated silicate rocks: Evidence from a global database. *Canadian Mineralogist*, 46(4), 741–752.

Zepf, V. (2013). *Rare Earth Elements: A New Approach to the Nexus of Supply, Demand and Use: Exemplified along the Use of Neodymium in Permanent Magnets*. Springer, Berlin.

Zhou, B., Li, Z., Chen, C. (2017). Global potential of rare earth resources and rare earth demand from clean technologies. *Minerals*, 7(11), 203.

Zozulya, D.R., Bayanova, T.B., Eby, G.N. (2005). Geology and age of the late Archean Keivy alkaline province, northeastern Baltic Shield. *J. Geol.*, 113, 601–608.

Zozulya, D.R., Bayanova, T.B., Serov, P.N. (2007). Age and isotopic geochemical characteristics of Archean carbonatites and alkaline rocks of the Baltic Shield. *Doklady Akademii Nauk*, 415, 383–388.

Zozulya, D.R., Lyalina, L.M., Eby, N., Savchenko, Y.E. (2012). Ore geochemistry, zircon mineralogy, and genesis of the Sakharjok Y-Zr deposit, Kola Peninsula, Russia. *Geology of Ore Deposits*, 54, 81–98.

Zozulya, D.R., Eby, N., Bayanova, T., Lyalina, L. (2013). Archean alkaline magmatism of OIB-type on the Baltic Shield. *30th International Conference on "Ore Potential of Alkaline, Kimberlite and Carbonatite Magmatism"*. GEOKHI RAS, Moscow.

2

The Cornubian Batholith: Post-Collisional Variscan Granites and Resources

Robin SHAIL[1] and Beth SIMONS[2]

[1] *Camborne School of Mines, University of Exeter, United Kingdom*

[2] *St Agnes, Cornwall, United Kingdom*

2.1. Introduction

The Early Permian igneous rocks of SW England, and the associated granite-related magmatic-hydrothermal mineralization, constitute one of Europe's major magmatic and mineralization provinces (Figure 2.1). It exhibits broad similarities with other Variscan massifs, but there are also contrasts in that: (1) magmatic activity occurred entirely within a post-collisional extensional tectonic regime, and (2) SW England is located to the north of the Rheic-Rhenohercynian suture.

The main expression of felsic magmatism is the 250 km long Cornubian Batholith. It comprises composite peraluminous granite plutons, predominantly of crustal origin, that were emplaced into Variscan-deformed low-grade regionally metamorphosed Devonian and Carboniferous successions. It is associated with numerous rhyolite-microgranite dykes ("elvans") and the remnants of formerly more extensive rhyolite lavas. Contemporaneous mantle-derived lamprophyric and basaltic lavas occur at or near the base of post-Variscan Permian "red-bed" sedimentary successions and lamprophyre dykes are hosted by the pre-Permian Variscan basement.

For a color version of all figures in this chapter, see www.iste.co.uk/decree/metallic1.zip.

Metallic Resources 1,
coordinated by Sophie DECRÉE. © ISTE Ltd 2023.

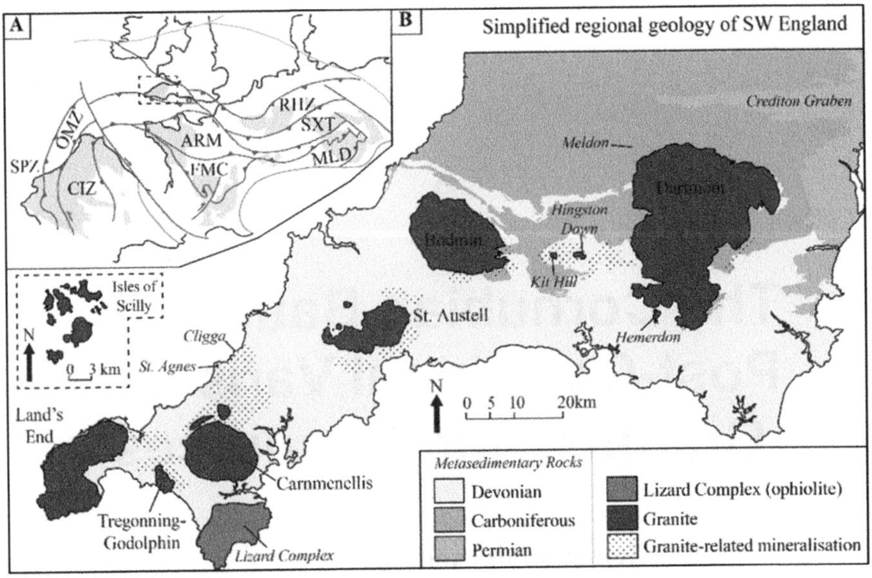

Figure 2.1. *(a) Tectonic map of the wider European Variscan belt showing the setting of SW England within the Rhenohercynian Zone (RHZ). ARM: Armorican Zone, CIZ: Central Iberian Zone, FMC: French Massif Central, MLD: Moldanubian Zone, OMZ: Ossa-Morena Zone, SPZ: South Portuguese Zone, SXT: Saxothurigan Zone. Modified after Franke (2000). (b) Simplified geological map of SW England, modified after Shail and Leveridge (2009) and showing the principal granite plutons, areas of granite-related mineralization and the Lizard Complex*

The Cornubian Batholith is associated with a wide range of metallic, industrial mineral and deep geothermal resources. Almost all of these have been controlled by structurally controlled fluid migration through steeply inclined joint and/or fault systems.

Granite-related mineralization occurred during the Early- to Mid-Permian, contemporaneous with batholith construction and generated a giant W-Sn-Cu-Zn-As-Zn orefield. Mineralization was primarily a consequence of the migration and mixing of magmatic-hydrothermal fluids with meteoric ± basinal fluids within fracture systems in the granites and their host rocks.

Subsequent and subordinate post-granite mineralization episodes occurred throughout the Mesozoic, most notably in the Mid-Triassic. These are related to the tectonically induced, fracture-controlled, migration of basinal brines from Permo-Triassic and younger sedimentary successions into the basement. The resulting "cross-course" veins have been exploited primarily for lead, zinc, silver, fluorite and barite within Devonian and Carboniferous successions and locally for iron within the granites.

Parts of the Cornubian Batholith constitute a poorly understood basement hydrocarbon reservoir that was charged during long-range fault-controlled migration in the Early Cretaceous. "Five-element" vein assemblages, comprising variable associations of cobalt, nickel, bismuth, arsenic, silver and uranium, may have also formed around this time. Multiple episodes of fluid–rock interaction, some related to weathering, have resulted in localized extensive kaolinization of the granites.

The history of tin mining stretches back over at least 4,000 years with major expansion of copper production during the 1700s industrial revolution. The last deep metal mine, South Crofty, closed in 1998 following a long period of low tin prices. However, increased metal prices from the mid-2000s have brought about renewed interest, primarily for tin, tungsten and, to a lesser extent, copper around known deposits such as Treliever (Sn), Redmoor (W-Sn ± Cu), United Downs (Sn-Cu) and Wheal Vor (Sn). The Hemerdon tungsten-tin deposit, one of the largest in the world, underwent substantial development and was in production during 2015–2018. There is ongoing work to dewater and redevelop South Crofty Mine and a continuing interest in offshore cassiterite placer deposits.

The St Austell granite represents one of the largest inferred hard rock lithium resources in Europe (Gourcerol et al. 2019) and exploration is focused on the occurrence of lithium-bearing micas. Recent granite-meteoric fluid reactions within the batholith have resulted in the development of fracture-hosted, lithium-enriched brines that are also being evaluated as a lithium source. The granites have high geothermal gradients, on account of their thickness and elevated contents of Th, U and K. Projected temperatures at a depth of 5 km are around 200°C in most of the plutons and SW England is the most prospective region in the UK for the development of deep geothermal energy (Beamish and Busby 2016). There is potential for the coproduction of power, heat and lithium brines.

While the region is best known for metalliferous mineralization, the mining of granite-related industrial minerals has generated significantly more revenue, and supported many more jobs, since the late 19th century. Localized extensive kaolinization of the granites formed world-class kaolinite deposits that have been worked since 1746. Quarrying of granite for dimension stone, armor stone and aggregate continues at multiple locations.

There is a published record of substantive geological research in SW England, relating to granites and mineralization, that stretches back over 250 years. It has been enhanced by data collected during the long history of surface and underground mining (e.g. Halls et al. 1985) and by the investigation of the hot dry rock (HDR) geothermal energy potential during the 1980s (e.g. Richards et al. 1991). Increased commercial activity directly and indirectly supports research and there has been substantial recent funding for both mineral and deep geothermal exploration across SW England from the European Union, private investors and the UK government. A substantial quantity of data have been generated and are starting to appear in a "new wave" of publications related to granites and resources (e.g. Simons et al. 2016, 2017; Dehaine et al. 2019; Yeomans et al. 2020).

In this chapter, we provide an overview of the (1) geological processes that have contributed to the development of granites of the Cornubian Batholith and the associated metallic, industrial mineral and deep geothermal resources, and (2) resurgence in exploration and development activity in SW England, in response to increased commodity prices and new opportunities associated with the production of metals deemed "critical" for low carbon technologies and meeting the needs of the energy transition.

2.2. Tectonic context of magmatism and resources

2.2.1. *Variscan tectonics*

SW England is an Upper Palaeozoic Massif surrounded by Permian and Mesozoic sedimentary basins (Evans 1990). It preserves a record of rifting and the development of a narrow Devonian Rhenohercynian Ocean basin on the northern margin of, or as a successor basin to, the Rheic Ocean (Franke 2000). Pre-Devonian basement is poorly constrained; while an Avalonian provenance is commonly assigned, other peri-Gondwana terranes may have been accreted at the northern margin of the Rheic Ocean prior to Rhenohercynian rifting (Nance et al. 2015; Dijkstra and Hatch 2018).

Variscan convergence in SW England occurred from the late Devonian to the late Carboniferous. In the south (Gramscatho Basin and Lizard Complex), peak-convergence-related (D1) regional metamorphism occurred during the late Devonian to earliest Carboniferous (ca. 370–360 Ma). It was related to accretionary processes accompanying the southwards subduction of oceanic lithosphere and the formation of the Rhenohercynian suture to which SW England forms the lower plate (e.g. Shail and Leveridge 2009; Alexander et al. 2019). Farther north, on the SW England proximal "passive" margin, rift-related magmatism and subsidence occurred throughout much of the Early Carboniferous on the southern margin of the Culm Basin (Leveridge and Hartley 2006). Here, Variscan primary deformation (D1) brought about a "thick-skinned" inversion of the Devonian and Carboniferous successions during the Visean to Serpukhovian (ca. 345–325 Ma) (Leveridge and Hartley 2006). Late Carboniferous Variscan (D2) convergence during the Moscovian to Kazimovian (ca. 315–305 Ma) occurred across the region; it brought about NNW-directed thrusting of the southern accretionary complex/Gramscatho Basin, and SSE-directed backthrusting of the Culm Basin, across the proximal passive margin (Leveridge and Hartley 2006).

Regional metamorphism is generally very low grade (anchizone-epizone), correlated with stratigraphical age and compatible with the burial of proximal passive margin successions during syn-rift subsidence and basin inversion (Warr et al. 1991). In the Gramscatho Basin, metamorphism is of a similar grade but related to thrust stacking during accretion; D2 thrusting at the southern margin of the Culm Basin may have generated local greenschist facies conditions (Warr et al. 1991). Overall, the predominance of very low-grade regional metamorphism suggests that Variscan convergence-related crustal thickening was modest.

2.2.2. *Post-Variscan tectonics*

A change in plate-boundary stresses occurred across NW Europe during the late Carboniferous as Variscan convergence was replaced by a dextral transtensional regime, manifested by major NW-SE faults (e.g. Ziegler and Dèzes 2006). In SW England, dextral movements on NW-SE faults were accompanied by the extensional reactivation of ENE-WSW to E-W Variscan thrust and transfer faults in the basement, including the Rheic-Rhenohercynian suture (Shail and Leveridge 2009; Alexander et al. 2019). The NNW-SSE extensional tectonic regime controlled the development of Early Permian "red-bed" sedimentary basins, bimodal magmatism (lamprophyres-basalts and granites) and most of the granite-related magmatic-hydrothermal mineralization (Shail and Wilkinson 1994). The latest magmatic-hydrothermal mineralization persisted into the Mid-Permian and was influenced by evolving stress regimes that included episodes of intraplate shortening (Shail and Alexander 1997).

Triassic ENE-WSW extension brought about the reactivation and further development of NNW-SSE fault systems and migration of brines from Permo-Triassic sedimentary basins into the basement to generate epithermal "cross-course" mineralization (Scrivener et al. 1994; Shail and Alexander 1997). Substantial rift-related uplift and exhumation of SW England occurred during the Early Cretaceous and was associated with the migration of basinal brines (e.g. Psyrillos et al. 2003) and hydrocarbons (Baba et al. 2019) into basement fault systems. Following Late Cretaceous subsidence and deposition of the Chalk, SW

England was once again exhumed during early Cenozoic uplift accompanying Atlantic opening (Evans 1990). Episodes of Cenozoic N-S intraplate shortening have brought about strike-slip reactivation of NW-SE and subordinate NNE-SSW fault systems as well as the localized development of pull-apart sedimentary basins in Devon (e.g. Holloway and Chadwick 1986).

2.3. Lamprophyres and basalts

Early Permian mafic volcanic rocks occur throughout SW England, including the presently offshore sedimentary basins. Onshore, there are two principal occurrences: (1) lamprophyric and basaltic lavas, collectively termed the "Exeter Volcanic Rocks", that occur at or near the base of post-Variscan Permian successions in the east of the region, and (2) intrusive lamprophyre sheets hosted by low-grade regionally metamorphosed Devonian and Carboniferous successions to the west. Permian mafic rocks are volumetrically subordinate, near-surface, to the felsic magmatism represented by the Cornubian Batholith and associated rocks but provide evidence of a link between the mantle and lower/mid-crustal melting and therefore the processes controlling Early Permian magmatism and mineralization. At locations in the Crediton Graben, where both lamprophyre and basaltic lavas occur together, the lamprophyres usually predate basalt and have provided $^{40}Ar/^{39}Ar$ biotite ages of 290.8 ± 0.8 (2σ) Ma (Killerton) and 281.8 ± 0.8 (2σ) Ma (Knowle Hill) (Edmonds et al. 1968; Edwards and Scrivener 1999). $^{40}Ar/^{39}Ar$ phlogopite ages for lamprophyric dykes range from 292.15 ± 1.5 (2σ) Ma to 284.38 ± 1.07 (2σ) Ma (Leat et al. 1987; Dupuis et al. 2015).

Basalts are laterally discontinuous lava flows with peperitic margins. They are typically chlorite-, carbonate- and clay-altered, with iddingsite pseudomorphs after olivine, rare plagioclase feldspar and quartz xenocrysts in a fine-grained groundmass of alkali feldspar, apatite, plagioclase and Fe-Ti oxides. The lamprophyric lavas are up to 50 m thick with peperitic margins and carbonate veins, like the basalts. Lamprophyre dykes are NE-SW to NW-SE striking and moderately- to steeply-dipping (30–70°) (Figure 2.2(a)).

Lamprophyre dykes and lamprophyric lavas have textural and mineralogical variations but are predominantly minettes using the classification of Rock (1991), with golden phlogopite mica, and more abundant alkali feldspar than plagioclase in a groundmass that includes iron oxides and apatite (Figure 2.2(b)). Clinopyroxene is rarely preserved, more commonly replaced by carbonates.

Figure 2.2. *(a) Exposure of the Mawnan lamprophyre dyke, Cornwall, hosted by Devonian metasedimentary rocks of the Portscatho Formation, looking east. (b) Photomicrograph showing a typical lamprophyre dyke texture – phlogopite phenocrysts (Phl) set in a fine-grained groundmass of alkali feldspar, apatite and Fe-Ti oxides. Adapted from Simons (2015)*

Basalts are tholeiitic in composition, while lamprophyre dykes are calc-alkaline and lamprophyre lavas can be either (Simons 2015; Dijkstra and Hatch 2018); all are high-K ($K_2O/Na_2O > 1$) or (ultra)potassic ($K_2O/Na_2O > 2.2$). On trace element spider diagrams, relative to normal mid-ocean ridge basalt (N-MORB), all rocks are enriched in light rare earth elements (REEs) and large ion lithophile elements (LILEs) relative to heavy REE, with distinctive Nb and Ti anomalies (Thorpe et al. 1986; Leat et al. 1987). Lamprophyre lavas have a wide range of Sr isotopic composition, from 0.7043 to 0.717 (Thorpe et al. 1986), with a more constrained range for the dykes of 0.7056 to 0.7062 (Dupuis et al. 2016). The Nd isotopic data have a narrow range, falling in the range of the enriched mantle, between EMI and EMII (Thorpe et al. 1986; Dupuis et al. 2016).

The notable depletion in Nb and Ti, combined with LREE and LILE enrichment and coupled with isotopic data, have been interpreted to reflect mantle refertilization processes (e.g. Wilson et al. 2004). Additionally, the steep heavy REE slopes for the lamprophyres are interpreted as representing a mantle source containing garnet, at a depth of >60–85 km, close to the base of the lithosphere (Dijkstra and Hatch 2018). Mixing and mingling of mafic and felsic melts, as well as crustal contamination, have also been proposed as important processes influencing the whole-rock geochemistry of basalts and lamprophyres in SW England (Edmonds et al. 1968; Hall 1982; Grimmer and Floyd 1986).

The lamprophyres and basalts in SW England have commonalities with similar rocks across the Variscan belt, including those in the French Massif Central, Armorica and Iberian Zones; the dominance of sedimentary geochemical fingerprints in refertilized mantle sources has been emphasized (Turpin et al. 1988; Orejana et al. 2008; Abdelfadil et al. 2013).

2.4. The Cornubian Batholith and associated felsic igneous rocks

The Cornubian Batholith (Figure 2.1) is exposed, from west to east, in six major plutons, the Isles of Scilly, Land's End, Tregonning-Godolphin, Carnmenellis, St Austell, Bodmin Moor and Dartmoor granites, plus smaller stocks including the Seven Stones, St Michael's Mount, Carn Brea, Carn Marth, St Agnes, Cligga Head, Castle-an-Dinas, Belowda Beacon, Kit Hill, Hingston Down and Gunnislake granites. The occurrence of rhyolite/microgranite sheets, contact metamorphism, magmatic-hydrothermal mineralization and host-rock alteration provides an approximate indication of the subsurface roof zone between the exposed plutons and stocks. Direct observations of the batholith, in outcrop and mines, have generally been within a vertical elevation range of ±600 m. However, the Carnmenellis granite has been proven to a depth of over 900 m below surface in the deepest mines. In 2019, the batholith close to the Carn Marth granite was drilled to a depth of 5,057 m at the United Downs Deep Geothermal Power Project. Knowledge of the subsurface geometry and extent of most of the batholith is indirect and entirely dependent upon the interpretation of geophysical data.

2.4.1. *Geophysical data*

2.4.1.1. *Gravity anomalies*

A negative gravity anomaly, coincident with the Cornubian Batholith (Bott et al. 1958), can be traced NE-SW across the Cornubian peninsula and platform for almost 300 km; its width increases from 20 to 50 km, and magnitude from −10 to −40 mGal, NE toward Dartmoor (British Geological Survey 2006). Combined onshore and offshore gravity anomaly datasets and 3D modeling indicate that all the major plutons are connected by ridges of granite that are generally less than 2 km below the surface (Willis-Richards and Jackson 1989).

Gravity-derived estimates of batholith thickness are dependent upon assumptions about the geometry and depth of the intrusion(s), the mean density contrast with the host rock and the regional background Bouguer anomaly. Density contrasts of 100–150 $kg{\cdot}m^{-3}$ have resulted in a mean modeled batholith thickness of 10–15 km and an estimated volume of 68,000 km^3 (Willis-Richards and Jackson 1989). The thickness to width ratio is anomalously high relative to many batholiths worldwide (e.g. Petford et al. 2000). A re-appraisal by Taylor (2007) modeled the short wavelength near-surface gravity anomaly and indicated a variable batholith thickness of ca. 3–5 km across the Bodmin, St Austell and Carnmenellis plutons, but ca. 9 km across the Dartmoor pluton. Nevertheless, other than around Dartmoor, the model requires additional deeper granite bodies to account for the longer wavelength gravity anomaly and the total thickness of granite in the crust remains similar to earlier estimates (Taylor 2007).

2.4.1.2. *Seismic reflection/refraction studies*

A crustal refraction survey along the length of the Cornubian Batholith interpreted the granite to form a 10–12 km thick upper seismic layer with a reasonably constant seismic velocity of 5.85 km·s^{-1} (Bott et al. 1970; Holder and Bott 1971). A subsequent crustal refraction/wide angle reflection survey identified a reflector (R1), only within the gravity-determined limits of the batholith, at depths of ca. 8 km in the southern Carnmenellis granite, ca. 7 km in the southern part of the Bodmin granite and, less well constrained, at ca. 10 km in the SW part of the Dartmoor granite (Brooks et al. 1984). Tentative suggestions for the geological significance of the R1 reflector included an internal granite boundary, a zone of thrusting through the granite, or stoped blocks (Brooks et al. 1984). The R1 reflector is interpreted here as the horizontal base of the batholith.

Deep offshore seismic reflection profiles, across the offshore continuation of the Cornubian Batholith (Klemperer and Hobbs 1991), and two onshore normal incidence seismic reflection profiles have been run across the Carnmenellis granite (Jones 1991). A marked shallowing and increase in the amplitude of lower crustal reflectivity occurs in a broad zone below the Cornubian Batholith. The most detailed investigation, using deep down-hole seismic sources and receivers, suggested that the reflectivity was a consequence of sharp lithological contrasts over vertical distances of 10 m or less (Warner et al. 1994). These features are interpreted here to represent mafic sills.

2.4.2. *Field occurrence*

2.4.2.1. *Relations with host rock deformation*

Contact relations with host rocks are most completely displayed in the coastal exposures of the Land's End, St Michael's Mount, Tregonning-Godolphin and Cligga Head granites, but also occur in inland exposures farther east. Contacts have also been proven throughout the region in mine, quarry and china clay workings and by exploration drilling; several unexposed near-surface stocks or culminations have been recognized.

Felsic igneous rocks always post-date Variscan convergence-related D1 and D2 deformation in the Devonian and Carboniferous host rocks (section 2.2.1). Post-Variscan ductile D3 structures in the host rocks, comprising SSE- or S-verging folds and associated cleavages, and/or fault systems displaying a top-to-the-south sense of shear, are truncated by pluton margins, cut by undeformed granite and aplite sheets, and included within host rock enclaves (Stone 1971). These relations are consistent with ductile D3 deformation, controlled by the post-Variscan extensional reactivation of thrust faults, occurring prior to the emplacement of the younger granites (Alexander and Shail 1995, 1996; Hughes et al. 2009). Magmatic- and solid-state fabrics in the older Scilly, Carnmenellisand Bodmin Moor G1 granites indicate coupling with host rock during ductile D3 deformation.

D3 and earlier structures were tilted away from the plutons during emplacement, particularly along their northern margins, to define broad open antiforms (Rattey and Sanderson 1984; Hughes et al. 2009). Bedding, metabasic rocks and the trace of S1 cleavage locally define marginal synclines around the northern parts of Carnmenellis, St Austell, Bodmin Moor and Dartmoor granites interpreted to reflect emplacement-related roof uplift (Taylor 2007). NW-dipping faults along the northern margin of the batholith have also been invoked as accommodating pluton growth (Alexander and Shail 1995).

2.4.2.2. *Granite contacts and fabrics*

2.4.2.2.1. Contacts with host rocks

Granite boundaries with host rocks reflect a mixture of intrusive contacts, influenced by pre- and syn-emplacement fractures, and post-emplacement faulting. Intrusive contacts are usually sharp and generally defined by:

– joints and/or faults parallel to bedding and/or S1 foliation, dominant in granite roof contacts, originally ENE–WSW to E–W striking and gently dipping but locally reorientated during emplacement;

– joints and/or faults that are inclined at a high angle to bedding/S1 foliation, dominant in sidewall contacts, typically ENE–WSW, ESE–WNW, NW–SE and NNE–SSW striking and steeply dipping.

Roof and sidewall boundaries typically comprise both gently and steeply inclined elements. The geometry of pluton-host rock contacts, and the distribution of granite mineralogical/textural facies, suggests that surface exposures of the mainland plutons are, vertically, within a few hundred meters of their original roof zones.

2.4.2.2.2. Internal contacts

The Cornubian Batholith is composite and internal contacts between both contrasting and similar mineralogical and/or textural granite types are common. Most represent the margins of sheet-like bodies, the majority of which are horizontal or gently inclined; some are steeply inclined, and a few have irregular forms (Powell et al. 1999; Salmon and Shail 1999) (Figure 2.3). The nature of these contacts reflects the rheological contrasts between the two granites at the time of emplacement. They range from planar fracture-controlled boundaries that truncate magmatic state foliations in the host granite, thus interpreted to have been essentially solid, through to co-existing "blobs" of mingling magma with irregular boundaries.

Figure 2.3. *(a) Sharp contact between coarse-grained porphyritic (G1a) and fine-grained (G1c) two mica granites on the Isles of Scilly. Field of view approximately 0.8 m across. (b) Sheet-like bodies of coarse-grained porphyritic (G3a) and coarse-grained poorly porphyritic (G3b) biotite granites within the Dartmoor pluton*

The fine-grained two-mica (G1) and biotite (G3) granites of, respectively, the Isles of Scilly and Land's End plutons have a variety of forms, ranging from subhorizontal sheet-like bodies, possibly >1 km^3 and interpreted as either intrusions or enclaves (Goode and Taylor 1988), to smaller bodies <1 m^3 that co-existed as enclaves of mingled magma within the coarse-grained biotite granite (Salmon 1994; Powell et al. 1999). Bodies within the Land's End granite occur as subhorizontal intrusive sheets sourced from a dyke, and possibly as large enclaves several hundred meters long (Salmon 1994; Müller et al. 2006a). Biotite-rich schlieren with variable rhythmic-graded and trough morphologies occur in both granites. Complex and gradational mineralogical and textural variations occur on the scale of up to tens of meters below internal contacts (e.g. Müller et al. 2006a).

China clay mining in the St Austell granite has progressively exposed contacts between the biotite (G3), tourmaline (G4) and topaz (G5) granites and their subtypes (Bray and Spooner 1983; Hill and Manning 1987; Manning et al. 1996). Contacts are commonly marked by the presence of a unidirectional solidification texture (UST) such as *stockscheiders*. Pegmatitic "comb-layered" feldspars have grown, often in branching forms, from a nucleation surface on solidified or largely solidified granite, into the melt-rich environment formed by the later magma batch. The textural variations within the tourmaline granites in part reflect multiple intrusive episodes; contacts may be sharp or diffuse and marked by magma mingling. Contacts between the fine-grained tourmaline granite and other types are often marked by zones up to 3 m wide with quartz-tourmaline ± feldspar vugs (Manning et al. 1996). Multiple smaller-scale unidirectional solidification textures are developed within the stocks of G2 granite, e.g. St Michael's Mount and Cligga Head granites, and may be related to episodic rapid exsolution and loss of magmatic-hydrothermal fluids accompanying the formation of greisen-bordered sheeted vein systems (Halls 1987).

2.4.2.2.3. Magmatic state fabrics

Magmatic state fabrics, comprising a foliation and/or lineation, develop in the presence of sufficient melt to allow the re-orientation of mineral grains without internal deformation (e.g. Paterson et al. 2019). They are variably developed in all Cornubian plutons and are defined by the alignment of K-feldspar and micas (Figure 2.4a). An early investigation of the Carnmenellis granite documented magmatic lineations defined by feldspar long axes and their dominant NNW-SSE orientation close to pluton-host rock and internal contacts (Ghosh 1934). The fabrics were interpreted to be generated by emplacement-related magma flow, initially steeply upwards in the SW and south, then spreading laterally northwards.

The application of anisotropy of magnetic susceptibility (AMS) techniques in the Carnmenellis granite by Mintsa Mi Nguema et al. (2002) confirmed the common occurrence of the NW-SE magma stretch. Broadly similar conclusions were obtained from an analysis of AMS fabrics in the Bodmin granite; fabrics, including a weak solid-state component, were generally more intense towards the pluton margins, and a NNW-SSE AMS lineation was interpreted as the direction of magma stretch (Bouchez et al. 2006). Integrated field mapping and traditional/AMS fabric studies along the north-western margin of the Land's End granite has indicated considerable complexity. The feldspar fabric preserves evidence for magma emplacement but the AMS fabric was strongly influenced by the growth of groundmass minerals during post-emplacement deformation (Kratinova et al. 2010). The former was influenced by the interaction with host rock boundaries and stoped blocks, and the latter by regional D3 NNW-SSE extension.

2.4.2.2.4. Solid-state fabrics

Solid-state fabrics are less common and reflect high temperature plastic deformation of minerals in melt-poor or melt-free rocks (e.g. Paterson et al. 2019). Significant crystal plastic deformation and the development of a solid-state foliation and mylonitic shear zones is restricted to localized areas of the older coarse-grained two-mica (G1) Isles of Scilly, Carnmenellis and Bodmin Moor granites (Dangerfield and Hawkes 1981). Along the NW margin of the Isles of Scilly granite, the coarse-grained two-mica granite possesses a steeply NW-dipping magmatic state foliation that is overprinted by a similarly orientated solid-state foliation characterized by the plastic deformation of quartz. The associated magmatic/solid-state lineation is typically steeply plunging within the foliation. Occasional kinematic indicators such as tiling of feldspar phenocrysts suggest these fabrics developed during a top sense-of shear to the SE. Gently SE-dipping extensional solid-state ductile shear zones are common and kinematic indicators are usually consistent with a top-sense-of-shear to the SE (Figure 2.4(b)). Coeval oblique- and strike-slip displacement occurred on less common NW-SE striking, steeply inclined, ductile transfer shear zones. The kinematics are compatible with the regional D3 tectonic episode and suggest coupling of granite and host rock deformation during the development of magmatic and solid-state fabrics.

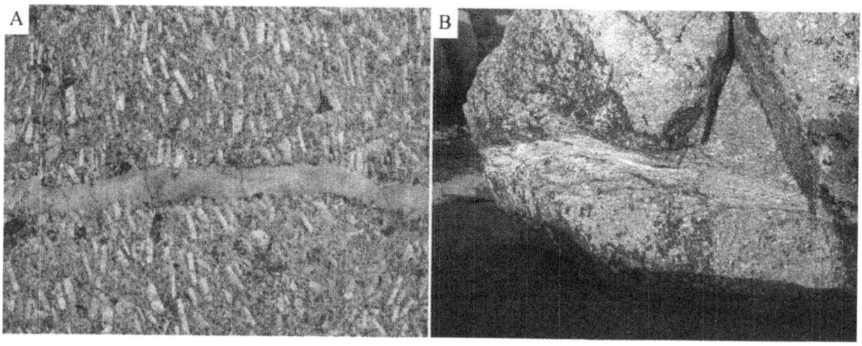

Figure 2.4. *(a) Alkali feldspar phenocrysts defining magmatic state foliation in coarse-grained porphyritic biotite granite (G3a) in the Land's End pluton, crosscut by a fine-grained aplitic sheet. Field of view approximately 0.75 m. (b) Mylonitic shear zone developed in coarse-grained porphyritic two mica (G1a) granite along the northern edge of the Isles of Scilly pluton. Field of view approximately 4 m*

2.4.2.2.5. Enclaves

Inclusions of rock fragments and mineral aggregates within granite plutons/stocks and, to a lesser extent, rhyolite/microgranite dykes are widespread; they are classified as igneous or non-igneous enclaves based on their mineralogy and texture. Igneous enclaves are typically ovoid, fine-grained and enriched in biotite and plagioclase relative to their host. One of the first substantive descriptions of enclaves globally was from the biotite granites of St Austell and Dartmoor (Phillips 1880).

A systematic investigation into the occurrence, mineralogy and whole-rock and mineral chemistry of igneous enclaves across the Cornubian Batholith was provided by Stimac et al. (1995). Igneous enclaves were estimated to constitute <1% by volume of the granites. Examples were found in all plutons, except the Carnmenellis granite, ranging in composition from monzogranite to tonalite, usually comprising a microgranular groundmass of biotite and plagioclase with minor quartz, ilmenite, alkali feldspar and pinitized cordierite. Larger crystals of alkali feldspar, plagioclase, biotite and quartz, similar in appearance and composition to phenocrysts in the host granite, were variably present. Almost all enclaves are mafic microgranular enclaves ("MME"; e.g. Vernon 1984) and interpreted as quenched hybrid magmas produced by mixing between biotite granite and unidentified mafic, probably basaltic, magmas (Stimac et al. 1995). MME are notably more abundant in the younger St Austell, Land's End and Dartmoor granites (Figures 2.5(a) and (b)).

Non-igneous enclaves, inclusions of metasedimentary rock, typically 20–50 mm in diameter but up to 1 m, sometimes elongate and possess ragged margins, occur widely (Figure 2.5(c)). Stimac et al. (1995) estimated that their disaggregation products, in the form of individual minerals or their aggregates, might account for up to 5% by volume of the Carnmenellis granite. Non-igneous enclaves commonly display compositional layering, some with leucosomes; the dominant mineralogy is biotite, andalusite, plagioclase, cordierite and quartz (Stimac et al. 1995). Rare sillimanite-hercynite-corundum-biotite assemblages probably have a restite origin, whereas the more common andalusite-bearing assemblage, in which sillimanite is absent, represent re-equilibrated restite or thermally metamorphosed aureole rocks from deeper levels than presently exposed near surface (Stimac et al. 1995). "Arrested" non-igneous enclaves, unequivocally reflecting host rocks incorporated by stoping at pluton roof and sidewall contacts, or by incorporation of "broken bridges" during the emplacement of granite sheets (Figure 2.5(d)), are relatively common.

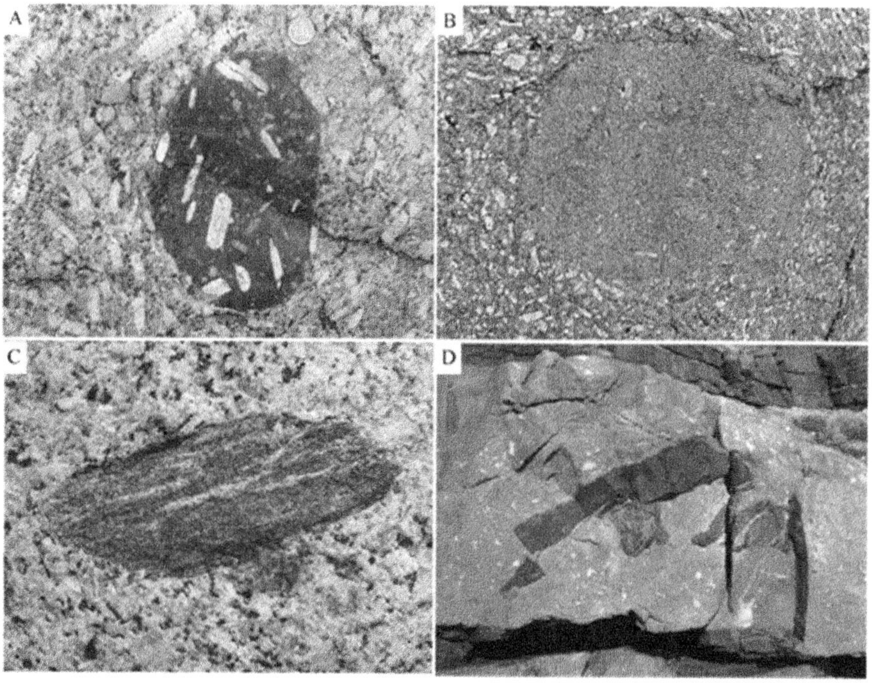

Figure 2.5. *(a) Mafic microgranular enclave (MME) approximately 0.2 m across with biotite and plagioclase in coarse-grained porphyritic biotite granite (G3a) of the Land's End pluton. (b) Fine-grained granite igneous enclave approximately 0.4 m across in coarse-grained porphyritic two mica (G1a) granite of the Isles of Scilly. (c) Foliated and partially disaggregated non-igneous enclave approximately 0.15 m across in the coarse-grained porphyritic biotite granite (G3a) of Land's End pluton. (d) Non-igneous host rock xenoliths attaining 1.5 m across representing "broken bridges" formed during emplacement of granite sills (G5) associated with the Tregonning pluton*

2.4.2.2.6. Post-emplacement mineralogical and textural evolution

Tourmaline-rich "orbicules" described from the roof zone of the Land's End granite probably formed from an immiscible boron-rich melt in the latest stages of crystallization, forming a rounded enclave predominantly composed of tourmaline with minor quartz, feldspar and mica. The orbicule tourmaline shows a higher Cr, Mg and V content, and has a late-stage overgrowth rich in Sr and Sn (Drivenes et al. 2015). Boron-rich fluids associated with an increased abundance of Sn share isotopic and chemical characteristics with evolved tourmaline granites in the Land's End roof zone; these granites are therefore interpreted as the magmatic precursors to Sn mineralization (Müller et al. 2006; Drivenes et al. 2015).

Vugs and aureoles of extensive tourmalinization are developed around topaz granites, suggesting the widespread separation of B-rich volatiles from their magmas (Manning and Hill 1990). Additionally, occurrences of massive quartz-tourmaline (MQT) rock occur, most notably at Roche Rock, north of the St Austell granite, and in smaller volumes in the roof zone of the Land's End granite. The origin of MQT from replacement of a granite by boron-rich fluids, highly evolved silicate liquids or magmatic-hydrothermal fluids have all been proposed (e.g. London and Manning 1995). The tourmaline within the MQT is similar in composition to that in the granites, with overgrowths that exhibit higher Sn contents, providing evidence for mixing with more oxidizing Fe-rich formation waters (Williamson et al. 2000). Luxullianite, a rock comprising almost entirely black tourmaline, pink K-feldspar and quartz, is also found around the St Austell granite. This rock is further evidence for movement of boron-rich fluids in the region, in this case replacing the original granite, precipitating tourmaline and growth of secondary K-feldspar (Alderton 1979).

2.4.2.3. *Intrusive rhyolite-microgranite sheets and hydrothermal breccias*

Rhyolite-microgranite sheets, often termed "elvans", or quartz ± feldspar porphyries in older literature, occur widely throughout SW England (Figure 2.6(a)). They are hosted by the Devonian-Carboniferous successions and most granite plutons; some sheets pass between the two. The majority occur as moderately to steeply inclined dykes that are 1–30 m thick and have typical strike-lengths of 1–2 km. However, local thicknesses can be greater and strike-lengths of up to 10 km do occur. Gravity anomaly data suggest that most of those hosted by Devonian-Carboniferous successions occur within 2 km of the batholith roof (Willis-Richards and Jackson 1989). Moderately to steeply inclined rhyolite/microgranite sheets are unknown from all but the south-western extremities of the Dartmoor granite (Hawkes 1982) and do not occur in the Land's End granite.

Intrusive elvan contacts are usually sharp and controlled by a similar range of pre- and syn-emplacement faults and joints as the granite plutons (Goode 1973). Dyke margins are often fine-grained and can exhibit flow foliations that may be folded and boudinaged (Goode 1973); they may contain abundant enclaves of wallrock and, occasionally, granite. Most dykes strike ENE-WSW to E-W, with subordinate ESE-WNW, NW-SE and N-S trends, and are usually parallel to nearby magmatic-hydrothermal mineral veins which they typically predate (e.g. Hosking 1964); many are displaced by or truncate against NW-SE striking faults. Paragenetically early mineral veins are occasionally cut by elvan dykes (e.g. Bray and Spooner 1983).

Intrusive hydrothermal breccias are usually hosted by the same fractures as rhyolite-microgranite sheets (Goode 1973); clasts, sometimes rounded, of host rock and granite occur in a matrix of comminuted rock that may be tourmalinized or chloritized. Rhyolite-microgranite can occur as angular or rounded clasts and, in rare instances, contemporaneous partially consolidated magmas that envelop other clasts (Goode and Taylor 1980; Allman-Ward et al. 1982) (Figure 2.6(b)).

Figure 2.6. *(a) Elvan (quartz porphyry) dyke hosted by metasedimentary rocks at St Agnes, looking towards the east. (b) Block from the Wheal Remfry Breccia, St Austell granite with sub-angular and rounded clasts of granite and coeval rhyolite in a tourmaline breccia*

2.4.2.4. *Rhyolite lavas*

Approximately 0.5 km^2 of rhyolite lava, in a succession up to 100 m in total thickness, occurs in SE Cornwall. It is bounded by the NW-SE trending Cawsand Fault and has gently inclined flow banding. The Kingsand Rhyolite Formation crops out at three small locations in SE Cornwall and provides the only in situ evidence for felsic volcanism associated with the Cornubian Batholith (Floyd et al. 1993; Leveridge et al. 2002). However, the abundance of rhyolite clasts in Permian sedimentary successions indicates that felsic volcanism was considerably more extensive (Cosgrove and Elliott 1976).

2.4.2.5. *Contact metamorphism and metasomatism*

The emplacement of the granite plutons generated contact metamorphic aureoles within the Devonian and Carboniferous host rocks that overprinted Variscan low-grade regional metamorphic assemblages. These are characterized by mineral spotting of the slates, typically cordierite + biotite ± andalusite. Aureoles extend up to 2.5 km from steep granite margins, with wider aureoles in the calcareous rocks. More extensive epizone and high anchizone metamorphic aureoles, not apparent in the field, are revealed by illite crystallinity studies (Warr et al. 1991). Exotic assemblages, such as cordierite-anthophyllite, were locally developed in altered pillow lavas and have formed the basis for a recent P-T estimate for the Land's End granite aureole of 1.5 ± 1.0 kbar and 615 ± 50°C (Pownall et al. 2012).

Metasomatism is widespread within the aureoles and can overprint earlier contact metamorphic assemblages (e.g. Alderton 1993). Carbonate-rich aureole rocks are uncommon but "skarns" are developed in calc-silicate and mafic igneous rocks on the NW margins of the Land's End, Carnmenellis, St Austell and Dartmoor granites (e.g. el-Sharkawi and Dearman 1966). Mineral assemblages include garnet (grossular-andradite), magnetite, epidote, diopside, hedenbergite, tourmaline, axinite, vesuvianite, wollastonite and Mn-silicates.

2.4.3. *Mineralogical and textural variation*

The granite plutons and stocks of the Cornubian Batholith comprise monzogranites, syenogranites and alkali feldspar granites with considerable variation, down to the cm scale. Classifications of the five main different granite types (G1-G5) are based on a combination of schemes using dominant ferromagnesian mineral (biotite or tourmaline), mean matrix grainsize, the size and abundance of alkali feldspar phenocrysts or rounded quartz (Figures 2.7 and 2.8) (Exley and Stone 1964; Dangerfield and Hawkes 1981; Manning et al. 1996; Simons et al. 2016). Mean groundmass grainsize may be coarse (>2 mm), medium (1-2 mm) or fine-grained (<1 mm). If alkali feldspar phenocrysts comprise >10% of the granite, it is considered porphyritic, <10% phenocrysts are moderately porphyritic, 1–5% are sparsely porphyritic and <1% are poorly porphyritic. This mineralogical and textural classification, when combined with radiometric dating, variations in mineral chemistry and whole-rock geochemistry and geochemical modeling, reflects two distinct stages of felsic magmatism, with differing metal budgets and implications for granite-related mineralization (Manning et al. 1996; Chappell and Hine 2006; Müller et al. 2006a; Simons et al. 2016, 2017).

The two-mica (G1) granites are represented by the Isles of Scilly, Carnmenellis and Bodmin plutons, and the muscovite granites (G2) by the granite stocks of St Michael's Mount, Carn Marth, Cligga, Kit Hill, Hingston Down and Hemerdon. Biotite (G3) and tourmaline (G4) granites are represented by the Dartmoor, St Austell and Land's End plutons. Topaz (G5) granites form the majority of the Tregonning pluton, small stocks within the St Austell granite and the Meldon Aplite dyke, NW of Dartmoor. Minor aplites and pegmatites, some mineralized, are associated with all granite types.

2.4.3.1. *Two-mica granite (G1)*

Two-mica granites (G1) are fine- to coarse-grained, typically porphyritic with perthitic orthoclase phenocrysts (<25 mm). *Quartz* is usually anhedral and can have late overgrowths. *Biotite group micas* attain up to 10% of the modal mineralogy and are Mg siderophyllite or siderophyllite (Figures 2.8(a) and (b)). *Alkali feldspar* phenocrysts are perthitic, euhedral to subhedral, often with irregular margins and Carlsbad twinning. In G1 and G2 granites, the phenocrysts are typically <25 mm in length, smaller than in the later granite variants. *Plagioclase* is typically euhedral to subhedral, with albite, Carlsbad or percline twinning. Normal zoning in plagioclase is common, with core and rim compositions of, respectively, An_{25-30} and An_{8-15} (Exley and Stone 1982). G1 granites have up to 6% modal muscovite, which can be both primary euhedral muscovite and secondary anhedral overgrowths on biotite, 1% modal tourmaline (schorl) and accessory minerals including zircon, andalusite, apatite, fluorite, ilmenite and topaz (Charoy 1986; Simons et al. 2016).

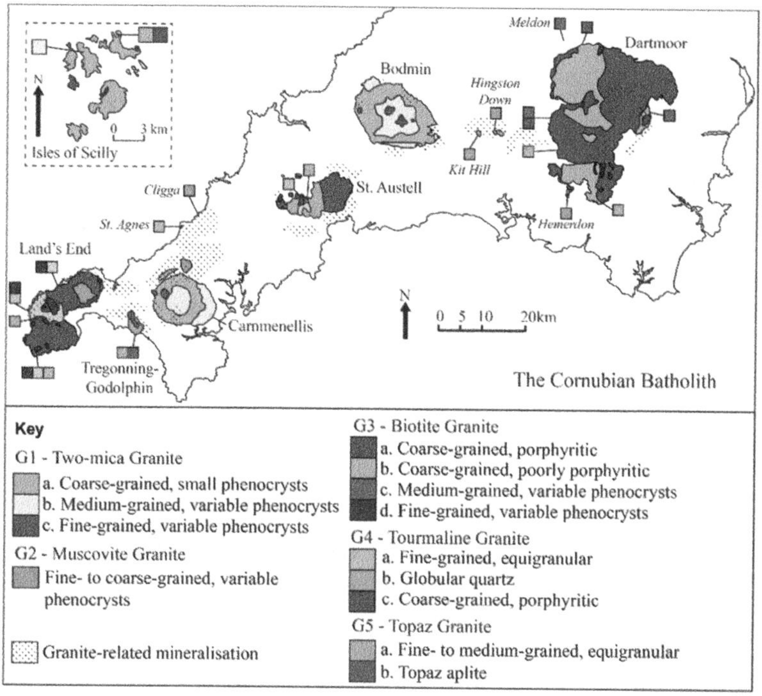

Figure 2.7. *Geological map of the principal mineralogical and textural variations in the Cornubian Batholith and subdivisions of G1-G5 granites with main areas of granite-related mineralization. Compiled from Ghosh (1927), Dangerfield and Hawkes (1981), Exley and Stone (1982), Manning et al. (1996), Müller et al. (2006a) and British Geological Survey Map Data NERC 2016. Modified from Simons et al. (2016)*

2.4.3.2. *Muscovite granite (G2)*

Muscovite (G2) granites (Figure 2.8(c)), like G1, are fine- to coarse-grained with variable sizes and abundances of alkali feldspar phenocrysts. G2 granites comprise up to 10% modal muscovite with minor *Li siderophyllite* to *zinnwaldite* biotite group micas and euhedral to subhedral *plagioclase* close to pure albite. *Quartz* is anhedral, but within G2 granites proximal to mineral deposits may contain radial inclusions of mica and tourmaline. *Muscovite* has two forms – primary, Fe-poor muscovite forming euhedral grains up to 3 mm across intergrown with biotite group micas; secondary, anhedral ferroan muscovite forming overgrowths and alterations of feldspars and biotite. Accessory minerals include andalusite, topaz, fluorite and Nb-bearing rutile. Tourmaline (schorl) attains up to 1.5% modal abundance, although with higher Li than tourmaline within G1 granites (Simons et al. 2016).

2.4.3.3. *Biotite granite (G3)*

Biotite (G3) granites are fine- to coarse-grained, with varying proportions of alkali feldspar phenocrysts (Figure 2.8(d)). Like G1 granites, *quartz* is anhedral with late microcrystalline quartz overgrowths. Mg siderophyllite and siderophyllite *biotite group micas* attain up to 10% modal abundance, and unlike G1-G2 granites, *muscovite* is a minor, anhedral, secondary accessory mineral. Within G3 granites, *alkali feldspar* phenocrysts are typically greater than 25 mm, distinguishing them from G1 granites. These phenocrysts are perthitic, euhedral to subhedral, often with irregular margins and Carlsbad twinning. Trace element chemistry shows variations in Ba and Sr, which decrease from core to rim, as Rb increases (Müller et al. 2006a). *Plagioclase* is subhedral to euhedral and normally zoned, but slightly more calcic than G1 granites. *Cordierite*, which is frequently pinitized, forms subhedral to euhedral coarse crystals within all G3 granite types. G3 granites have up to 1.5% modal tourmaline (schorl) with accessory zircon, monazite, xenotime, apatite and ilmenite.

2.4.3.4. *Tourmaline granite (G4)*

Tourmaline (G4) granites are characterized by tourmaline as the dominant ferromagnesian mineral. They are usually monzogranites and syenogranites and display considerable textural variation down to the sub-meter scale. They comprise fine- to coarse-grained variants, some with globular quartz (Figure 2.8(e)) whilst others appear similar in texture to porphyritic G1 granites, with alkali feldspar phenocrysts <25 mm in length. G4 granites are associated with G3 granites throughout the batholith, often forming gradational contacts or crosscutting sheets.

Biotite group minerals are anhedral to subhedral and trend toward lithium-rich varieties such as Li siderophyllite and zinnwaldite (Henderson et al. 1989). Like G2 granites, *plagioclase* is euhedral, with albite twinning and trends toward albite in composition (<An_7). The accessory mineral assemblage is diverse, with apatite, monazite, zircon, topaz, fluorite and ilmenite. Rutile contains elevated levels of Nb and Ta with micro inclusions of columbite-tantalite. Very fine-grained cassiterite (<10 μm) occurs in G4 granites in the St Austell pluton (Manning et al. 1996; Scott et al. 1998; Müller et al. 2006a; Simons et al. 2016).

2.4.3.5. *Topaz granite (G5)*

The topaz (G5) granites are medium-grained, equigranular and aphyric in all exposures, with very rare alkali feldspar phenocrysts (Figure 2.8(f)). Equigranular aplites are commonly associated with the granites. *Biotite group micas* are subhedral to euhedral and lithium-rich, attaining zinnwaldite or lepidolite in composition; there is almost no muscovite. Euhedral albite-twinned *plagioclase* is almost pure albite (<An_5) in composition with little zoning and, consequently, the topaz granites are classified as alkali feldspar granites. *Topaz* is a key mineral, attaining up to 3% modal abundance, forming euhedral disseminated crystals. *Tourmaline* is enriched in lithium, occasionally attaining elbaite composition with pink-green watermelon tourmaline found in the Tregonning granite (London and Manning 1995). Accessory minerals are diverse and comprise unusual phosphates and disseminated magmatic mineralization including amblygonite, zircon, Nb-(Ta)-rutile, Mn-ilmenite, columbite-tantalite and cassiterite, Mn-ilmenite and fluorite (Stone and George 1983; Manning et al. 1996; Scott et al. 1998; Simons et al. 2016).

2.4.3.6. *Rhyolite intrusive sheets (elvans) and (lavas)*

Elvans are fine- to medium-grained with euhedral *alkali feldspar* and/or rounded *quartz* phenocrysts in a fine-grained matrix. Some show grainsize variation from fine-grained margins to coarse-grained centers, with local areas of flow banding. Euhedral Carlsbad-twinned *alkali feldspar* often shows flow alignment parallel to the margins of the sheets. Fine-grained aggregates of *muscovite mica* and *tourmaline* are common (Stone 1968; Goode 1973). Rhyolite lavas crop out at very few localities in SW England, all restricted to SE Cornwall. At Kingsand, the rhyolite is fine-grained, with gently inclined flow banding. It has a devitrified quartz-feldspar groundmass with glass and spherulites and sanidine and quartz phenocrysts (Cosgrove and Elliot 1976).

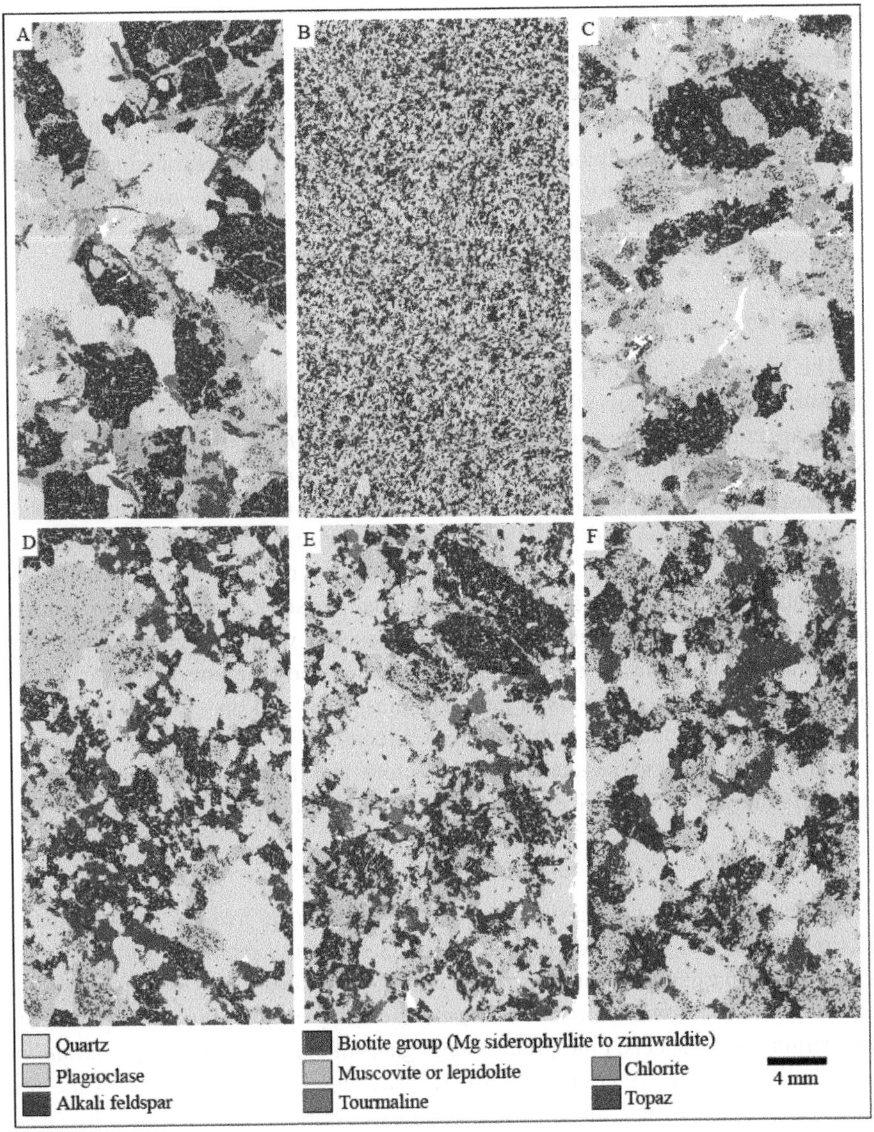

Figure 2.8. *QEMSCAN false color images showing predominant granite mineralogy. (a) Coarse-grained two-mica (G1a) granite (Bodmin). (b) Fine-grained two-mica (G1c) granite (Isles of Scilly). (c) Muscovite (G2) granite (Cligga). (d) Coarse-grained biotite granite (Dartmoor). (e) Globular quartz tourmaline (G4b) granite (St Austell). (f) Topaz (G5a) granite (St Austell). All images are at the same scale. Note that muscovite and lepidolite in these samples cannot be distinguished by QEMSCAN*

2.4.4. *Mineral chemistry*

The composition of biotite group micas and plagioclase feldspar effectively tracks granite geochemical evolution. The least evolved G1 and G3 granites have more calcic plagioclase (average $An_{15\text{-}30}$), with lower Ca in the more evolved G2 ($An_{<10}$) and G4 ($An_{<8}$) granites. G5 topaz granites have plagioclase compositions of almost pure albite ($An_{<5}$) (Simons et al. 2016). For micas, using the Mg-Li diagram of Tischendorf et al. (1997), a clear continuum between the different granite types is identified. The least evolved G1 and G3 granites have the highest Mg and lowest Li, through to the G2 and G4 granites, which have the highest Li; G5 topaz granites plot toward the highest Li and lowest Mg (Figure 2.9).

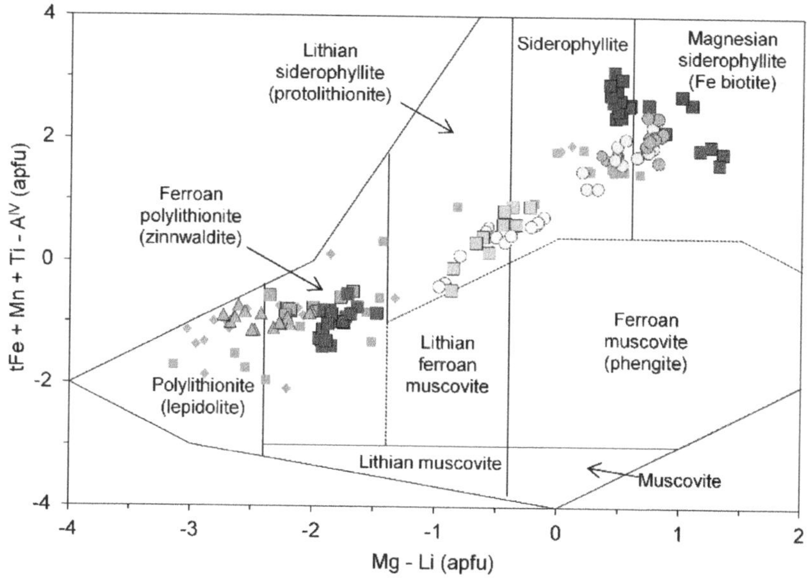

Figure 2.9. *The range of biotite group mica compositions in the different granite types using the Mg-Li diagram of Tischendorf et al. (1997); after Simons et al. (2016). Color key as per Figure 2.7. Additional data from Henderson et al. (1989) – gray diamonds and Stone et al. (1988) – gray squares*

Biotite and muscovite micas host several key metals including Li, Sn and W. Biotite group micas are the principal host of Li, reaching a maximum of 22,650 ppm in micas of G5 granites (Simons et al. 2017). Secondary muscovite and cordierite also host Li. Muscovite is the predominant host of W in G1, G2 and G4 granites, whereas Ga is distributed evenly across feldspars, micas and tourmaline. Tin is predominantly hosted in biotite group micas within G1, and likely partitions into Fe-Ti oxides in G5 (e.g. Scott et al. 1998).

2.4.5. *Whole-rock geochemistry*

The granites are all peraluminous, with A/CNK [(Al_2O_3/(CaO + Na_2O + K_2O)] values of approximately 1.25 for G1-G4 granites and 1.35 for G5 granites. The mineralogical and geochemical characteristics of the batholith are similar to "S-type" granites worldwide (Chappell and White 2001), but they have higher Al, K, Rb and lower Ca, Sr, Ba, Ti and Fe (Chappell and Hine 2006). The term "S-type" is best avoided in SW England as the G3 granites have a minor mantle component in their source. The Cornubian Batholith is also notably enriched in F, Li and B, reflected in high tourmaline and Li content of micas.

Trace element data for Nb, Zr and Ga support the mineralogical classification of five broad granite types G1-G5 (Figure 2.10) (Manning et al. 1996). Additionally, a range of geochemical characteristics distinguish the older G1 granites from the younger G3; the younger granites tend to have high Sr, Ba, FeO, MgO and CaO as well as higher abundances of Sn (Stone and Exley 1989; Stone 1997; Chappell and Hine 2006; Simons et al. 2017). The geochemistry of elvans is variable; some are comparable to G1 and G3 granites, with higher Zr (e.g. Hall 1970) but some represent G4 and G5 compositions with high Nb and low Zr (London and Manning 1995; Penfound-Marks and Shail 2015).

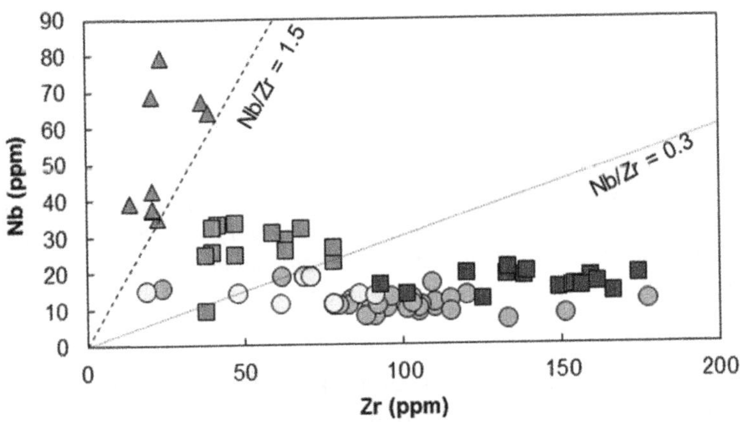

Figure 2.10. *Example geochemical plot of Zr versus Nb showing distinct groupings for G1, G2 and G3 versus G4 versus G5 supporting granite classification after Manning et al. (1996). Data from Simons et al. (2017). Color key as per Figure 2.7*

On the majority of major element and trace element plots there is geochemical continuum between G1-G2 and G3-G4 granites but G5 granites are dictinct. Using various differentiation indicators, including SiO_2, $1/tFe_2O$, Rb/Sr, Zr and Nb/Ta, G2 and G4 are the most evolved granites, with G5 granite displaying unique, evolved characteristics that differ from the G1-G4 granite types (e.g. Chappell and Hine 2006; Müller et al. 2006a; Simons et al. 2016).

REE data are available for all major and minor granite plutons (Figure 2.11). REE chemistry is strongly influenced by plagioclase, biotite, apatite, monazite, xenotime and zircon. Chondrite-normalized REE profiles for G1 and G3 granites show similar steep LREE enriched patterns, with Eu anomalies. The more evolved granites have flatter REE profiles, lower overall REE abundance and marked negative Eu anomalies (Stone and Exley 1989; Chappell and Hine 2006; Müller et al. 2006a; Simons et al. 2016).

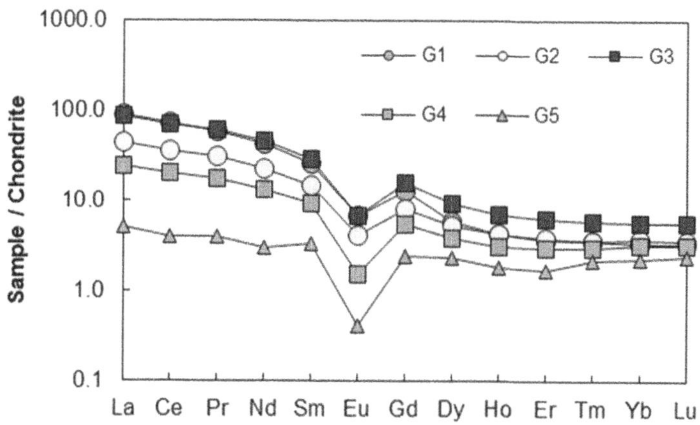

Figure 2.11. *Average chondrite-normalized REE patterns for G1-G5 granites using data from Chappell and Hine (2006), Müller et al. (2006a) and Simons et al. (2016). Normalizing values from McDonough and Sun (1995)*

2.4.6. *Geochronology*

U-Pb monazite/xenotime dating indicates that the granite plutons were emplaced during the Early Permian (Chen et al. 1993; Chesley et al. 1993). Dating of primary muscovite from the same samples using the ^{40}Ar-^{39}Ar method yields ages that are typically 3-5 Ma younger, and these have therefore been used as proxy indicators for other smaller granites where no U-Pb determinations are available, as well as to infer cooling rates (Chen et al. 1993; Chesley et al. 1993; Clark et al. 1993).

Granite magmatism occurred over a period of at least 20 Ma, from 293 Ma (Carnmenellis granite) to 274 Ma (Land's End granite) (Chen et al. 1993; Chesley et al. 1993). The two-mica (G1) granites of the Isles of Scilly, Carnmenellis and Bodmin plutons are ≥290 Ma, while the biotite granites (G3) of the Land's End, St Austell and Dartmoor plutons are ≤286 Ma, with most <282 Ma. The muscovite (G2) granite stocks of Cligga Head, St Michael's Mount, Kit Hill and Hingston Down were emplaced early in the second magmatic episode (≤286 Ma), for example: Cligga Head granite, U-Pb zircon age of 285.2 ± 0.1 Ma (Tapster and Bright 2020). An early short-lived episode of two-mica (G1) granite magmatism (293–290 Ma) was therefore followed by a longer episode of muscovite (G2), biotite (G3), tourmaline (G4) and topaz (G5) granite magmatism (286–274 Ma). A two-stage batholith construction history has also been recognized within the composite Crownhill stock; U-Pb ages of zircon rims in biotite (G3) and tourmaline (G4) granites are 10–12 Ma younger than the cores that are interpreted to be inherited from earlier two-mica (G1) and muscovite (G2) granites (Smith et al. 2019).

Rhyolite/microgranite "elvan" dykes are invariably altered and have low zircon abundance so there are few high precision dates. Low precision Rb-Sr whole rock ages include 282 ± 6 Ma (Wherry elvan, near Penzance) and 270 ± 9 Ma (Brannel elvan, near St Austell) (Darbyshire and Shepherd 1985). Indirect dating using field relations and dating of other geological features has provided further constraints. For instance, greisen-bordered veins in the St Austell granite have ^{40}Ar-^{39}Ar muscovite ages of 276.8 ± 1.2 Ma and are cut by rhyolite dykes; the Wheal Remfry tourmaline breccia has an ^{40}Ar-^{39}Ar muscovite age of 270.3 ± 0.8 Ma and includes fragments of coeval rhyolite magma (Figure 2.6(b)) implying that granite magmatism persisted beyond the 274 Ma recorded by the Land's End granite (Chesley et al. 1993). However, recent advances in U-Pb zircon methodology have demonstrated that high precision dating is possible and have yielded an age of 283.21 ± 0.45 Ma for an elvan dyke cutting the Cligga granite (Tapster and Bright 2020). A single, province-wide emplacement episode between 280 and 270 Ma, as proposed by Darbyshire and Shepherd (1985), is unlikely. Elvans always post-date two-mica (G1) granites and are likely to reflect the age range of their parental G3, G4 and G5 magmas ≤286 Ma.

2.4.7. *Synthesis*

2.4.7.1. *Tectonic controls on granite magmatism*

Granite magmatism in SW England contrasts with that in most other Variscan massifs in that it (i) is located north of the Rheic-Rhenohercynian suture, (ii) occupies a lower plate setting, and (iii) is exclusively post-collisional and Early Permian in age. Such distinctiveness has resulted in the proposal of "Cornwall-type" granite magmatism and mineralization (Romer and Kroner 2014, 2016).

Variscan convergence brought about moderate thickening of lower plate continental lithosphere during the Carboniferous (340–305 Ma). Resultant peak metamorphic temperatures likely pre-dated all but possibly the initial short-lived episode of two-mica (G1) granite magmatism (293–290 Ma). While the thermal effects of Variscan thickening of the lithosphere would have contributed to high lower and mid-crustal temperatures, additional heat sources are required to account for the extensive partial melting and batholith construction over a ca. 20 Ma interval during the Early Permian.

Thinning and exhumation of the lower plate continental lithosphere during the Early Permian extensional reactivation of the Rhenohercynian suture and other Variscan thrust faults (e.g. Alexander et al. 2019) brought about the decompression partial melting of lithospheric, and maybe asthenospheric, mantle (Leat et al. 1987). Lamprophyres and basalts span both the early episode of two-mica granite (G1) magmatism and the initial stages of biotite granite (G3) magmatism (e.g. Dupuis et al. 2015). Anomalous lower crustal reflectivity below SW England is interpreted to represent the presence of Early Permian mafic sills. High temperature mantle-derived melts are therefore interpreted to have been emplaced in a lower and middle crust that was already anomalously hot, following convergence-related thickening, and provided the additional heat necessary for high degrees of partial melting to generate the granites of the Cornubian Batholith (Simons et al. 2016).

2.4.7.2. *Granite petrogenesis*

The G1 granites were generated by partial melting of a greywacke source at moderate temperatures (730–800°C) and pressures (>5 kbar) and are linked to the G2 granites through fractionation of biotite, feldspars, monazite, zircon, allanite and apatite (Chappell and Hine 2006; Simons et al. 2016). G3 granites were generated by higher temperature (765–850°C) and lower pressure (<5 kbar) partial melting of a similar greywacke source and are linked to G4 granites through fractionation of biotite, feldspars and minor monazite, zircon, allanite and apatite (Simons et al. 2016). G4 granites often show considerable geochemical variation which has been interpreted as being due to fluid overprint (Müller et al. 2006b).

The brief hiatus in lamprophyric-basaltic and granite magmatism between ca. 289 and 287 Ma corresponds to the fundamental change from G1 to G3 granites as the dominant pluton-forming components. The younger (≤286 Ma) G3 granites are distinguished from the older (≥290 Ma) G1 granites by their textures, abundant mafic enclaves, less aluminous compositions, higher femic elements, less negative ε_{Nd} values, gentle REE patterns and biotite:muscovite ratios >1 (Stone 1997; Simons et al. 2016). These changes are consistent with the onset of higher temperature biotite-dominated partial melting in source regions that had previously undergone lower temperature muscovite-dominated partial melting, and greater mixing and mingling with mantle-derived melts (Simons et al. 2016). Higher temperature/lower pressure partial melting is likely controlled by the renewed emplacement of substantial volumes of mantle-derived melt into a lower and middle crust that was undergoing continued extensional exhumation (Simons et al. 2016). G2 granite stocks, for example, Kit Hill, Cligga and St Michael's Mount are one of the earliest expressions of the second episode of magmatism (≤286 Ma) and may result from muscovite-dominated partial melting of shallower greywacke sources that were unaffected by the initial magmatic episode.

The G5 granites do not exhibit the geochemical characteristics or trends of the G1-G2 or G3-G4 granites. The continued emplacement of mantle-derived melts may have contributed to flux-induced granulite facies partial melting of lower crustal restite biotite (Stone 1992; Simons et al. 2016, 2017). The model is like that inferred for the petrogenesis of the Beauvoir granite, French Massif Central (Cuney and Barbey 2014) in which F, Li, Nb, Ta and Sn are released into a melt from source biotite and Fe-Ti oxides.

2.4.7.3. *Granite emplacement*

The transport of magma from lower and middle crust source regions to the upper crust was most likely achieved via NW-SE faults; many have surface horizontal trace lengths that are consistent with a vertical persistence through much of the crust. Some of the earliest expressions of magmatic activity, such as eruption of the Kingsand Rhyolite Formation and the emplacement of the Hemerdon dyke, are controlled by the NW-SE and subordinate NNE-SSW faults. G2 muscovite granites and G5 topaz granites are commonly localized along, or close to, NW-SE faults. Magma transport in the upper crust, as indicated by intrusive sheets including elvans, was primarily controlled by ENE-WSW, E-W and NW-SE faults.

Field relations in well-exposed areas suggest that individual magma batches were commonly added to plutons as horizontal or gently inclined sheet-like intrusions (e.g. Powell et al. 1999; Müller et al. 2006a). The horizontal R1 reflector (Brooks et al. 1984), interpreted here as the base of the batholith, indicates a laccolithic geometry; emplacement of individual plutons was accommodated by roof uplift that brought about localized folding and/or tilting of host rocks (Taylor 2007; Hughes et al. 2009). The roof zones of the Land's End, Tregonning-Godolphin and Carnmenellis plutons are very close to the projected position of the extensionally reactivated Carrick Thrust and it may have controlled the magma emplacement level. Pre-existing faults and joints, some parallel to foliation, determined the geometry of contacts and contributed to minor stoping. AMS studies indicate that most magmatic- and solid-state fabrics within plutons are consistent with NNW-SSE regional extension during and after emplacement (Mintsa Mi Nguema et al. 2002; Bouchez et al. 2006; Kratinova et al. 2010).

2.4.7.4. *Textural variations*

Groundmass grainsize in most medium and coarse-grained porphyritic granites is related to undercooling following emplacement and steady-state growth of minerals from interstitial melt. Given the composite nature of the plutons, individual melt batches can have varying cooling histories influenced by earlier and later intrusive episodes. However, the loss of volatiles is also a significant control on groundmass grainsize, particularly in B-enriched melts. Liquidus and solidus temperatures are significantly reduced by volatiles but, on their loss, the melt is very significantly undercooled and very high nucleation to growth ratios result in a very fine-grained groundmass. Such "decompression quenching" is a very important control in the large range of textural variations exhibited by the larger composite bodies of tourmaline granite, in particular the globular quartz G4 facies (Manning et al. 1996). "Two-phase granites" or "disequilibrium texture granites" resulting from these processes have been recognized in all the major plutons (Cobbing and Clarke 2006). Complex banded pegmatite-aplite sheets are associated with the G5 topaz granites, most notably in the Tregonning granite at Megiliggar Rocks (e.g. Breiter et al. 2018).

2.4.7.5. *Implications for mineralization*

Peraluminous granites around the world are a major source of magmatic and magmatic-hydrothermal ore deposits. Elements described as "lithophile" are often associated with peraluminous granites, for example, Li, Nb, Be, Sn and W (e.g. Černý et al. 2005). Relative to average upper and lower continental crust, the Cornubian Batholith has a strong enrichment in Li and a moderate enrichment in Sn, Ta, W and Bi (Figure 2.12). G5 granites and aplites are notably enriched in Li, Nb, Ta, Sn and W (Stone 1992; Simons et al. 2017). While the region is enriched in Cu, it is low in the granites (e.g. Müller et al. 2006a; Simons et al. 2017).

Source composition, melting temperatures and subsequent fractionation exert the first-order control on the prospectivity of the G1-G4 granites for magmatic-hydrothermal mineralization. At lower temperatures, muscovite is one of the first minerals to undergo melting within crustal sources and would have been one of the main minerals that melted to form the G1 granites (Simons et al. 2016). Although the source in SW England is not exposed, the distribution of metals within the granites indicates that a primary muscovite hosts significant proportions of W. Biotite micas, which melt at higher temperatures and would have likely undergone more substantial source melting to form the later G3 granites, contain high proportions of Li and Sn. Metals such as Nb and Ta are known to be hosted in Fe-Ti oxides (e.g. Ballouard et al. 2016).

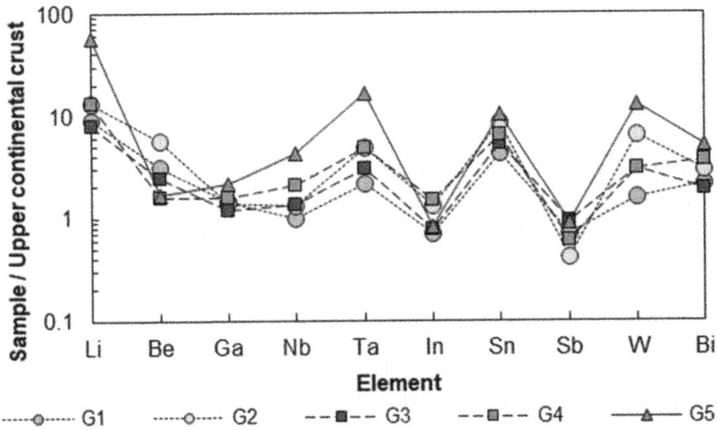

Figure 2.12. *Multi-element plot showing average metal abundances normalized to the upper continental crustal (UCC) values of Rudnick and Gao (2004), showing strong Li, Nb and Ta enrichment in G5 granites. Figure from Simons et al. (2017). Color key as per Figure 2.7*

The presence of P and F in granitic melts has an implication for the retention and fractionation of metals including Nb, Ta, Sn and W within peraluminous granite melts, which have been demonstrated across SW England. Increased F within fractionated granite melts is common, along with B and P. The presence of these elements lowers the temperature of the granitic melt (e.g. Černy et al. 2005) and promotes retention within residual melts/fluids of Nb, Ta and W, as these metals partition in favor of the melt during the crystallization of minerals such as plagioclase, micas and quartz. If F partitions into topaz, Nb, Ta and W will partition into micas and oxides, such as rutile, if the melt is saturated with respect to these minerals (Keppler 1993; Linnen 1998). The Cornubian Batholith G4 and G5 granites have the highest F and also Nb and Ta, which is manifested as disseminated magmatic mineralization (section 2.5.1).

Phosphorus also increases with increasing fractionation in the Cornubian Batholith. Within the more evolved G4 and G5 granites, the high-P and low-Ca, termed the "Pedrobernardo-type" trend by Bea et al. (1992), causes P to behave incompatibly during fractionation and concentrate in residual fluids. This, in turn, ensures that the critical threshold of P is reached to crystallize phosphate minerals such as amblygonite, documented in the G5 granites (Stone and George 1978). The combination of high F and P within evolved granites supports retention of metals such as Sn and W within residual melts, so that they can later partition into magmatic-hydrothermal fluids. This is also demonstrated for B-rich melts, where G4 granites are interpreted as the magmatic precursors to mineralization in the Land's End granite (Müller et al. 2006a).

Metagreywackes in the lower and middle crust (section 2.4.7.2) were the primary sources of the Sn, W, Li, Nb and Ta within the granite and associated mineralization (Simons et al. 2017). Sn and W were concentrated through a combination of early Palaeozoic peri-Gondwanan chemical weathering and, possibly, pre-Devonian accretion to the Avalonian margin before they underwent Early Permian partial melting (Romer and Kroner 2014; Dijkstra and Hatch 2018).

There are suggestions that lamprophyres affect metallogenetic processes for metals such as Sn, W and U within mineral deposits associated with Variscan granites, particularly in Germany (Seifert 2008; Štemprok and Seifert 2011). A very minor mantle component is supported by contemporaneous mantle- and crustal-derived magmatism, particularly for the younger granites which contain MME. The initial $^{87}Sr/^{86}Sr$ ratios for the major plutons vary between 0.709 and 0717, indicating a dominant crustal source (Darbyshire and Shepherd 1994).

The occurrence of substantial magmatic-hydrothermal Cu mineralization associated with peraluminous granites is potentially problematic since the reducing conditions that favor the retention of Sn in a residual granite melt usually result in the incorporation of Cu into crystallizing igneous minerals and hence offer little prospect of concentration. Rare melt inclusions hosted by quartz in porphyritic Li-siderophyllite granite in the southern part of the St Just district contain drop-like micro-inclusions of chalcopyrite and rare galena that have been interpreted as remnants of an exotic sulfide melt that co-existed with the granite magma (Müller et al. 2006a).

2.5. Granite-related mineralization

The metallic resources of SW England are primarily related to the migration and mixing of granite-derived magmatic-hydrothermal fluids with meteoric ± basinal fluids within fracture systems in the granites and their host rocks. Less substantial mineralization episodes pre-date and post-date granite-related mineralization. Post-granite mineralization is briefly described as it is often recorded during exploration for granite-related mineralization, primarily within NW-SE to N-S striking "cross-course" fault zones that segment earlier lodes (see Box 2.1 for mineralization definitions), but also as late lode infills. These same cross-course fault zones influence the development of kaolinite deposits and are targets for deep geothermal and lithium brine exploration.

Terminology of mineralization in SW England

Carbonas are irregularly-shaped bodies of highly-altered tin-bearing granite. Tabular or pipe-like geometries are common and are typically structurally controlled by intersecting fractures and/or dilational jogs.

Cross-courses are lodes orientated approximately perpendicular to the granite-related Sn-Cu lodes, typically NW-SE to N-S.

Fluccans are any clay or gouge-infilled fault. They may include lodes, slides or cross-courses.

Gossans are zones of oxidation and leaching in the upper parts of sulfide-rich lodes that are characterized by the presence of iron oxides such as goethite or "limonite" and, occasionally, manganese oxides.

Greisens are vein wall rock alteration assemblages comprising secondary white mica, quartz ± topaz

Lodes comprise mineral veins and adjacent wall rock alteration that are persistent, usually >100 m in strike length, and generally controlled by faults or fault zones rather than joints. The more persistent lodes have been worked over strike lengths >1 km, e.g. Great Flat Lode (6 km) and vertical intervals of many 100s m, occasionally up to almost 1 km (e.g. Dolcoath Main Lode). Lode widths vary markedly; most are 1–2 m but a few lodes were worked persistently over widths >10 m.

Sheeted veins are closely spaced (usually <1 m) sub-parallel veins that result from the mineralization of a single joint set.

Slides are faults that strike approximately parallel to a lode but intersect and displace it (their dip direction may be the same or opposite to that of the lode).

Stockworks comprise closely spaced mineralized fractures in which more than one set, and/or numerous non-systematic fractures of limited persistence, are present. Most occurrences are dominated by one fracture set and are therefore closer in style to sheeted veins.

Box 2.1. *Terminology used to describe mineralization styles in SW England*

Early- to Mid-Permian granite-related mineralization is diverse, including disseminated magmatic mineralization (Li, Nb, Ta, Sn), mineralized pegmatites, sulfide and magnetite "skarns" (replacement deposits) and greisen-bordered sheeted vein complexes (W-Sn). However, most historical production has been from fault-controlled quartz-tourmaline (Sn) and polymetallic sulfide (Cu, Zn, Sn, As) lodes (Scrivener 2006 and references therein).

Post-granite mineralization occurred throughout the Mesozoic, but most notably in the Mid-Triassic when most of the cross-course infills formed (Pb, Zn, Ag, Fe, fluorite and barite in host rocks; Fe in granites). Cross-courses have also controlled the long-range migration of hydrocarbons into the batholith and localized development of "five-element" vein assemblages (variable combinations of Ni, Co, Bi, Ag, U and As). These post-granite mineralization episodes reflect structurally controlled fluid flow in the basement (granites and host rocks), driven by episodes of rifting and inversion that are recognized in adjacent post-Variscan sedimentary basins.

Multiple episodes of fluid-rock interaction, some related to weathering, have resulted in localized extensive kaolinization of the granites and the development of china clay. Supergene enrichment, associated with the development of gossans in the upper parts of sulfide-bearing lodes and a huge variety of complex secondary minerals, is widespread. Cassiterite eroded from mineralized areas formed substantial alluvial and shallow marine placer deposits.

2.5.1. *Disseminated magmatic mineralization*

Low-grade, disseminated, rare metal mineralization occurs within the Nanpean G5 topaz granite and within G3 biotite and G4 granites of the St Austell granite. Metallurgical studies of the micaceous residue generated during china clay extraction has shown fine-grained (<100 μm) heavy minerals including columbite-tantalite, Nb-Ta rutile, cassiterite and apatite, together with subordinate zircon, monazite and Mn-ilmenite (Dehaine et al. 2019); these are likely to have a magmatic origin (Scott et al. 1998). Although unexploited to date, these represent one of the few significant regional occurrences of niobium- and tantalum-bearing minerals and have the potential for extraction as a by-product (Dehaine et al. 2019). Disseminated cassiterite and a variety of tin sulfides including stannite, stannoidite ($Cu_8(Fe,Zn)_3Sn_2S_{12}$) and mawsonite ($Cu_6Fe_2SnS_8$), together with bornite, digenite and tennantite occur locally within the Hensbarrow G5 granite in the St Austell pluton (Manning 1983).

2.5.2. *Pegmatites*

Pegmatites ± aplites, located within plutons or at their host rock contacts, are occasionally mineralized, particularly those associated with topaz G5 granites. Other occurrences that have been referred to as "pegmatites" are more appropriately classified as quartz- and feldspar-bearing hydrothermal veins (Hosking 1954) and are described in section 2.5.4. Pegmatite-aplite sheets in the roof zone of the Tregonning (G5) granite host löllingite, arsenopyrite, molybdenite, bismuth and wolframite with traces of columbite, Sn-bearing rutile, cassiterite and uraninite (Bromley 1989) (Figure 2.13); many of these minerals are also identified in the Meldon topaz (G5) aplite (e.g. Floyd et al. 1993). The Carnmenellis granite (G1) hosts internal pegmatites that comprise quartz, albite, microcline, orthoclase, tourmaline, muscovite, Li-mica, topaz, apatite, fluorite and bertrandite ($Be_4Si_2O_7(OH)_2$), but which also include cassiterite, stokesite ($CaSnSi_3O_9{\bullet}2H_2O$), sphalerite, pyrite and galena (Hosking 1964). The pegmatite assemblages have not been commercially exploited but have strong affinities with vein systems exhibiting potassic or greisen alteration and provide a linkage between magmatic and magmatic-hydrothermal processes (Bromley 1989).

Figure 2.13. *(a) Pegmatite with large alkali feldspars at Megiliggar Rocks associated with topaz G5 granite. (b) Layered pegmatite-aplite zone at Porthmeor Cove, Land's End granite*

2.5.3. *Replacement deposits*

2.5.3.1. *Skarns*

Calc-silicate mineral assemblages developed during contact metamorphism ± metasomatism of Ca-bearing lithologies in the granite aureoles can undergo further metasomatism as they interacted with magmatic-hydrothermal fluids. In some instances, these have resulted in pervasive Sn or Cu mineralization, comprising cassiterite and sulfides. Fe-rich protoliths may be transformed to magnetite-bearing assemblages that, in a few cases, have been worked as a magnetite ore. Overall mineral production from skarns was minor, but a few high tonnage bodies have been exploited and it has attracted exploration attention in the 20th century, given the potential for large tonnage, lower grade deposits. However, complex mineralogies, often accompanied by fine-grained cassiterite, create mineral processing challenges.

Limestones have a limited occurrence in the granite aureoles and true skarns are largely restricted to the northern margin of the Dartmoor granite. However, calc-silicate assemblages are developed in other metasedimentary rocks and altered metabasic igneous rocks, in which Ca is derived from feldspar and/or pyroxene. Replacement mineralization is usually related to the infiltration of magmatic-hydrothermal fluids containing B, Sn, Fe, As and Cu. Axinite ((Ca, Fe, Mn)$_3$Al$_2$BO$_3$Si$_4$O$_{12}$OH) is the dominant boron-bearing mineral.

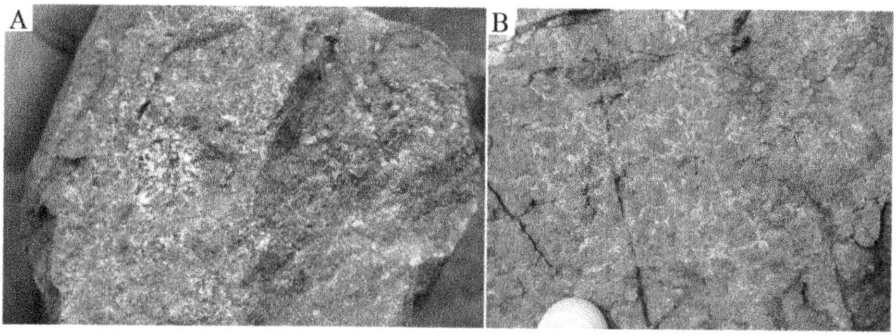

Figure 2.14. *(a) Replacement deposit at Red-a-Ven, Dartmoor granite comprising fine-grained pyrrhotite, pyrite, sphalerite and löllingite in bands and lenses with minor chalcopyrite. (b) Hydrous grossular-andradite garnets in the replacement deposits at Botallack, north of the Land's End granite*

Replacement deposits are widely developed around the northern margin of the Dartmoor (G3) granite within the cherts and limestones of the Meldon Shale and Quartzite Formation (el-Sharkawi and Dearman 1966). Contact metamorphism and metasomatism at Red-a-Ven produced a complex assemblage of grossular, vesuvianite, wollastonite, diopside and Mn-silicates that was variably replaced by hedenbergite, andradite, Fe-wollastonite, axinite and datolite, together with pyrrhotite, pyrite, chalcopyrite, sphalerite, löllingite, arsenopyrite, bornite, Bi, Co and Ni sulfides and scheelite (el-Sharkawi and Dearman 1966; Edmonds et al. 1968; Alderton 2018). Sn occurs in malayaite, grossular and andradite (Figure 2.14(a)). It is possible that some of the metals may have been derived from pre-granite stratiform mineralization (Beer and Scrivener 1982).

Around Botallack, on the NW margin of the Land's End granite, regionally and contact metamorphosed mafic lavas and tuffs, located within 100 m of the granite contact, are variably replaced. Transgressive zones up to 15 m thick and later veins contain an assemblage that includes garnet (grossular-andradite), magnetite, epidote, axinite, diopside and tourmaline (Jackson and Alderton 1974; van Marcke de Lummen 1985; Alderton 2018) (Figure 2.14(b)). Tin is hosted by cassiterite, malayaite, epidote, titanite and amphibole. Cassiterite has also been worked from a series of tourmalinite "floors" developed by selective replacement of hornblende-biotite horizons in metabasic rocks (Jackson 1974). A similar magnetite-hornblende-cassiterite assemblage, developed in metabasic sills on the north-eastern margin of the Carnmenellis (G1) granite, was worked for Sn at the Magdalen Mine (Dines 1956). Calc-silicate rocks, comprising axinite, epidote, garnet, hedenbergite and diopside, occur in the extensive metamorphic aureole to the north of the St Austell (G3-G4) granite, and together with other metasedimentary rocks, they are partially replaced, commonly along bedding and/or cleavage, by quartz, tourmaline, cassiterite and sulfides (Camm and Dominy 1997).

Bedding-parallel ore bodies of massive intergrown fine-grained magnetite and hornblende were worked from metasedimentary rocks close to the south-eastern margin of the Dartmoor granite at Haytor Iron Mine. The assemblage includes andradite, axinite, siderite, calcite and apatite and traces of arsenopyrite, pyrite, sphalerite and chalcopyrite (Scrivener et al. 1987), and likely results from the metamorphism and metasomatism of diagenetic siderite concretions.

2.5.3.2. *Carbonas*

Although wall rock alteration is a common aspect of lode and vein development, and the associated replacement mineralization may contribute significantly to the overall resource, it is not usually distinguished as a separate style. However, carbonas represent more extensive irregularly-shaped granite-hosted replacement bodies, typically with a cylindrical or tabular form (Dominy et al. 1996). These are often developed on, or in the vicinity of, existing lodes due to favorable structural controls such as fracture intersections or dilational jogs. Carbonas are best known from the Land's End and Carnmenellis granites and host greisen-, tourmaline- or chlorite-dominated cassiterite ± sulfide assemblages like those in adjacent lodes (Dominy et al. 1996).

2.5.4. *Mineralization associated with greisen and potassic alteration*

Greisen wall rock alteration, characterized by secondary muscovite or Li-mica, occurs around sheeted veins, stockworks and some lodes, both within the granite (endogranitic) or its host rock (exogranitic). Potassic wall rock alteration, characterized by secondary alkali feldspar, is associated with some veins and lodes, commonly at deeper structural levels within granites. Pervasive potassic alteration or greisening of granite in the vicinity of its external or internal intrusive contacts is rare. These styles, together with the tourmaline-dominated sheeted veins and breccias, typically represent the earliest occurrence of significant fracture-controlled magmatic-hydrothermal mineralization, and occasionally pre-date the emplacement of rhyolite/microgranite sheets and later episodes of pluton construction.

2.5.4.1. *Greisen-bordered sheeted veins*

Occurrences of limited extent are widely developed within, or immediately adjacent to, all plutons and there is an overlap with types in which wall rock tourmalinization is dominant. Endogranitic sheeted veins, with horizontal dimensions up to several hundred meters, occur in the G2 granite stocks of St Michael's Mount (Halls et al. 2000), Cligga Head (Moore and Jackson 1977) and Hemerdon, where more than one set occurs (Figure 2.15). In the larger plutons, they have been described from areas of combined tin and china clay production, in the Land's End (Jackson et al. 1982) and Goonbarrow in the St Austell granites (Bray and Spooner 1983).

Figure 2.15. *(a) Sheeted greisen-bordered vein complex hosted by muscovite (G2) granite at Cligga Head. (b) Greisen-bordered quartz-tourmaline vein with wolframite in muscovite (G2) granite at St Michael's Mount. Field of view approximately 0.5 m*

The typical occurrence is a series of closely spaced quartz ± tourmaline veins up to 0.1 m wide that may contain topaz, muscovite, alkali feldspar and apatite, plus coarse-grained wolframite, cassiterite, stannite, arsenopyrite and other sulfides, some of which may have been introduced during later mineralization episodes. Veins usually have a symmetrical 0.1–0.3 m wide greisened margin that is dominated by quartz and secondary white mica/Li mica plus tourmaline, topaz, fluorite and chlorite.

Significant occurrences of sheeted veins are found SE of the Tregonning (G5) granite at Wheal Fortune, and to the south of the St Austell (G4) granite at Carclaze. In these exogranitic greisens, coarse secondary muscovite occurs within the vein while intense tourmalinization of the wall rock may sometimes be associated with disseminated fine-grained cassiterite and rutile. Intense argillic alteration of the rock between the margins of the sheeted veins is common. The potential for near-surface, large tonnage/low grade deposits associated with this style of mineralization generated further exploration during the 1970s–1980s and the discovery of a major deposit, Redmoor, to the west of the Kit Hill (G2) granite.

2.5.4.2. *Feldspathic/greisen veins and lodes*

Discontinuous subhorizontal or gently-inclined quartz-alkali feldspar veins, known as "*Quartz Floors*", occur in the Carnmenellis (G1) granite at South Crofty Mine, and host wolframite, arsenopyrite/löllingite ± cassiterite, chalcopyrite and fluorite (Figure 2.16). These veins are post-dated by steeply inclined lodes that exhibit the same assemblages as well as elvan dykes and lodes with quartz-tourmaline and polymetallic sulfide parageneses. Similar discontinuous quartz-tourmaline ± feldspar veins with greisened margins containing wolframite, cassiterite, arsenopyrite and occasional chalcopyrite and chlorite are also widely developed around the NE margins of the Bodmin (G1) granite (Dines 1956). These veins have mylonitic fabrics and are cut by rhyolite/microgranite dykes (Hosking 1964). Extensional ductile shear zones in the Isles of Scilly (G1) granite locally host mylonitized quartz-tourmaline veins containing wolframite and cassiterite (Grant and Smith 2012).

Feldspathic and/or or greisen assemblages are widely developed as early lode components elsewhere (Beer and Scrivener 1982), but are commonly overprinted or obscured by subsequent tourmaline- and chlorite-dominated episodes of mineralization and wall rock alteration (Jackson et al. 1989). At Wheal Jane Mine to the east of the Carnmenellis (G1) granite, an initial greisen assemblage, comprising wolframite-löllingite-bismuth-cassiterite, set in vein quartz with a coarse muscovite border and tourmalinized wall rock, was cut by tourmalinite breccia mineralization; this was then overprinted by a second greisen assemblage exhibiting a progressive change to a quartz-chlorite-fluorite-sulfide paragenesis (Bromley and Holl 1986). Castle-an-Dinas Mine, to the north of the St Austell (G3-G4) granite, worked a single greisen-bordered quartz-wolframite ± löllingite lode that was locally cut by veins of topaz (G5) granite (Dines 1956).

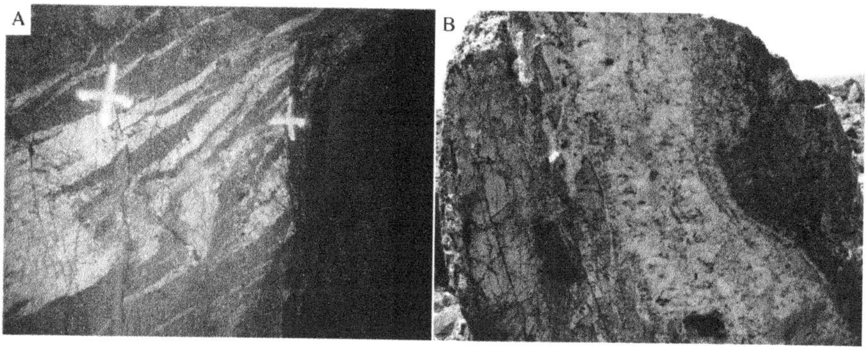

Figure 2.16. *(a) Gently-inclined Quartz Floor (quartz ± alkali feldspar ± wolframite) veins in top-to-SSE brittle-ductile shear zone hosted by Carn Brea (Carnmenellis) G1 granite; stope to N marks later steeply dipping quartz-tourmaline-cassiterite lode (looking WSW). (b) Quartz-alkali feldspar-tourmaline vein, Levant/Geevor, Land's End (G4) granite*

2.5.4.3. *Pervasive greisens*

Pervasive greisens are locally developed in the upper parts of stocks or culminations in the roof of the batholith, but not to the extent found in Variscan granites elsewhere such as the Erzegebirge. The Bosworgey (G2) granite, identified by drilling north of the Godolphin (G1) granite, has a 20 m thick greisened zone, overprinted by tourmalinization and silicification, dominated by chalcopyrite, cassiterite and tourmaline, with minor apatite, pyrite, arsenopyrite, bismuth, columbite, wolframite and uraninite (Ball and Basham 1984).

2.5.5. *Tourmaline-dominated veins, lodes and intrusive breccias*

Tourmaline-dominated mineralization, characterized by quartz, tourmaline ± cassiterite ± rutile ± specular hematite, occurs widely in granites and the immediately adjacent host rocks (Durrance et al. 1982; Bromley and Holl 1986) (Figures 2.17(a) and (b)). Joint-controlled mineralization occurs as sheeted veins/stockworks, containing quartz and/or tourmaline, that may host or grade into intrusive (or hydrothermal) breccias. Fault-controlled mineralization usually results in persistent lodes that exhibit brecciated textures resulting from both fluid pressure and frictional processes.

2.5.5.1. *Tourmaline-dominated sheeted veins and stockworks*

Tourmaline-dominated sheeted vein/stockwork and breccia systems hosted by metasedimentary rocks are very common to the north of the St Austell (G3-G4) granite, including Treliver (Bromley and Holl 1986; Camm and Dominy 1997). Cassiterite can occur in sheeted veins, along bedding- and cleavage-parallel fractures, in a breccia matrix, or as a replacement of calc-silicate bodies intersected by sheeted veins (Bromley and Holl 1986). Examples of closely spaced, quartz-tourmaline-cassiterite, joint-controlled veins occur at Ding Dong (Land's End (G3-G4) granite) and Beam (St Austell (G3-G4) granite).

2.5.5.2. *Tourmaline-dominated lodes*

Lodes with a primary quartz-tourmaline-cassiterite ± rutile ± arsenopyrite ± fluorite ± specular hematite assemblage have been a significant source of Sn production in granites and, to a lesser extent, metasedimentary rocks (Figure 2.16). Examples include the Dolcoath Main Lode, the Great Flat Lode and lodes at South Crofty Mine (all Carnmenellis (G1) granite). The tourmaline-dominated paragenesis may overprint earlier potassic/greisen mineralization and be post-dated by cassiterite-hematite, quartz-chlorite-fluorite-sulfide and later assemblages, often as distinct vein generations (Dines 1956; Hosking 1969).

Lode infills in granite usually comprise clasts of black schorl ± cassiterite and rutile that are cemented by a fine-grained quartz-blue tourmaline matrix containing cassiterite ± rutile; in metasedimentary rocks, the tourmaline may be blue-green (Bromley and Holl 1986; LeBoutillier 2002). Tourmalinization is the dominant style of wall rock alteration; in granites, it may be accompanied by secondary micas or feldspar, while in metasedimentary rocks, it may be accompanied by quartz and rutile; cassiterite can be present in both (Bromley and Holl 1986). Multiple episodes of hydrothermal and fault-related brecciation usually occur and the cassiterite often displays complex growth, zoning and cataclastic textures (Halls 1987; Farmer et al. 1991). Lodes in the central Dartmoor granite area are controlled by narrow brittle shear zones, comprising discontinuous quartz-tourmaline-dominated veins and local intrusive breccias (Shepherd et al. 1985). In central and eastern Dartmoor, the tourmaline-cassiterite assemblage is post-dated by a quartz-hematite ± chlorite assemblage; micaceous hematite was worked at several sites (Beer and Scrivener 1982).

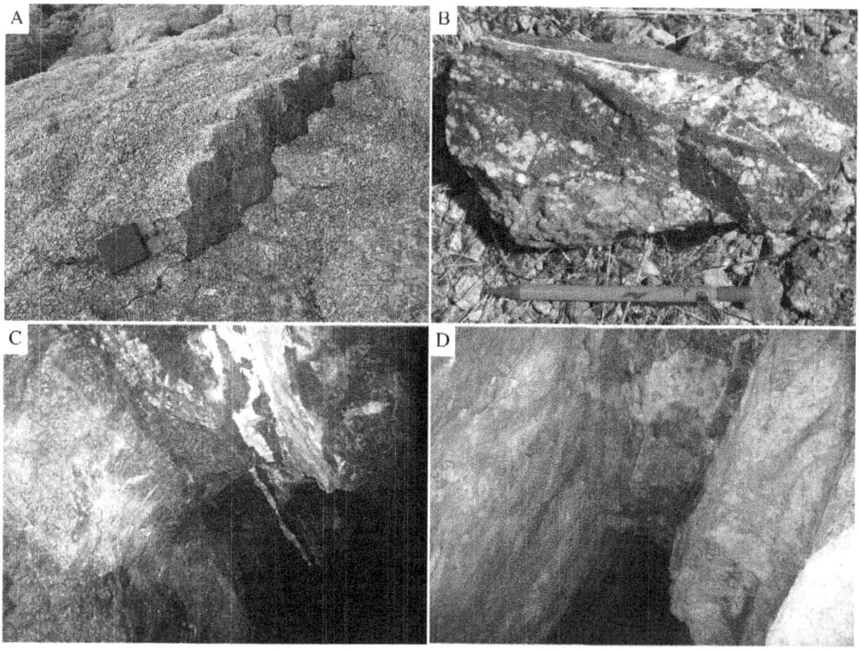

Figure 2.17. *Examples of tourmaline-dominated veins and lodes. (a) Steeply dipping tourmaline-quartz veining within Isles of Scilly G1 two-mica granite. (b) Tourmaline-quartz vein from Wall's Lode, Birch Tor, Dartmoor. (c) The footwall contact of the tourmaline-dominated Roskear A Lode (0.8 m wide), South Crofty Mine, Carn Brea G1 granite. (d) Red lode in Rosevale Mine, Land's End G3 biotite granite. The lode is approximately 0.6 m wide*

2.5.5.3. *Tourmaline-dominated intrusive breccias*

Irregular tabular or pipe-like bodies of breccia in granite or its host rock, that contain angular to rounded clasts of granite, rhyolite/microgranite and/or host rock up to 2 m in diameter, usually in a tourmaline-rich matrix, occur across the province (Goode and Taylor 1980). The breccias are often spatially and temporally associated with intrusive rhyolite/microgranite dykes and tourmaline-dominated sheeted veins/stockworks and may utilize the same fracture systems. A spectacular example, with minimum horizontal dimensions approximately 600 m (N-S) and up to 100 m (E-W), controlled principally by a subvertical NNW-SSE trending joint set, has been exposed whilst china clay mining at Wheal Remfry in the western part of the St Austell G3-G4 granite (Allman-Ward et al. 1982). Coeval rhyolite magma was included within the breccia and the main matrix minerals are tourmaline of magmatic origin (brown), inherited from the granite, and magmatic-hydrothermal origin (blue or pale green) formed in situ (Figure 2.6(b)). Accessory rutile is the principal host for Sn and W, as well as containing elevated Fe, Nb and V (Müller and Halls 2005).

2.5.6. *Polymetallic sulfide lodes*

Lodes in which quartz and/or chlorite and/or fluorite form the principal gangue minerals and which host variable combinations of cassiterite, chalcopyrite, arsenopyrite, sphalerite, galena and sporadic wolframite were the main source of Cu and As production, as well as substantial Sn; monazite is also common (LeBoutillier et al. 2003) (Figure 2.18). Indium, predominantly hosted within chalcopyrite and sphalerite, occurs widely in this paragenesis (Andersen et al. 2016). Uranium mineralization is sometimes hosted by sulfide-rich lodes, commonly as a later infill (Ball et al. 1982). In many lodes, it is possible to identify multiple infill and wall rock alteration episodes, and these may be superimposed upon, or incorporate, earlier potassic/greisen and tourmaline-dominated assemblages (e.g. Kettaneh and Badham 1978). Cassiterite, arsenopyrite and tourmaline can occur in veins in which chlorite, fluorite and quartz are the dominant gangue mineral(s) and which may or may not include chalcopyrite and other sulphides. Quartz ± chlorite ± fluorite assemblages hosting chalcopyrite ± sphalerite ± galena, but lacking cassiterite and tourmaline, are also common.

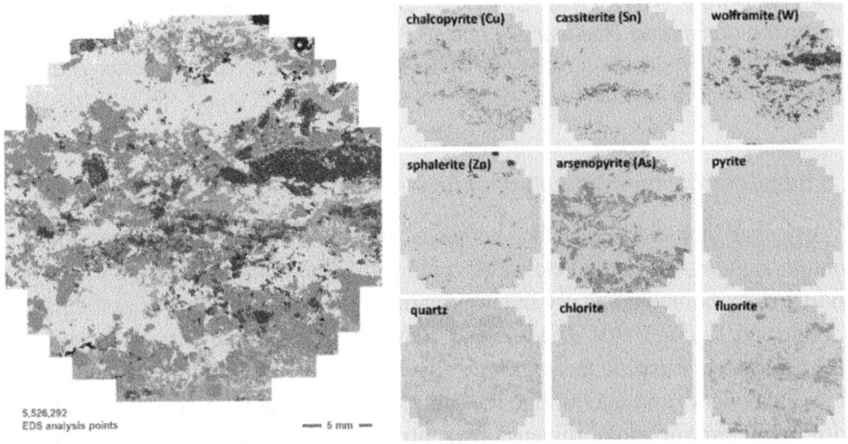

Figure 2.18. *QEMSCAN fieldscan image of polymetallic vein sample from underground exploration diamond drillcore in the Dolcoath section of South Crofty Mine. Composite shows distribution of each mineral*

The polymetallic quartz-chlorite-fluorite sulfide assemblage is the dominant lode paragenesis in host rocks surrounding the granites, but also occurs within the granites. In some lodes, Cu production from near surface quartz-chlorite-fluorite sulfide assemblages was replaced at depth by Sn production from tourmaline-dominated assemblages. The transition from “copper zone” to “tin zone” typically occurs in the vicinity of the granite–host rock boundary but can be somewhat above or below and include a zone of combined Cu and Sn production.

In the Camborne-Redruth district (Carnmenellis G1, Carn Brea G1 and Carn Marth G2 granites), granite-hosted tourmaline-dominated lodes include later quartz-chlorite-fluorite segments, often best-developed along E-W oriented structures, which suggests that tourmaline-cassiterite mineralization was postdated by quartz-chorite-fluorite-chalcopyrite mineralization (Farmer and Halls 1993). However, it is also possible that, in some lodes, the upwards transition from tourmaline-cassiterite to shallower polymetallic quartz-chlorite sulfide (± cassiterite) assemblages reflects a single episode of mineralization (LeBoutillier 2002) that was post-dated by other quartz ± chlorite ± fluorite sulfide events. Similar zonation occurs in the St Agnes-Perranporth (St Agnes and Cligga granites) district.

Lode and wallrock textures are highly variable and include crackle, mosaic and chaotic breccias, cataclasites and banded infills. At Wheal Jane, the quartz-chlorite-fluorite assemblage includes cassiterite-arsenopyrite-(stannite)-chalcopyrite-sphalerite and overprints quartz-tourmaline-cassiterite and greisen mineralization (Bromley 1989). Later oxidative alteration of chlorite- and some tourmaline-dominated lodes may result in secondary hematite-cassiterite assemblages in the Camborne-Redruth and St Just districts (Bromley 1989).

2.5.7. *Mineralization models*

2.5.7.1. *Tectonic and structural controls*

Granite magmatism and related mineralization spanned a 30 Ma interval during the Early to Mid-Permian (ca. 290–260 Ma), during which fracture formation and reactivation, in the granite and their host rocks, was controlled by the evolving post-Variscan tectonic regime (e.g. Shail and Alexander 1997). Fractures facilitated the migration and mixing of magmatic, meteoric ± basinal fluids and were fundamental to the mineralization process. The release of large volumes of magmatic fluids at suprahydrostatic pressures also contributed to fracturing and may locally be associated with decompression quenching and multiple generations of unidirectional solidification texture within their source granites (e.g. Halls 1987; Halls et al. 2000). Variations in the regional orientation and mutual relations of steeply dipping lodes/veins and rhyolite-microgranite dykes have been documented since the 1800s (e.g. Carne 1822; Hosking 1964; Moore 1975). The dominant trend is ENE-WSW to E-W, with subordinate ESE-WNW and NW-SE trends. Host rocks usually exhibit more complex fracture systems than the granites, reflecting their longer tectonic history, but veins/lodes are usually similar in orientation to those in the adjacent granite.

Analysis of post-Variscan fracture systems in west Cornwall, remote from mineralized areas, suggests that three successive Permian tectonic regimes can be defined; each is associated with fractures, both new and reactivated, and sometimes by the localized development of steeply inclined cleavages in metasedimentary rocks (Shail and Alexander 1997). These fracture systems encompass the range of vein/lode orientations typically observed in strongly mineralized areas and suggest a common origin, involving changes in Early- to Mid-Permian regional stresses (Figure 2.19).

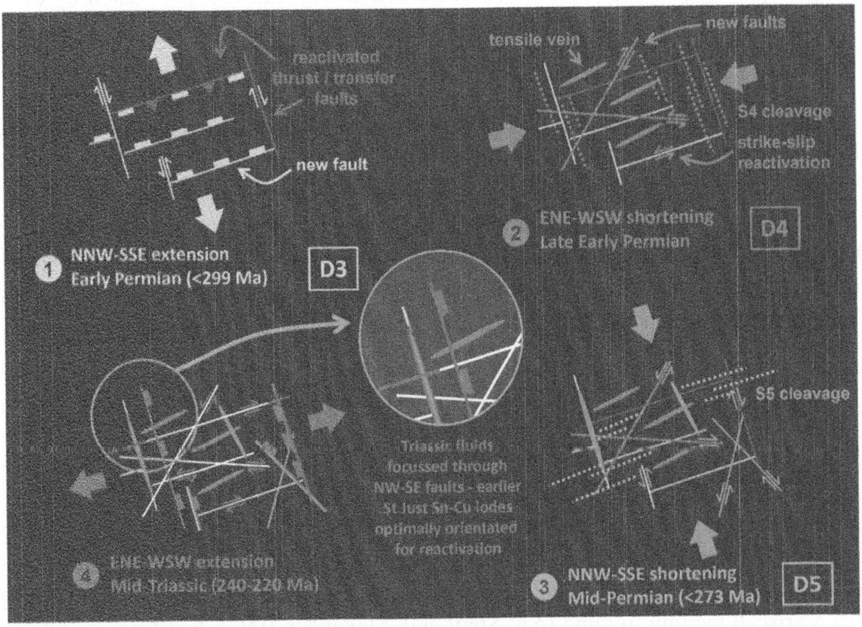

Figure 2.19. *Schematic map representation of the progressive development and reactivation of fracture systems controlling Early- to Mid-Permian granite-related and post-granite Mid-Triassic basinal brine "cross-course" mineralization (Shail and Alexander 1997)*

The initial regime corresponds to Early Permian D3 NNW-SSE extension. Its earliest expression occurs locally in two-mica G1 plutons where potassic/greisen "quartz floor" veins containing wolframite ± cassiterite formed within subhorizontal or gently inclined extensional ductile shear zones that exhibit a top sense-of-shear to the SSE. These represent coupling of deformation in two mica G1 plutons with extensional reactivation of thrust faults in the host rocks. The ENE-WSW to E-W striking, steeply dipping, vein/lode network represents the ubiquitous later brittle expression of this regime and can host all parageneses. NW-SE faults, formed in the host rocks as strike-slip transfer faults during Variscan convergence, were also active as transfer faults during extension and sometimes form the boundary between blocks with contrasting lode geometries and distributions.

The two subsequent regimes involved episodes of strike-slip with ENE-WSW (D4) and then NNW-SSE (D5) shortening directions that were compatible with the formation of new ESE-WSW and NW-SE striking lodes/veins as well as the selective reactivation of earlier structures. Some of the deeper ENE-WSW striking lodes in South Crofty Mine that formed as extensional faults indicate distinct episodes of strike- or oblique-slip reactivation (Farmer and Halls 1993). Quartz-chlorite-fluorite sulfide parageneses are often localized in E-W striking dilational jogs and that are coeval with the development of E-W to ESE-WNW trending "caunter" lodes. The latest veins/lodes probably formed in the Mid-Permian (<273 Ma) and are NW-SE striking in the St Just district (Land's End granite) and N-S striking around the St Austell granite, suggesting that regional stresses had changed substantially by this time.

2.5.7.2. *Geochronology*

The principal challenges of establishing the age of mineralization have been (i) use of mineral/isotope systems with sufficient analytical precision to resolve spatial age variations within and between mineralization styles, (ii) correct linkage of dated minerals to a paragenesis, particularly important in composite lodes and (iii) resetting the isotope systems of earlier paragenetic stages during later episodes of mineralization. Recent developments in the U-Pb dating of cassiterite indicate that high precision ages are possible and will significantly enhance understanding of the mineralization system (Moscati and Neymark 2020; Tapster and Bright 2020). The key findings from geochronological studies are as follows:

1) Granite-related mineralization occurred over a period of 30 Ma during the Early to Mid-Permian and is "diachronous and independent" (Chen et al. 1993), at least for greisen/potassic and quartz-tourmaline parageneses.

2) Replacement deposits in aureole rocks represent the earliest mineralization associated with each pluton, for example, garnet-pyroxene-amphibole-magnetite

skarns have an identical Ar-Ar muscovite age to the U-Pb monazite emplacement age of the adjacent Dartmoor granite (Chesley et al. 1993).

3) Ar-Ar dates for secondary muscovites in greisen-bordered sheeted veins and quartz-(feldspar)-wolframite ± cassiterite lodes overlap with muscovite cooling ages for the host/adjacent granite and suggest formation within 2–3 Ma of emplacement (Chen et al. 1993; Chesley et al. 1993). High precision U-Pb dating of cassiterite within greisen-bordered sheeted veins, and zircons from the host Cligga Head G2 granite, confirms mineralization occurred soon after emplacement (Tapster and Bright 2020).

4) Preliminary U-Pb cassiterite dating (Moscati and Neymark 2020), allowing for systematically older cassiterite ages (Tapster and Bright 2020), suggests that quartz-tourmaline-cassiterite veins/lodes within the Land's End (G4), St Austell (G4/G5) and Dartmoor (G3) plutons develop shortly after emplacement. However, U-Pb cassiterite ages for quartz-tourmaline-cassiterite veins/lodes hosted by the Carn Brea (Carnmenellis) G1 granite are ca. 10 Ma younger than the granite U-Pb monazite emplacement ages (Chen et al. 1993; Chesley et al. 1993) and consistent with wall rock secondary muscovite Ar-Ar ages (Chen et al. 1993). It is likely that the major lode systems, developed within G1 granites and adjacent host rocks of the Camborne-Redruth mining district, are unrelated to those granites. Instead, they were formed from magmatic-hydrothermal fluids released from underlying biotite G3 and tourmaline G4 granites that also sourced elvan dykes.

5) Polymetallic sulfide lodes are poorly constrained. Those associated with fluorite, or barren fluorite veins, have yielded fluorite Sm-Nd and Rb-Sr fluid isochron ages, and secondary muscovite Ar-Ar ages, in the interval 269–259 Ma (Shepherd and Darbyshire 1986; Chesley et al. 1991; Chen et al. 1993), although it is possible that more than one paragenetic episode has been dated (Chen et al. 1993; Clark et al. 1993). These mark the final contribution of magmatic-hydrothermal fluids to mineralization in the region.

2.5.7.3. *Nature and origin of mineralizing fluids*

Research into mineralizing fluids has been dominated by the application of fluid inclusion and stable isotope techniques, together with the integrated grain-scale evaluation of mineral textures, zonation and chemistry. Inclusion fluid homogenization temperature (Th) approximates the temperature of trapping and the freezing temperature indicates the salinity; these characteristics, together with the presence of variable liquid, vapor and solid components, allow inference of the fluid composition. Substantive reviews of the fluid and stable isotope characteristics of the major paragenetic types have been presented by Jackson et al. (1989), Alderton and Harmon (1991) and Alderton (1993). Some studies have utilized bulk crush-leach methods to release the fluids and allow their chemistry to be determined directly (Bottrell and Yardley 1988). Stable isotopes, particularly variations in H- and O-isotopic compositions, have been used to assess the relative contribution of magmatic and meteoric fluids. The occurrence of mantle-derived He within inclusion fluids from a range of parageneses implies that devolatization of mantle-derived magmas emplaced in the lower and/or middle crust occurred throughout mineralization and may account for the anomalous association of magmatic-hydrothermal Cu with peraluminous granites (Shail et al. 2003). Crush-leach and laser ablation studies of inclusion fluids in the earliest parageneses derived wholly from magmatic fluids indicate significant contents of Cu, Zn and Pb (Wilkinson et al. 1994). Sulfur isotope data are limited but indicate that a magmatic component is usually dominant within greisen/potassic and quartz-chlorite-fluorite sulfide vein and lode parageneses (Jackson et al. 1989; Shail et al. 2003).

2.5.7.3.1. Granites, pegmatites and replacement deposits

Fluid inclusions are common in magmatic quartz within unmineralized granites; they comprise several populations (Th <70°C to >550°C and salinities 0 to >50 wt%) that indicate multiple fluid flow episodes, some of which post-date granite-related mineralization (Rankin and Alderton 1983). The earliest high temperature (Th >500°C) inclusions are common and suggest the exsolution of large volumes of variable salinity aqueous fluids, including NaCl and $CaCl_2$ (Drivenes et al. 2016). This is supported by analyses of melt inclusions in the Land's End granite that indicate immiscible hypersaline fluids co-existed with the silicate melt, most likely due to volatile oversaturation of the magma (Müller et al. 2006b). H- and O-isotope data for mineralized pegmatites and the Botallack replacement deposits are similar to those for the early high temperature granite-derived fluids, but often indicate some mixing with meteoric fluids (Jackson et al. 1982; van Marcke de Lummen 1985).

2.5.7.3.2. Greisen and potassic alteration

At Cligga Head (G2 granite), mineralizing fluids had temperatures of 400–350°C, pressures of 0.7–1.7 kbar and salinites of 5–10 wt%; crush leach analysis indicated that they were Na-Ca-K-Fe chloride solutions with minor Mn, Cu, Li and Mg; major anions were B, F and SO_4, with Br/Cl ratios similar to other magmatic-derived fluids (Smith et al. 1996). Some samples are consistent with vapor separation from the fluid; this may have been triggered by pressure reductions during vein propagation and dilation. Earlier work at Hemerdon (G2 granite) suggested that a similar mechanism caused moderately saline magmatic hydrothermal fluids to separate into a high salinity (>29 wt%) component associated with Sn mineralization and a low salinity (12 wt%), CO_2-enriched component

(Th 450–375°C) associated with W mineralization (Shepherd et al. 1985). A slightly cooler CO_2-rich fluid component (Th 380–300°C) was identified in the early quartz-wolframite paragenesis present within composite quartz-chlorite-fluorite-sulfide lodes of the Kit Hill-Gunnislake area (Shepherd et al. 1985), and tentatively within the Complex Lode at South Crofty (Scrivener et al. 1986). Fluids hosted by wolframite-bearing Quartz Floors at South Crofty are moderately saline (15–27 wt%), high temperature (Th 320–430°C) and CO_2 rich (Bromley 1989). Initial laser ablation analyses show that these fluids contain modest W and Sn, and high Cu, Zn and Pb (Wilkinson et al. 1994). Geochemical modeling indicates that potassic (feldspar) alteration represents a higher temperature and/or pH equivalent of greisen; as fluid temperatures and/or pH decrease, micas rather than feldspar will be formed in the wall rock (e.g. Burt 1981). A temperature control is consistent with the high Th (430–350°C) of K-feldspar-cassiterite lodes in the St Just district (Jackson et al. 1982) and the more frequent occurrence of potassic alteration at depth.

2.5.7.3.3. Tourmaline-dominated veins, lodes and intrusive breccias

The fluids responsible for tourmaline-cassiterite mineralization at Wheal Jane had high homogenization temperatures (Th of 450–380°C) and high salinity (>30 wt %) (Holl 1990). Changes in the chemistry of hydrothermal tourmalines indicate that the fluids from which they precipitated commonly became more oxidizing with time and this probably reflects greater degrees of mixing with meteoric water (e.g. London and Manning 1995; Williamson et al. 2000; Duchoslav et al. 2017). Cassiterite inclusion fluids from South Crofty tourmaline-dominated lodes have a high but variable salinity and exhibit a decrease in homogenization temperature, from those co-existing with black tourmaline (Th 375–360°C) to those coexisting with blue tourmaline (Th 340–280°C), accompanied by progressively lower Mg/Fe and Ti/Fe ratios (Farmer et al. 1991). Laser ablation analyses of fluid inclusions from the early quartz-tourmaline veins at Lee Moor (Dartmoor) indicate Na, K, Ca and Fe as major components, and the presence of Cu, Zn and Mn in a moderately to highly saline fluid whose salinity variation may be caused by mixing with meteoric water (Wilkinson et al. 1994).

2.5.7.3.4. Polymetallic sulfide lodes

Fluid inclusion analyses of Pb-Zn ores hosted by ENE-WSW to E-W trending polymetallic lodes indicated that they could be distinguished, based on their lower salinities (0–13 wt%) and higher temperatures (Th 350–170°C), from subsequent cross-course Pb-Zn mineralization (Alderton 1975). Little systematic variation in sulfur isotopic composition has been detected, between polymetallic lode Sn-Cu parageneses close to the granites and Pb-Zn parageneses at greater distances (Gavelin et al. 1960). Inclusion fluids in fluorite from polymetallic veins in the Kit

Hill-Gunnislake area (Th 215–305°C) have significant $CaCl_2$ contents, and the fluorites have REE characteristics similar to those in nearby lower temperature cross-course veins; these data have been interpreted to indicate a basinal brine component (Shepherd et al. 1985). A $CaCl_2$ fluid component has also been recognized within fluids in South Crofty lodes (Scrivener et al. 1986). Inclusion fluids within South Crofty cassiterites co-existing with Fe-chlorite and subordinate tourmaline have homogenization temperatures of 340–260°C, like those for the blue tourmaline-dominated stage (Th 340–280°C), but salinity data indicate greater mixing with meteoric water and an increased tourmaline Ca content (Farmer et al. 1991). Although a basinal brine component has often been inferred from high $CaCl_2$ contents, it also forms a significant component in magmatic-hydrothermal fluids (Drivenes et al. 2016). Inclusion fluids hosted by quartz-fluorite-chlorite assemblages at Wheal Jane have generally lower homogenization temperatures (Th 310–180°C) and lower salinities (3–7 wt%) than the earlier tourmaline-dominated paragenesis, although in some cases there is a continuum and sulfide precipitation might be caused by mixing with cooler, less saline groundwaters (Holl 1990). It has been proposed that the occurrence of the polymetallic sulfide lodes in SW England and elsewhere is related to the incursion of mantle-derived Cu-bearing oxidized mafic melts into highly reduced residual granite magmas (Clark et al. 1993). Another possibility is that there was continual devolatization of mafic melts in the lower and middle crust and Cu did not occur as a significant component within the silicate melt but as a coexisting exotic sulfide melt (Müller et al. 2006a) or in a volatile phase.

2.6. Post-granite mineralization

2.6.1. *Cross-course mineralization*

A diverse range of mineralization is hosted by lodes that typically trend NW-SE to NNE-SSW and consistently cut granite-related Sn-Cu-Zn-Pb lodes. The gangue mineralogy typically comprises one or more of quartz, chalcedony, carbonate, fluorite, hematite and clay, often with banding and vuggy cavities; many structures host solely barren quartz, chalcedony hematite or clay. Several broad paragenetic associations can be defined: (i) Pb-Zn ± F, Ba, (ii) U ± Co-Ni-Bi-As-Ag (considered in section 2.6.2) and (iii) Fe (e.g. Jackson et al. 1989; Alderton 1993).

2.6.1.1. *Lead-zinc ± fluorite and barite mineralization*

This most common type of cross-course mineralization occurs throughout the region, usually hosted by slates, and is associated with the production of significant Pb, Ag and Zn, and occasionally fluorite. Major lodes persist for some distance along strike, although their deepest workings are usually shallower than those associated with granite-related mineralization, for example, Menheniot (1.5 km lode strike length, ca. 600 m depth), Tamar Valley (6 km, 465 m), Mary Tavy (5 km, 310 m) and Teign Valley (8 km, 265 m) (Beer and Scrivener 1982; Alderton 1993). The ore mineralogy is dominated by galena (often argentiferous) and sphalerite, with minor marcasite, arsenopyrite-löllingite, pyrite, chalcopyrite, tetrahedrite and Ag minerals, including native Ag, argentite and pyrargyrite (e.g. Andersen et al. 2016). Wall rock alteration is not usually extensive and comprises argillization and subordinate hematization. The only significant barite production has been to the east of the Dartmoor granite, where predominantly banded or botryoidal barite with some quartz, pyrite, galena, sphalerite and tetrahedrite was worked (Beer and Ball 1977).

2.6.1.2. *Iron mineralization*

There has been substantial production of Fe from cross-courses, largely in and around the St Austell granite. The mineralization is represented by quartz-hematite veins that predate the pervasive kaolinization of the granites and are characterized by several cycles of precipitation; early hematite, kaolin and illite are post-dated by chlorite, kaolin and illite assemblages (Psyrillos et al. 2001).

2.6.1.3. *Age and origins*

Pb-Zn-fluorite ± barite mineralization in the Tamar Valley cross-course system has been dated using the Rb-Sr method on inclusion fluids hosted by quartz and fluorite, at 236 ± 3 (2σ) Ma, Mid-Triassic (Scrivener et al. 1994). Cross-course Pb-Zn inclusion fluids exhibit lower temperatures (Th 175–115°C) and higher salinities (22–26 wt%) than those in Pb-Zn ores hosted by ENE-WSW to E-W trending "granite-related" polymetallic lodes (Alderton and Harmon 1991). The link between cross-course mineralization and the evolution and migration of basinal fluids has been confirmed by their similarities with veins hosted by Permo-Triassic successions of the Plymouth Bay Basin (Gleeson et al. 2000). The cross-course fracture system is dominated by the formation and reactivation of NW-SE to N-S striking faults in response to Triassic ENE-WSW rifting (Shail and Alexander 1997) (Figure 2.18). However, the cross-course parageneses of quartz, chalcedony, hematite and kaolinite also occur within reactivated/dilated segments of granite-related lode systems. In the St Just district (Land's End granite), the NW-SE lodes were optimally oriented for reactivation and resulted in common Triassic K-Ar illite and feldspar wall rock alteration ages that are unrelated to earlier granite-related Sn-Cu mineralization (Halliday 1980).

2.6.2. *Five-element (Co-Ni-Bi-As-Ag) and uranium mineralization*

Five-element (Co-Ni-Bi-As-Ag) mineralization occurs locally in cross-courses and Sn-Cu lodes and can be associated with uranium mineralization (Ball et al. 1982). There has been minor production of all these metals (Dines 1956). The St Austell area has notable occurrences of this style of mineralization in N-S structures such as at South Terras Mine, where the parageneses are complex with multiple precipitation episodes. The lode is dominated by banded quartz-hematite-siderite and includes uraninite associated with chalcopyrite and secondary U minerals zippeite, torbernite and autunite (Ball et al. 1982). U-Pb uraninite ages (pre-1977 constants) on these parageneses hosted by N-S cross-courses include 225 ± 5 (2σ) Ma (Triassic) and 47 ± 2 (2σ) (Paleogene) at South Terras Mine and 165 ± 4 (2σ) Ma (Jurassic) at Wheal Bray (Bodmin Moor) (Pockley 1964; Darnley et al. 1965). Similar parageneses hosted by Sn-Cu lodes at Botallack and Geevor (Land's End) have, respectively, a U-Pb uraninite age of 58 ± 3 (2σ) Ma and a chemical age of 45 Ma (Pockley 1964; Darnley et al. 1965). There is little fluid inclusion data available, although Alderton (1993) describes a study in which low temperature, low salinity fluids containing compressed gases (CO_2 or low-order hydrocarbons) were inferred. While these data are incomplete, they support a model of episodic hydrothermal fluid flow that is correlated with periods of regional Triassic and later rifting and inversion rather than long-lived radiogenically induced convection about the granites.

2.6.3. *Hydrocarbons*

In South Crofty Mine, hydrocarbons seep from otherwise barren NW-SE joints and bitumen occurs with paragenetically late kaolin ± pyrite/marcasite infills within quartz-tourmaline-cassiterite lodes (Harrison and Beer 1977). Elsewhere, there is an association between hydrocarbons and uranium mineralization in some cross-courses (Ball et al. 1982). Biomarkers in hydrocarbons from South Crofty Mine indicate a Lower Jurassic source (Baba et al. 2019). NW-SE cross-course fault zones have probably acted as migration pathways for hydrocarbons generated from source rocks that were subsequently eroded during late Jurassic–Early Cretaceous inversion. Migration of basinal brines and hydrocarbons into basement fault zones may also have contributed to kaolinization (e.g. Psyrillos et al. 2003) and low-temperature hydrothermal Co-Ni-Bi-As-Ag ± U mineralization.

2.7. China clay

Minerals of the kaolin sub-group ($Al_2Si_2O_5(OH)_4$), predominantly kaolinite, formed during the pervasive alteration of granite, primarily due to breakdown of plagioclase feldspar, and occur in tabular or funnel-shaped zones that may be more than 250 m deep and 1 km in diameter (Figure 2.20). Extensive kaolinization is also locally developed in metasedimentary host rocks surrounding the St Austell granite (Bristow and Scott 1998). The alteration process affects G3, G4 and G5 granites; some of the highest quality clays are derived from G5 due to its low biotite and high Na-feldspar content (Bristow and Exley 1994). Two main textural types of kaolin occur: (i) finely crystalline, typically <5 μm in mean diameter, present within feldspar dissolution porosity, and (ii) coarsely crystalline vermiform aggregates, up to 100 μm or more in length, that are closely associated with precipitation of kaolinite along hydrothermal mica cleavage planes, resulting in expanded micas (Psyrillos et al. 1999).

Figure 2.20. *(a) View over the Karslake complex, a number of smaller china clay pits merged to form one mining area operated by Imerys. (b) China clay extraction in Treviscoe Pit (G5 granite, St Austell)*

There has been considerable debate regarding the origins of china clay, with end-member models reflecting either hydrothermal or weathering processes. There is a strong correlation between zones of intense kaolinization and the occurrence of quartz-tourmaline and greisen-bordered sheeted veins, and it has been inferred that kaolin was formed during this event (Bray and Spooner 1983). Others have suggested origins during later stages of hydrothermal circulation (Alderton and Rankin 1983). There is also an apparent association in the St Austell district between kaolinization and the NW-SE to N-S (sporadically ENE-WSW) trending quartz-hematite "cross-course" veins. However, O- and H-isotope data, indicating that kaolin equilibrated with low temperature meteoric fluids, have been used to support an exclusively or predominantly supergene weathering origin (Sheppard 1977; Jackson et al. 1982). Attempts to reconcile these data have involved multi-stage models in which early high temperature mineralization resulted in the development of permeable fracture networks and the removal of biotite. Significant argillic alteration, initially comprising a transitional smectite and illite assemblage, did not commence until cross-course mineralization; this was followed by extensive kaolinization, including recrystallization, related to persistent convective flow of Mesozoic meteoric fluids and Palaeogene weathering (Durrance et al. 1982; Alderton 1993; Bristow and Exley 1994).

Pervasive kaolinization in the St Austell granite was dominated by the fracture-controlled migration of hydrothermal fluids, locally manifested by networks of kaolin veins (<10 mm wide) that post-dated the development of quartz-hematite "cross-course" veins (Psyrillos et al. 2003). Previous correlation of pervasive kaolinization with the quartz-hematite and earlier sheeted vein parageneses is likely to reflect utilization of their intrinsic fracture permeability during reactivation and is consistent with the occurrence of kaolin veins in the center of, and locally cross-cutting, these earlier veins (Psyrillos et al. 1998). O- and H-isotopic data from inclusion fluids have been used to infer that kaolinization occurred at temperatures between 50°and 100°C, and involved high salinity fluids that were either highly evolved meteoric fluids circulating in the granite host rocks or basinal brines derived from the Plymouth Basin (Psyrillos et al. 2003). While the latter fluids are similar to those responsible for the Triassic cross-course mineralization, it has been tentatively suggested that kaolinization was related to a major episode of fluid expulsion, controlled by major regional uplift and basin inversion during the late Jurassic to Early Cretaceous (Psyrillos et al. 2003).

2.8. The past, present and future of the resources sector in SW England

Elevated commodity prices, increased awareness of security-of-supply, environmental, social and governance compliance, and the requirements of the energy transition all present opportunities for the development of metal mining and deep geothermal energy in SW England. There has been substantial recent funding for both mineral and deep geothermal exploration across SW England from the European Union, private investors and the UK government. The principal challenge is whether or not resources can be extracted for a profit and in a manner that is compliant with contemporary environmental and societal expectations. This final section provides an overview of these recent developments.

2.8.1. *W, Sn, Cu*

Tin mining in SW England has occurred for at least 4,000 years and peak annual production, accounting for 50% of the global supply, was achieved around 1870

(Dines 1956). Copper mining expanded substantially during the industrial revolution, accounting for up to 40% of annual global production in the early 1800s and peak production was achieved around 1860 (Dines 1956). However, there was a rapid and substantial decline in the mining of Cu from the 1860s, and Sn in the 1890s, as cheaper sources became available. The industry went through a series of boom-and-bust cycles for much of the first half of the 20th century (Buckley 2007).

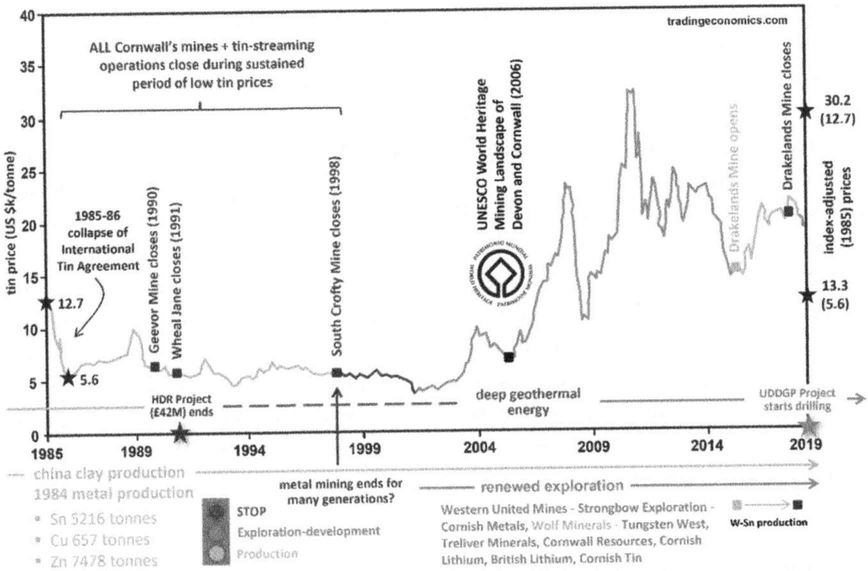

Figure 2.21. *The SW England resources sector and global tin price 1985–2019. A prolonged period of low tin prices following the collapse of the International Tin Agreement in 1985–1986 caused the closure of all metal mines by 1998. Increased metal prices from the early 2000s generated renewed exploration activity and W-Sn production from Drakelands Mine 2015–2018. China clay production has continued throughout, and there has been renewed exploration for deep geothermal energy*

A period of stable and relatively high tin prices during the 1960s–early 1980s, underpinned by the International Tin Agreement, resulted in extensive mineral exploration, primarily for Sn and W, as well as the opening of Wheal Jane, Mount Wellington, Pendarves and Wheal Concord (Buckley 2007). Tin production in 1984 was 50% of that at its peak in 1870 but was achieved from just four mines – Geevor, Pendarves, South Crofty and Wheal Jane. However, the collapse of the International Tin Agreement in 1985–1986 marked the beginning of a sustained period of low tin prices and the closure of all SW England metal mines by 1998 (Figure 2.21).

Increased global metal prices from the mid-2000s have brought about renewed interest, primarily for Sn, W and, to a lesser extent, Cu (Figure 2.21).

2.8.1.1. *Hemerdon W-Sn deposit*

The Hemerdon W-Sn deposit is centered upon the subvertical, NNE-SSW striking, 100+ m wide Hemerdon granite dyke (G2) hosted by Devonian metasedimentary and metavolcanic rocks (Figure 2.22). Mineralization is associated with moderately to steeply NW-dipping greisen-bordered quartz-ferberite ± cassiterite sheeted veins. The deposit had been permitted for open-pit mining in 1985–1986 but never progressed, due to the collapse in metal prices. It was re-developed from 2007 by Wolf Minerals and in 2015, as Drakelands Mine, it became the first metal mine to open in England for 40 years. The 2015 granite-hosted resource estimate was 145.2 Mt at 0.15% WO_3 and 0.02% Sn. Production during 2015–2018 was 3735 tonnes of WO_3 and 693 tonnes of Sn and, despite significant mineral processing challenges and the mine ceasing operations before the year's end, it resulted in the UK becoming the fifth biggest global tungsten producer in 2018 (Brown et al. 2019). It is currently being re-developed as Hemerdon Mine by Tungsten West who have substantially increased the resource estimate, largely through the re-inclusion of mineralized host rocks.

Figure 2.22. *Drakelands W-Sn Mine open pit and processing plant looking SW (2018); the Hemerdon (G2) granite dyke is clearly visible. The mine was operated by Wolf Minerals (2015–2018), and the UK was the fifth biggest global tungsten producer in 2018 due solely to production from this location. It is currently being re-developed, as Hemerdon Mine, by Tungsten West*

2.8.1.2. *Redmoor W-Sn-Cu deposit*

The Redmoor W-Sn-Cu deposit, developed in metasedimentary host rocks above Kit Hill (G2) granite, was evaluated by South West Minerals Ltd between 1979 and 1985 in a 35 hole diamond drilling program totaling 12,146 m, but was later abandoned due to low metal prices. Since 2017, Cornwall Resources have undertaken over 14,000 m of diamond drilling with primary focus on a sheeted vein system. The 2019 inferred mineral resource is 11.7 Mt at 0.56% WO_3, 0.16% Sn and 0.50% Cu. Preliminary plans are for mining by long-hole stoping, with paste back-fill, accessed via a decline with a potential mine life of 12 years.

2.8.1.3. *Treliver Sn deposit*

A substantial exploration project, focused on metasediment-hosted Sn to the north of the St Austell granite, was undertaken during 2013–2015 by Treliver Minerals. The aim was to follow-up on drilling undertaken by Billiton in the 1980s (Camm and Dominy 1997) with the objective of identifying near-surface tin mineralization suitable for open pit mining. A total of 3,685 m of diamond drilling (31 holes) was completed but the persistence of grade and intersection thicknesses in the target calc-silicate horizons was generally lower than expected from the Billiton campaign. The 2015 inferred resource was 4.52 Mt at 0.15% Sn, at the lower end of what was anticipated, and exploration ended that year.

2.8.1.4. *South Crofty Mine (Sn ± W, Cu, Zn)*

South Crofty Mine was the last metal mine to close in SW England in 1998 and worked multiple ENE-WNW lodes within the Carn Brea (Carnmenellis) G1 granite and its overlying host rocks. It was owned by Baseresult/Western United Mines from 2001 to 2013 who undertook >30,000 m of diamond drilling and development work, largely in the upper mine section west of the Great Cross-course. It is presently owned by Cornish Metals who are undertaking a feasibility study to inform a decision on future production. A surface drilling campaign in 2020 proved the continuations of known lodes at depth, including high-grade intersections, for example, No. 4 lode, 974.2–976.8 m (true thickness 2.6 m) at 10.33% Sn, as well as five new lodes. The 2021 indicated mineral resource for the lower mine (tin only) was 2.08 Mt at 1.59% Sn and for the upper mine (polymetallic) was 0.277 Mt at 0.67% Sn, 0.78% Cu and 0.57% Zn. Underground mining is fully permitted to 2071 (covering 1,490 hectares incorporating 26 former mines), there is planning permission to construct a new processing plant and a discharge permit from the Environment Agency to dewater the mine.

2.8.1.5. *United Downs (Sn, Cu ± W)*

Cornish Metals have commenced exploration drilling in the United Downs area, to the east of Carn Marth granite, where earlier Cornish Lithium diamond drill holes for Li brine discovered new zones of polymetallic Cu-Sn mineralization over depth intervals of ca. 750 m in a previously unmined area between the historic United Mine and Consolidated Mines. The mineralization style is similar to that mined at the Wheal Jane and Mount Wellington mines, located 1.5 km along strike to the east.

2.8.1.6. *Wheal Vor (Sn)*

In 2020, Cornish Tin announced plans to commence exploration for Sn around Wheal Vor, located between the Tregonning (G5) and Carnmenellis (G1) granites.

2.8.1.7. *Placer tin deposits*

There is ongoing exploration interest in placer Sn deposits. Marine minerals have undertaken detailed studies of placer tin in the shallow marine sands off the north coast of Cornwall between St Ives and Perranporth. In St Austell Bay, on the south coast, placer tin is present within gravels in drowned river systems; offshore exploration was initially undertaken by Billiton in the 1980s and MOCEAN are undertaking a technical feasibility study.

2.8.2. *Lithium micas (St Austell granite)*

The St Austell granite and its derived china clay waste is recognized as one of the largest inferred hard rock Li resources in Europe (Hawkes et al. 1987; Gourcerol et al. 2019). Of particular interest are the fractionated tourmaline (G4) and topaz (G5) granites that host zinnwaldite (ferroan polylithionite) and lepidolite (polylithionite) micas with up to 5.5 wt% Li and very little Fe (Henderson et al. 1989; Simons et al. 2017). Exploration in the St Austell granite, in sites of former china clay extraction, has been undertaken by two companies, British Lithium and Cornish Lithium. Presently, there is no commercial-scale extraction of Li from micas anywhere globally, so establishing appropriate energy-efficient and environmentally acceptable mineral processing and extraction methodologies is critical. British Lithium is developing proprietary Li-Sep technology for mica separation and a proposed production of ca. 21,000 tpa of battery grade lithium carbonate from a 100 Mt mineral resource estimate. Cornish Lithium has completed two drilling programs at its Trelavour prospect and has licensed use of the Lepidico L-Max and LOH-Max technologies for lithium hydroxide production.

2.8.3. *Deep geothermal energy*

Radiogenic heat generation, due to elevated Th, U and K content, means that temperatures of 200°C are projected at a depth of ca. 5 km (Figure 2.23(a)) for most of the Cornubian Batholith, and SW England is consequently the most prospective region in the UK for the development of deep geothermal energy (Beamish and Busby 2016). The current exploration model involves targeting permeable fault zones in granite at depths >4,500 m to ensure temperatures appropriate for power generation.

2.8.3.1. *Camborne School of Mines Geothermal Energy Project*

Camborne School of Mines ran a pioneering £42m "Hot Dry Rock" Geothermal Energy Project 1976–1991, funded by the UK Department of Energy and European Commission, at Rosemanowes Quarry in the Carnmenellis (G1) granite (Parker 1999). The objective was to determine if a geothermal reservoir could be developed in "background" jointed granite remote from fault zones. Three boreholes were drilled between depths of 2,000–2,600 m and large-scale hydraulic stimulations were undertaken. A very large reservoir was created but flow rates were low and water losses were high; microseismic events showed that reservoir growth was predominantly downwards and reached depths of more than 4 km. The focus of deep geothermal energy research was subsequently transferred to the Rhine Graben (Soultz) from 1992 (Parker 1999). A summary of the geological and rock mechanics research from the Hot Dry Rock Project is provided by Richards et al. (1991).

2.8.3.2. *United Downs Deep Geothermal Power Project*

Geothermal Engineering Limited drilled two deep geothermal wells UD-1 (5057 m) and UD-2 (2,214 m) into the granite-hosted segment of the NW-SE Porthtowan Fault Zone in 2018–2019 (Figure 2.23(b)). Final reservoir testing has been undertaken and the intention is for a 3 MW power plant to be operational by 2024. Lithium extraction from geothermal brines and heat usage for a gin distillery are also planned.

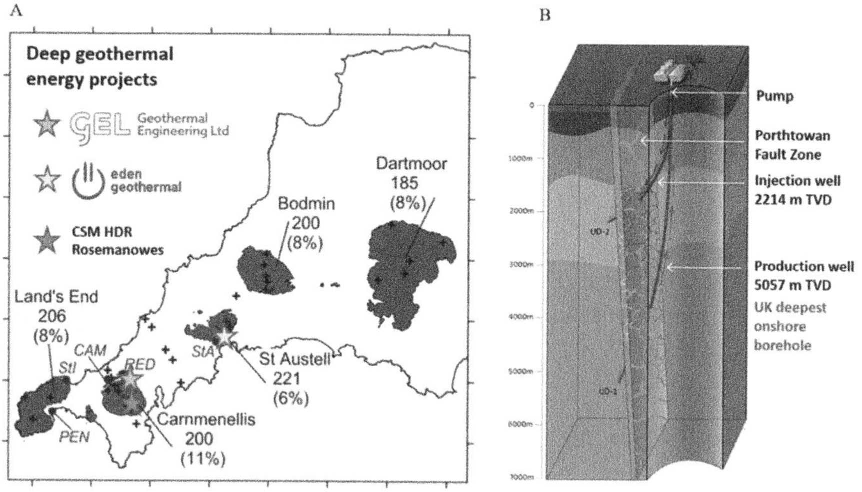

Figure 2.23. *(a) Location of deep geothermal energy projects superimposed on the revised predicted temperatures at 5 km depth within the Cornubian Batholith (from Beamish and Busby 2016; http://creativecommons.org/licenses/by/4.0/; Geothermal Energy). (b) Schematic representation of the target Porthtowan Fault Zone reservoir and wells UD-1 and UD-2 at the United Downs Deep Geothermal Power Project (Geothermal Engineering Limited)*

2.8.3.3. *Eden Geothermal Project*

Eden Geothermal Limited commenced drilling a single 4,500 m well EG-1, targeting the NW-SE Great Cross-course in the southern part of the St Austell (G3) granite, in May 2021. Following completion, there will be a 1-year period of heat production, using a co-axial circulation system, that will be used to provide direct heat for the facilities at the Eden Project. The intention is that a second well (EG-2) will then be drilled and a combined heat and power plant will be developed.

2.8.4. *Lithium brines*

Lithium was first discovered in "hot springs" in deep Cornish mines during the mid-19th century (Miller 1864). Fluid sampling of brines from deep mines and boreholes, and associated experimental studies, were undertaken in the 1980s as part of the Camborne School of Mines Geothermal Energy Project. Brines are developed through fracture-controlled granite-meteoric fluid reactions and lithium enrichment is consistent with biotite dissolution (Edmunds et al. 1985). Cornish Lithium are undertaking an extensive exploration and characterization program of groundwaters from shallow (1–2 km) and deeper (4–5 km) sources with the intention of extracting lithium from these brines. The former have been evaluated by two 1 km deep boreholes targeting sites of historical mine discharges in the United Downs area. The latter are being evaluated through the GeoCubed joint venture, between Cornish Lithium and Geothermal Engineering Limited at the United Downs Deep Geothermal Power Project, which will create a pilot-scale lithium extraction plant producing lithium hydroxide using Direct Lithium Extraction technologies. There is considerable future potential for the co-production of geothermal energy and low-carbon lithium.

2.8.5. *Industrial minerals, aggregates and building stones*

Granite-hosted kaolin (china clay) deposits have been historically worked in from all plutons but production is now solely from the western part of the St Austell granite (Imerys) and the SW part of the Dartmoor granite (Sibelco and Imerys). China clay extraction has occurred since 1746 and has been the dominant regional mining activity for well over a century; total kaolin production is estimated at ca. 200 million tonnes (Bristow et al. 2002; Bristow 2006). In Cornwall, Imerys produces approximately 1 million tpa of kaolin and employs around 800 people across 20 active quarry and industrial facilities.

Cornish kaolin is used in paints, sealants, adhesives, plastics, rubbers, cosmetics, pharmaceuticals, sanitaryware, tableware, tiles, paper and cardboard. Each tonne of kaolin generates approximately 9 tonnes of inert granite waste (stent, sand, mica). Some of this is used as secondary aggregates and the manufacture of concrete blocks. The potential for co-production of disseminated magmatic Sn-Nb-Ta-REE from china clay waste streams has been recently investigated in some detail (Dehaine et al. 2019). The commercial case for proceeding with co-recovery is not presently strong as concentrate grades are low and monazite, as the principal REE host, can potentially generate high levels of radioactivity due to its Th content (Dehaine et al. 2019).

Cornubian granites are still quarried for aggregates, building stone, monumental masonry and armor stone, although very much reduced from the scale of operations in the 1800s–1900s (Stanier 1999). De Lank Quarry (Bodmin G1 granite) maintains a full range of granite products and "De Lank silver-grey granite" has been used on many high-quality building projects across the region and in London. It has historically been used to construct bridges and monuments in London, including parts of the London Stock Exchange, Trafalgar Square and the Royal Opera House. Other larger operations include the Carnsew Quarry and Chywoon Quarry, which produce Carnmenellis (G1) granite, and the Castle-an-Dinas Quarry, which produces a fine-grained G2 granite in the Land's End pluton. There is an increased awareness of the importance and diversity of local building stones, of which granites and elvans are key components, in part to be used in the restoration of historical buildings (e.g. Bristow 2014; Historic England 2017).

2.9. Environmental, cultural and social impact of mining

Mining activity, from early tin streaming in the Bronze Age through to the present day, has extensively changed the landscape of SW England and contributed to unusual biodiversity in many locations. Abandoned mine buildings, equipment and spoil heaps are found throughout the region, from both hard rock and china clay extraction. The acidic waste grounds are biodiverse, containing unusual species, including a bryophyte (moss) that is one of the rarest in the world, growing only on copper-rich mine wastes (Callaghan 2018).

It has been estimated that up to one-third of metals were lost during processing, many into rivers and estuaries around the region (Thomas 1913). Mining has had a huge impact on estuarine sediments, with mine waste being released into the environment as particulate waste or in solution through acid mine drainage. The Fal, Fowey, Hayle, Camel, Gannel, Helford and Teign estuaries (Pirrie et al. 1997, 2000, 2002, 2003; Rollinson et al. 2007; Simons et al. 2011) all retain a geochemical and mineralogical record of the type and extent of mining in their respective catchments. The presence in these sediments of diagenetic Cu, Pb and Zn minerals is indicative that these elements were, and may still be, bioavailable.

Alongside the environmental legacy of mining, a social impact also persists. Throughout the 1980s, the low metal prices lead to mine closures and a loss of associated industries. There was widespread economic decline and deindustrialization in areas that had once been among the richest in the world. With a lack of industry to replace jobs, and a reliance on seasonal tourist and farming jobs in a rural location, Cornwall was recognized as one of the poorest regions of Western Europe and received significant investment from European Union Regional Development Funds from 2000 onwards. Several mines, including Wheal Martyn, Poldark and Geevor, are now open to tourists, and the redevelopment of a disused china clay pit into the Eden Project is one of the UK's major tourist attractions.

Despite the decline in mining, numerous mining-related industries remain, many working internationally, including mineral processing, petrographic services and mine remediation, and the Cornwall Mining Alliance has over 100 members.

Large parts of the countryside of SW England have protected status. Regional schemes include the Dartmoor National Park, the Cornwall and South Devon Areas of Outstanding Natural Beauty and the Cornish Mining World Heritage Site. In 2006, 10 mining landscapes across west Devon and Cornwall were designated as a UNESCO World Heritage Site, in celebration of their unique industrial and cultural heritage.

Also, although better known for its stately homes, the National Trust (a private charity) owns and manages large stretches of the SW England coastline as well as many inland areas. Many geological features are protected under site-specific schemes including Sites of Special Scientific Interest (SSSI), Geological Conservation Review (GCR) sites and County Geology Sites (CGS). Field-based evidence for most of these magmatic and mineralization processes is preserved in a network of GCR sites (Floyd et al. 1993; Bevins et al. 2010) and sites designated by the Cornwall Geoconservation Group and Devon RIGS Group.

2.10. References

Abdelfadil, K.M., Romer, R.L., Seifert, T., Lobst, R. (2013). Calc-alkaline lamprophyres from Lusatia (Germany) – Evidence for a repeatedly enriched mantle source. *Chemical Geology*, 353, 230–245.

Alderton, D.H.M. (1975). Fluid inclusion studies in SW England. *Proceedings of the Ussher Society*, 3, 214–217.

Alderton, D.H.M. (1979). Luxullianite in situ within the St Austell granite, Cornwall – A discussion. *Mineralogical Magazine*, 43(327), 441–442.

Alderton, D.H.M. (1993). Mineralization associated with the Cornubian Granite Batholith. In *Mineralization in the British Isles*, Pattrick, R.A.D. and Polya, D.A. (eds). Chapman and Hall, London.

Alderton, D.H.M. (2018). Skarn garnets from South-West England. *Geoscience in South-West England*, 14, 91–97.

Alderton, D.H.M. and Harmon, R.S. (1991). Fluid inclusion and stable isotope evidence for the origin of mineralizing fluids in South-West England. *Mineralogical Magazine*, 55, 605–611.

Alderton, D.H.M. and Rankin, A.H. (1983). The character and evolution of hydrothermal fluids associated with the kaolinised St Austell granite, SW England. *Journal of the Geological Society*, 140, 297–309.

Alexander, A.C. and Shail, R.K. (1995). Late Variscan structures on the coast between Perranporth and St Ives, Cornwall. *Proceedings of the Ussher Society*, 8, 398–404.

Alexander, A.C. and Shail, R.K. (1996). Late- to post-Variscan structures on the coast between Penzance and Pentewan, south Cornwall. *Proceedings of the Ussher Society*, 9, 72–78.

Alexander, A.C., Shail, R.K., Leveridge, B.E. (2019). Late Paleozoic extensional reactivation of the Rheic–Rhenohercynian suture zone in SW England, the English Channel and Western Approaches. *Geological Society Special Publications*, 470, 353–373.

Allman-Ward, P., Halls, C., Rankin, A., Bristow, C. (1982). An intrusive hydrothermal breccia body at Wheal Remfry in the western part of the St Austell granite pluton, Cornwall. In *Metallization Associated with Acid Magmatism*, Evans, A.M. (ed.). Wiley, Chichester.

Andersen, J.C.Ø., Stickland, R.J., Rollinson, G.K., Shail, R.K. (2016). Indium mineralisation in SW England: Host parageneses and mineralogical relations. *Ore Geology Reviews*, 78, 213–238.

Baba, M., Parnell, J., Bowden, S. (2019). The geochemistry of oil in Cornish granites. *Petroleum Geoscience*, 25, 298–305.

Ball, T.K. and Basham, I.R. (1984). Petrogenesis of the Bosworgey granitic cusp in the SW England Tin Province and its implications for ore mineral genesis. *Mineralium Deposita*, 19, 70–77.

Ball, T.K., Michie, U.M., Basham, I.R. (1982). Uraniferous vein occurrences of south-west England – Paragenesis and genesis. In *Vein-type and Similar Uranium Deposits in Rocks Younger than Proterozoic*. IAEA, Vienna.

Ballouard, C., Poujol, M., Boulvais, P., Branquet, Y., Tartèse, R., Vigneresse, J.L. (2016). Nb-Ta fractionation in peraluminous granites: A marker of the magmatic-hydrothermal transition. *Geology*, 44(3), 231–234.

Bea, F., Fershtater, G., Corretgé, L.G. (1992). The geochemistry of phosphorus in granite rocks and the effect of aluminium. *Lithos*, 29(1), 43–56.

Beamish, D. and Busby, J. (2016). The Cornubian geothermal province: Heat production and flow in SW England: Estimates from boreholes and airborne gamma-ray measurements. *Geothermal Energy*, 4(1), 4.

Beer, K.E. and Ball, T.K. (1977). Mineral investigations in the Teign Valley, Devon. Part 1 – Barytes. Mineral Reconnaissance Programme Report, Institute of Geological Sciences.

Beer, K.E. and Scrivener, R.C. (1982). Metalliferous mineralisation. In *The Geology of Devon, Durrance*, Laming, E.M. and Selwood, E.B. (eds). University of Exeter Press.

Bevins, R.E., Young, B., Mason, J.S., Manning, D.A.C., Symes, R.F. (2010). *Mineralization of England and Wales.* Joint Nature Conservation Committee, Peterborough.

Bott, M.H.P., Day, A.A., Masson-Smith, D. (1958). The geological interpretation of gravity and magnetic surveys in Devon and Cornwall. *Philosophical Transactions of the Royal Society, Mathematical and Physical Sciences*, 251, 161–191.

Bott, M.H.P., Holder, A., Long, R., Lucas, A. (1970). Crustal structure beneath the granites of south-west England. In *Mechanism of Igneous Intrusion*, Newall, G. and Rast, N. (eds). Wiley, New York.

Bottrell, S.H. and Yardley, B.W.D. (1988). The composition of a primary granite-derived ore fluid from S.W. England, determined by fluid inclusion analysis. *Geochimica et Cosmochimica Acta*, 52, 585–588.

Bouchez, J.L., Nguema, T.M.M., Esteban, L., Siqueira, R., Scrivener, R. (2006). The tourmaline-bearing granite pluton of Bodmin (Cornwall, UK): Magnetic fabric study and regional inference. *Journal of the Geological Society*, 163(4), 607–616.

Bray, C.J. and Spooner, E.T.C. (1983). Sheeted vein Sn-W mineralization and greisenization associated with economic kaolinization, Goonbarrow china clay pit, St Austell, Cornwall, England; geologic relationships and geochronology. *Economic Geology*, 78(6), 1064–1089.

Breiter, K., Ďurišová, J., Hrstka, T., Korbelová, Z., Vašinová Galiová, M., Müller, A., Simons, B., Shail, R.K., Williamson, B.J., Davies, J.A. (2018). The transition from granite to banded aplite-pegmatite sheet complexes: An example from Megiliggar Rocks, Tregonning topaz granite, Cornwall. *Lithos*, 302, 370–388.

Bristow, C.M. (2006). *China Clay: A Geologist's View.* Cornish Hillside Publications, St Austell.

Bristow, C.M. (2014). The geology of the building and decorative stones of Cornwall, UK. *Geological Society Special Publications*, 391, 93–120.

Bristow, C.M. and Exley, C.S. (1994). Historical and geological aspects of the china-clay industry of Southwest England. *Transactions of the Royal Geological Society of Cornwall*, 21, 247–314.

Bristow, C.M. and Scott, P.W. (1998). Kaolinized Devonian metasediments adjacent to the St Austell granite, Cornwall. *Proceedings of the Ussher Society*, 9, 255–262.

Bristow, C.M., Palmer, Q., Witte, G.J., Bowditch, I., Howe, J.H. (2002). The ball clay and china clay industries of Southwest England in 2000. In *Industrial Minerals and Extractive Industry Geology*, Scott, P.W. and Bristow, C.M. (eds). Geological Society of London.

British Geological Survey (2006). Gravity anomaly index map for 1:1 000 000 UTM Series. British Geological Survey.

Bromley, A.V. (ed.) (1989). *Field Guide – The Cornubian Orefield.* International Association of Geochemistry and Cosmochemistry, Columbus.

Bromley, A.V. and Holl, J. (1986). Tin mineralisation in Southwest England. In *Mineral Processing at a Crossroads: Problems and Prospects*, Wills, B.A. and Barley, R.W. (eds). Springer, Dordrecht.

Brooks, M., Doody, J.J., Al-Rawi, F.R.J. (1984). Major crustal reflectors beneath SW England. *Journal of the Geological Society*, 141, 97–103.

Brown, T.J., Idoine, N.E., Wrighton, C.E., Raycraft, E.R., Hobbs, S.F., Shaw, R.A., Everett, P., Deady, E.A., Kresse, C. (2019). *World Mineral Production 2015–19*. British Geological Survey, Nottingham.

Buckley, A. (2007). *The Story of Mining in Cornwall*. Cornwall Editions, Fowey.

Burt, D.M. (1981). Acidity-salinity diagrams – Application to greisen and porphyry deposits. *Economic Geology*, 76, 832–843.

Callaghan, D.A. (2018). Status, conservation and ecology of the exceptionally rare metallophyte Cornish Path-moss (Ditrichum cornubicum Paton). *Journal of Bryology*, 40(4), 358–370.

Camm, S. and Dominy, S. (1997). Metasediment-hosted tin mineralization in the Indian Queens area, mid-Cornwall. *Proceedings of the Ussher Society*, 9, 370–373.

Carne, J. (1822). On the relative age of the veins of Cornwall. *Transactions of the Royal Geological Society of Cornwall*, 2, 49–128.

Černý, P., Blevin, P., Cuney, M., London, D. (2005). Granite-related ore deposits. *Economic Geology*, 100, 1.

Chappell, B.W. and Hine, R. (2006). The Cornubian Batholith: An example of magmatic fractionation on a crustal scale. *Resource Geology*, 56(3), 203–244.

Chappell, B.W. and White, A.J.R. (2001). Two contrasting granite types: 25 years later. *Australian Journal of Earth Sciences*, 48(4), 489–499.

Charoy, B. (1986). The genesis of the Cornubian Batholith (South-West England): The example of the Carnmenellis pluton. *Journal of Petrology*, 27(3), 571–604.

Chen, Y., Clark, A.H., Farrar, E., Wasteneys, H.A.H.P., Hodgson, M.J., Bromley, A.V. (1993). Diachronous and independent histories of plutonism and mineralization in the Cornubian Batholith, southwest England. *Journal of the Geological Society*, 150(6), 1183–1191.

Chesley, J.T., Halliday, A.N., Scrivener, R.C. (1991). Samarium-neodymium direct dating of fluorite mineralisation. *Science*, 252, 949–951.

Chesley, J.T., Halliday, A.N., Snee, L.W., Mezger, K., Shepherd, T.J., Scrivener, R.C. (1993). Thermochronology of the Cornubian Batholith in southwest England: Implications for pluton emplacement and protracted hydrothermal mineralization. *Geochimica et Cosmochimica Acta*, 57(8), 1817–1835.

Clark, A.H., Chen, Y., Farrar, E., Wasteneys, H.A.H.P., Stimac, J., Hodgson, M.J., Willis-Richards, J., Bromley, A.V. (1993). The Cornubian Sn-Cu (-As, W) metallogenetic province: Product of a 30 m.y. history of discrete and concomitant anatectic, intrusive and hydrothermal events. *Proceedings of the Ussher Society*, 8, 112–116.

Cobbing, J. and Clarke, M. (2006). Textural features of South-West England granites: A reinterpretation. *Geoscience in South-West England*, 11, 229–242.

Cosgrove, M.E. and Elliot, M.H. (1976) Supra-batholithic volcanism of the South West England granites. *Proceedings of the Ussher Society*, 3, 391–401.

Cuney, M. and Barbey, P. (2014). Uranium, rare metals, and granulite-facies metamorphism. *Geoscience Frontiers*, 5(5), 729–745.

Dangerfield, J. and Hawkes, J.R. (1981). The Variscan granites of south-west England: Additional information. *Proceedings of the Ussher Society*, 5, 116–120.

Darbyshire, D.P.F. and Shepherd, T.J. (1985). Chronology of granite magmatism and associated mineralization, SW England. *Journal of the Geological Society*, 142(6), 1159–1177.

Darbyshire, D.P.F. and Shepherd, T.J. (1994). Nd and Sr isotope constraints on the origin of the Cornubian Batholith, SW England. *Journal of the Geological Society*, 151, 795–802.

Darnley, A.G., English, T.H., Sprake, O., Preece, E.R., Avery, D. (1965). Ages of uraninite and coffinite from south-west England. *Mineralogical Magazine*, 34, 159–176.

Dehaine, Q., Filippov, L.O., Glass, H.J., Rollinson, G. (2019). Rare-metal granites as a potential source of critical metals: A geometallurgical case study. *Ore Geology Reviews*, 104, 384–402.

Dijkstra, A.H. and Hatch, C. (2018). Mapping a hidden terrane boundary in the mantle lithosphere with lamprophyres. *Nature Communications*, 9, 3770.

Dines, H.G. (1956). *The Metalliferous Mining Region of South-West England*. Her Majesty's Stationery Office, London.

Dominy, S., Camm, G., Bussell, M. (1996). Characteristics and development of carbona-style replacement tin mineralization in west Cornwall. *Proceedings of the Ussher Society*, 9, 97–104.

Drivenes, K., Larsen, R.B., Müller, A., Sørensen, B.E., Wiedenbeck, M., Raanes, M.P. (2015). Late-magmatic immiscibility during batholith formation: Assessment of B isotopes and trace elements in tourmaline from the Land's End granite, SW England. *Contributions to Mineralogy and Petrology*, 169, 56.

Drivenes, K., Larsen, R.B., Müller, A., Sørensen, B.E. (2016). Crystallization and uplift path of late Variscan granites evidenced by quartz chemistry and fluid inclusions: Example from the Land's End granite, SW England. *Lithos*, 252, 57–75.

Duchoslav, M., Marks, M.A.W., Drost, K., McCammon, C., Marschall, H.R., Wenzel, T., Markl, G. (2017). Changes in tourmaline composition during magmatic and hydrothermal processes leading to tin-ore deposition: The Cornubian Batholith, SW England. *Ore Geology Reviews*, 83, 215–234.

Dupuis, N.E., Braid, J.A., Murphy, J.B., Shail, R.K., Archibald, D.A., Nance, R.D. (2015). 40Ar/39Ar phlogopite geochronology of lamprophyre dykes in Cornwall, UK: New age constraints on Early Permian post-collisional magmatism in the Rhenohercynian Zone, SW England. *Journal of the Geological Society*, 172(5), 566–575.

Dupuis, N.E., Murphy, J.B., Braid, J.A., Shail, R.K., Nance, R.D. (2016). Mantle evolution in the Variscides of SW England: Geochemical and isotopic constraints from mafic rocks. *Tectonophysics*, 681, 353–363.

Durrance, E.M., Bromley, A.V., Bristow, C., Heath, M., Penman, J. (1982). Hydrothermal circulation and post-magmatic changes in granites of SW England. *Proceedings of the Ussher Society*, 5, 304–320.

Edmonds, E., Wright, J., Beer, K., Hawkes, J., Williams, M., Freshney, E., Fenning, P. (1968). Geology of the country around Okehampton. Memoirs of the Geological Survey of Great Britain, Sheet 324 (England and Wales). Her Majesty's Stationery Office, London.

Edwards, R.A. and Scrivener, R.A. (1999). Geology of the area around Exeter: Memoir for 1: 50 000 Geological Sheet 325. British Geological Survey, London.

Evans, C.D.R. (1990). *The Geology of the Western English Channel and its Western Approaches*. British Geological Survey, Her Majesty's Stationery Office, London.

Exley, C. and Stone, M. (1964). The granitic rocks of south-west England. In *Present Views of Some Aspects of the Geology of Cornwall and Devon*, Hosking, K.F.G. (ed.). Royal Geological Society of Cornwall.

Exley, C. and Stone, M. (1982). Hercynian intrusive rocks. In *Igneous Rocks of the British Isles*, Sutherland, D.S. (ed.). Wiley–Blackwell, New York.

Farmer, C.B. and Halls, C. (1993). Paragenetic evolution of cassiterite-bearing lodes at South Crofty Mine, Cornwall, United Kingdom. In *Proceedings of the 8th Quadrennial IAGOD Symposium, Ottawa 1990*, Maurice, E. (ed.). E. Schweizerbart'sche Verlagsbuchhandlung, Stuttgart.

Farmer, C.B., Searl, A., Halls, C. (1991). Cathodoluminescence and growth of cassiterite in the composite lodes at South Crofty Mine, Cornwall, England. *Mineralogical Magazine*, 55(380), 447–458.

Floyd, P.A., Exley, C.S., Styles, M.T. (1993). *Igneous Rocks of South-West England*. Springer, Dordrecht.

Franke, W. (2000). The mid-European segment of the Variscides: Tectonostratigraphic units, terrane boundaries and plate tectonic evolution. *Geological Society Special Publications*, 179, 35–61.

Gavelin, S., Parwel, A., Ryhage, R. (1960). Sulfur isotope fractionation in sulphide mineralization. *Economic Geology*, 55, 510–530.

Ghosh, P.K. (1927). Petrology of the Bodmin Moor granite (eastern part). *Cornwall Mineralogical, Mineralogical Society*, 21(118), 285–309.

Ghosh, P.K. (1934). The Carnmenellis granite: Its petrology, metamorphism and tectonics. *Quarterly Journal of the Geological Society*, 90(1–4), 240–276.

Gleeson, S.A., Wilkinson, J.J., Shaw, H.F., Herrington, R.J. (2000). Post-magmatic hydrothermal circulation and the origin of base metal mineralization, Cornwall, UK. *Journal of the Geological Society*, 157, 589–600.

Goode, A.J.J. (1973). The mode of intrusion of Cornish elvans. Report, Her Majesty's. Stationery Office, London.

Goode, A.J.J. and Taylor, R.T. (1980). Intrusive and pneumatolytic breccias in south-west England. Report, Institute of Geological Sciences, Her Majesty's Stationery Office, London.

Goode, A.J.J. and Taylor, R.T. (1988). *Geology of the Country Around Penzance*. Her Majesty's Stationery Office, London.

Gourcerol, B., Gloaguen, E., Melleton, J., Tuduri, J., Galiegue, X. (2019). Re-assessing the European lithium resource potential – A review of hard-rock resources and metallogeny. *Ore Geology Reviews*, 109, 494–519.

Grant, J. and Smith, C.W.E.H. (2012). Evidence of tin and tungsten mineralisation in the Isles of Scilly. *Geoscience in South-West England*, 13, 65–70.

Grimmer, S. and Floyd, P. (1986). Geochemical features of Permian rift volcanism – A comparison of Cornubian and Oslo basic volcanics. *Proceedings of the Ussher Society*, 6, 352–259.

Hall, A. (1970). The composition of Cornish quartz-porphyry (Elvan) dykes. *Proceedings of the Ussher Society*, 2, 205–208.

Hall, A. (1982). The Pendennis peralkaline minette. *Mineralogical Magazine*, 45(337), 257–266.

Halliday, A.N. (1980). The timing of early and main stage ore mineralization in southwest Cornwall. *Economic Geology*, 75, 752–759.

Halls, C. (1987). A mechanistic approach to the paragenetic interpretation of mineral lodes in Cornwall. *Proceedings of the Ussher Society*, 6, 548–554.

Halls, C., Exley, C.S., Brunton, E.V. (1985). *A Bibliography of Magmatism and Mineralization in Southwest England*. Institution of Mining and Metallurgy, London.

Halls, C., Cosgrove, J., Camm, G. (2000). The influence of fluid pressure in governing fracture geometry and mineral textures in the pneumatolytic lode systems of South West England. *Proceedings of the Ussher Society*, 10, 58–63.

Harrison, R.K. and Beer, K.E. (1977). Elateritic hydrocarbon from South Crofty Mine, Cornwall: A preliminary note. *Journal of Microscopy*, 109, 165–169.

Hawkes, J.R. (1982). The Dartmoor granite and later volcanic rocks. In *The Geology of Devon*, Durrance, E.M., Laming, D.J.C., Selwood, E.B. (eds). University of Exeter Press.

Hawkes, J.R., Harris, P.M., Dangerfield, J., Strong, G.E., Davis, A.E., Nancarrow, P.H.A., Francis, A.D., Smale, C.V. (1987). The lithium potential of the St Austell Granite. Report, British Geological Survey, London.

Henderson, C.M.B., Martin, J.S., Mason, R.A. (1989). Compositional relations in Li-micas from S.W. England and France: An ion- and electron-microprobe study. *Mineralogical Magazine*, 53(372), 427–449.

Hill, P. and Manning, D. (1987). Multiple intrusions and pervasive hydrothermal alteration in the St Austell granite, Cornwall. *Geoscience in South-West England*, 6, 447–453.

Historic England (2017). Strategic stone study. A building stone atlas of Cornwall and the Isles of Scilly. Report, Historic England, Swindon.

Holder, A.P. and Bott, M.H.P. (1971). Crustal structure in the vicinity of south-west England. *Geophysical Journal International*, 23(5), 465–489.

Holl, C.J. (1990). A paragenetic study of the complex polymetallisation of South Lode, Wheal Jane Mine, Cornwall. PhD Thesis, Camborne School of Mines, Penryn.

Holloway, S. and Chadwick, R.A. (1986). The Sticklepath-Lustleigh fault zone: Tertiary sinistral reactivation of a Variscan dextral strike-slip fault. *Journal of the Geological Society*, 143(3), 447–452.

Hosking, K. (1954). The pegmatites of Trolvis Quarry, Carnmenellis, Cornwall. *Geological Magazine*, 91(4), 273–285.

Hosking, K. (1964). *Present Views of Some Aspects of the Geology of Cornwall and Devon*. Royal Geological Society of Cornwall, Penzance.

Hosking, K. (1969). The nature of the primary tin ores of the south-west of England. In *Second Technical Conference on Tin*. International Tin Council, London.

Hughes, S.P., Stickland, R.J., Shail, R.K., LeBoutillier, N.G., Alexander, A.C., Thomas, M. (2009). The chronology and kinematics of late Palaeozoic deformation in the NW contact metamorphic aureole of the Land's End Granite. *Geoscience in South-West England*, 12, 140–152.

Jackson, N.J. (1974). Gryll's Bunny, a "tin floor" at Botallack. *Proceedings of the Ussher Society*, 3, 186–188.

Jackson, N.J. and Alderton, D. (1974). Discordant calc-silicate bodies in the Botallack area. *Proceedings of the Ussher Society*, 3, 123–128.

Jackson, N.J., Halliday, A.N., Sheppard, S., Mitchell, J. (1982). Hydrothermal activity in the St Just Mining district, Cornwall, England. In *Metallization Associated with Acid Magmatism*, Evans, A. (ed.). Wiley, Chichester.

Jackson, N.J., Willis-Richards, J., Manning, D.A.C., Sams, M.S. (1989). Evolution of the Cornubian ore field, Southwest England. Part II: Mineral deposits and ore-forming processes. *Economic Geology*, 84(5), 1101–1133.

Jones, R. (1991). A seismic reflection survey as part of a geophysical investigation of the Carnmenellis granite. *Proceedings of the Ussher Society*, 7, 418–420.

Keppler, H. (1993). Influence of fluorine on the enrichment of high field strength trace elements in granitic rocks. *Contributions to Mineralogy and Petrology*, 114(4), 479–488.

Kettaneh, Y.A. and Badham, J.P.N. (1978). Mineralization and paragenesis at the Mount Wellington Mine, Cornwall. *Economic Geology*, 73(4), 486–495.

Klemperer, S.L. and Hobbs, R. (1991). *The BIRPS Atlas: Deep Seismic Reflections Profiles Around the British Isles*. Cambridge University Press.

Kratinová, Z., Jevzek, J., Schulmann, K., Hrouda, F., Shail, R., Lexa, O. (2010). Noncoaxial K-feld-spar and AMS subfabrics in the Land's End granite, Cornwall: Evidence of magmatic fabric decoupling during late deformation and matrix crystallization. *Journal of Geophysical Research*, 115, B09104.

Leat, P.T., Thompson, R.N., Morrison, M.A., Hendry, G.L., Trayhorn, S.C. (1987). Geodynamic significance of post-Variscan intrusive and extrusive potassic magmatism in SW England. *Earth and Environmental Science Transactions of the Royal Society of Edinburgh*, 77(4), 349–360.

LeBoutillier, N.G. (2002). The tectonics of Variscan magmatism and mineralisation in South West England. PhD Thesis, University of Exeter.

LeBoutillier, N.G., Shail, R.K., Jewson, C. (2003). Monazite in polymetallic chlorite-(tourmaline)-quartz-(fluorite)-cassiterite sulphide lodes and its potential for constraining the chronology of magmatic hydrothermal mineralisation in Cornwall. *Proceedings of the Ussher Society*, 10, 403–409.

Leveridge, B.E. and Hartley, A.J. (2006) The Variscan orogeny: The development and deformation of Devonian/Carboniferous basins in SW England and South Wales. In *The Geology of England and Wales*, Brenchley, P.J. and Rawson, P.F. (eds). The Geological Society, London.

Leveridge, B.E., Holder, A.P., Goode, A.J.J., Scrivener, R.A., Jones, N., Merriman, R. (2002). Geology of the Plymouth and South-east Cornwall area: Memoir for 1: 50 000 geological sheet 348. British Geological Survey, London.

Linnen, R.L. (1998). The solubility of Nb-Ta-Zr-Hf-W in granitic melts with Li and Li + F; constraints for mineralization in rare metal granites and pegmatites. *Economic Geology*, 93(7), 1013–1025.

London, D. and Manning, D.A.C. (1995). Chemical variation and significance of tourmaline from Southwest England. *Economic Geology*, 90(3), 495–519.

Manning, D.A.C. (1983). Disseminated tin sulphides in the St Austell granite. *Proceedings of the Ussher Society*, 5, 411–416.

Manning, D.A.C. and Hill, P.I. (1990). The petrogenetic and metallogenetic significance of topaz granite from the southwest England orefield. In *Ore-bearing Granite Systems; Petrogenesis and Mineralizing Processes*, Stein, H. and Hannah, J. (eds). Geological Society of America Special Papers, Boulder.

Manning, D.A.C., Hill, P.I., Howe, J. (1996). Primary lithological variation in the kaolinized St Austell Granite, Cornwall, England. *Journal of the Geological Society*, 153, 827–838.

van Marcke de Lummen, G. (1985). Mineralogy and geochemistry of skarn deposits in the Land's End aureole, Cornwall. *Proceedings of the Ussher Society*, 6, 211–217.

McDonough, W.F. and Sun, S. (1995). The composition of the Earth. *Chemical Geology*, 120, 223–253.

Miller, J.A. (1864). Chemical examination of a hot spring containing caesium and lithium in Wheal Clifford, Cornwall. *Chemical News and Journal of Physical Science*, 10, 181–182.

Mintsa Mi Nguema, T., Trindade, R.I.F., Bouchez, J.L., Launeau, P. (2002). Selective thermal enhancement of magnetic fabrics from the Carnmenellis granite (British Cornwall). *Physics and Chemistry of the Earth, Parts A/B/C*, 27(25), 1281–1287.

Moore, J.M. (1975). A mechanical interpretation of the vein and dyke systems of the S.W. England orefield. *Mineralium Deposita*, 10, 374–388.

Moore, J.M. and Jackson, N. (1977). Structure and mineralization in the Cligga granite stock, Cornwall. *Journal of the Geological Society*, 133(5), 467–480.

Moscati, R.J. and Neymark, L.A. (2020). U–Pb geochronology of tin deposits associated with the Cornubian Batholith of southwest England: Direct dating of cassiterite by in situ LA-ICPMS. *Mineralium Deposita*, 55(1), 1–20.

Müller, A. and Halls, C. (2005). Mineral deposit research: Meeting the global challenge rutile – The tin-tungsten host in the intrusive tourmaline breccia at Wheal Remfry, SW England. In *Proceedings of the Eighth Biennial SGA Meeting, Beijing, China, 18–21 August 2005*, Mao, J. and Bierlein, F.P. (eds). Springer, Berlin.

Müller, A., Seltmann, R., Halls, C., Siebel, W., Dulski, P., Jeffries, T., Spratt, J., Kronz, A. (2006a). The magmatic evolution of the Land's End pluton, Cornwall, and associated pre-enrichment of metals. *Ore Geology Reviews*, 28(3), 329–367.

Müller, A., Thomas, R., Wiedenbeck, M., Seltmann, R., Breiter, K. (2006b). Water content of granitic melts from Cornwall and Erzgebirge: A Raman spectroscopy study of melt inclusions. *European Journal of Mineralogy*, 18, 429–440.

Nance, R.D., Neace, E.R., Braid, J.A., Murphy, J.B., Dupuis, N., Shail, R.K. (2015). Does the Meguma terrane extend into SW England? *Geoscience Canada*, 42, 61–76.

Orejana, D., Villaseca, C., Billström, K., Paterson, B.A. (2008). Petrogenesis of Permian alkaline lamprophyres and diabases from the Spanish Central System and their geodynamic context within Western Europe. *Contributions to Mineralogy and Petrology*, 156(4), 477–500.

Parker, R. (1999). The Rosemanowes HDR project 1983–1991. *Geothermics*, 28(4–5), 603–615.

Paterson, S.R., Ardill, K., Vernon, R., Žák, J. (2019). A review of mesoscopic magmatic structures and their potential for evaluating the hypersolidus evolution of intrusive complexes. *Journal of Structural Geology*, 125, 134–147.

Penfound-Marks, L.R.G. and Shail, R.K. (2015). Early Devonian rift-related felsic igneous rocks in the western Looe Basin, SW England. *Geoscience in South-West England*, 13, 471–482.

Petford, N., Cruden, A.R., McCaffrey, K.J.W., Vigneresse, J.L. (2000). Granite magma formation, transport and emplacement in the Earth's crust. *Nature*, 408(6813), 669–673.

Phillips, J.A. (1880). On concretionary patches and fragments of other rocks contained in granite. *Quarterly Journal of the Geological Society*, 36(1–4), 1–22.

Pirrie, D., Camm, G.S., Sear, L.G., Hughes, S.H. (1997). Mineralogical and geochemical signature of mine waste contamination, Tresillian River, Fal Estuary, Cornwall, UK. *Environmental Geology*, 29, 58–65.

Pirrie, D., Power, M.R., Payne, A., Camm, G.S., Wheeler, P.D. (2000). Impact of mining on sedimentation; the Camel and Gannel estuaries, Cornwall. *Geoscience in South-west England*, 10, 21–28.

Pirrie, D., Power, M.R., Rollinson, G., Cundy, A., Watkins, D.C. (2002). Impact of mining on the sediment geochemistry and mineralogy of the Helford River, Cornwall. *Geoscience in South-West England*, 10, 323–328.

Pirrie, D., Power, M.R., Rollinson, G., Camm, G.S., Hughes, S.H. (2003). The spatial distribution and source of arsenic, copper, tin and zinc within the surface sediments of the Fal Estuary, Cornwall, UK. *Sedimentology*, 50, 579–595.

Pockley, R.P.C. (1964). Four new uranium-lead ages from Cornwall. *Mineralogical Magazine*, 33, 1081–1092.

Powell, T., Salmon, S., Clark, A., Shail, R. (1999). Emplacement styles within the Land's End Granite, west Cornwall. *Geoscience in South-West England*, 9, 333–339.

Pownall, J.M., Waters, D.J., Searle, M.P., Shail, R.K., Robb, L.J. (2012). Shallow laccolithic emplacement of the Land's End and Tregonning granites, Cornwall, UK: Evidence from aureole field relations and P-T modeling of cordierite-anthophyllite hornfels. *Geosphere*, 8(6), 1467–1504.

Psyrillos, A., Manning, D.A.C., Burley, S.D. (1998). Geochemical constraints on kaolinisation in the St Austell Granite, Cornwall, England. *Journal of the Geological Society*, 155, 829–840.

Psyrillos, A., Howe, J.H., Manning, D.A.C., Burley, S.D. (1999). Geological controls on kaolin particle shape and consequences for mineral processing. *Clay Minerals*, 34, 193–208.

Psyrillos, A., Manning, D.A.C., Burley, S.D. (2001). The nature and significance of illite associated with quartz-hematite hydrothermal veins in the St Austell pluton, Cornwall, England. *Clay Minerals*, 36, 585–597.

Psyrillos, A., Burley, S.D., Manning, D.A.C., Fallick, A.E. (2003). Coupled mineral-fluid evolution of a basin and high: Kaolinization in the SW England granites in relation to the development of the Plymouth Basin. In *Hydrocarbons in Crystalline Rocks*, Petford, N. and McCaffrey, K.J.W (eds). Geological Society Special Publications, London.

Rankin, A.H. and Alderton, D.H.M. (1983). Fluid inclusion petrography of SW England granites and its potential in mineral exploration. *Mineralium Deposita*, 18, 335–347.

Rattey, P.R. and Sanderson, D.J. (1984). The structure of SW Cornwall and its bearing on the emplacement of the Lizard Complex. *Journal of the Geological Society*, 141(1), 87–95.

Richards, H., Willis-Richards, J., Pye, J. (1991). A review of geological investigations associated with the UK Hot Dry Rock programme. *Geoscience in South-West England*, 7, 321–326.

Rock, N.M.S. (1991). *Lamprophyres*. Blackie and Son, Glasgow.

Rollinson, G.K., Pirrie, D., Power, M.R., Cundy, A., Camm, G.S. (2007). Geochemical and mineralogical record of historical mining, Hayle Estuary, Cornwall, UK. *Geoscience in South-west England*, 11, 326–337.

Romer, R.L. and Kroner, U. (2014). Sediment and weathering control on the distribution of Paleozoic magmatic tin–tungsten mineralization. *Mineralium Deposita*, 50, 327–338.

Romer, R.L. and Kroner, U. (2016). Phanerozoic tin and tungsten mineralization – Tectonic controls on the distribution of enriched protoliths and heat sources for crustal melting. *Gondwana Research*, 31, 60–95.

Rudnick, R.L. and Gao, S. (2004). Composition of the continental crust. In *Treatise on Geochemistry*, Holland, H.D. and Turekian, K.K. (eds). Elsevier, Amsterdam.

Salmon, S. (1994). Mingling between coexisting granite magmas within the Land's End granite – Preliminary observations. *Proceedings of the Ussher Society*, 8, 219–223.

Salmon, S. and Shail, R.K. (1999). Field excursion to examine the granites in the area between Cape Cornwall and Porth Nanven, West Penwith, 3rd January 1999. *Geoscience in South-West England*, 9, 391–393.

Scott, P., Pascoe, R., Hart, F. (1998). Columbite-tantalite, rutile and other accessory minerals from the St Austell topaz granite, Cornwall. *Geoscience in South-West England*, 9, 165–170.

Scrivener, R.C. (2006). Cornubian granites and mineralization of SW England. In *The Geology of England and Wales*, Brenchley, P. and Rawson, P. (eds). Geological Society of London.

Scrivener, R.C., Shepherd, T.J., Garrioch, N. (1986). Ore genesis at Wheal Pendarves and South Crofty Mine, Cornwall – A preliminary fluid inclusion study. *Proceedings of the Ussher Society*, 6, 412–416.

Scrivener, R.C., Cooper, B., George, M.C. (1987). Mineralogy and paragenesis of the Haytor iron ore deposit. *Proceedings of the Ussher Society*, 6, 558.

Scrivener, R.C., Darbyshire, D.P.F., Shepherd, T.J. (1994). Timing and significance of cross-course mineralization in SW England. *Journal of the Geological Society of London*, 150, 587–590.

Seifert, T. (2008). *Metallogeny and Petrogenesis of Lamprophyres in the Mid-European Variscides: Post-collisional Magmatism and its Relationship to Late-Variscan Ore Forming Processes in the Erzgebirge (Bohemian Massif)*. IOS Press, Amsterdam.

Shail, R.K. and Alexander, A.C. (1997). Late Carboniferous to Triassic reactivation of Variscan basement in the western English Channel: Evidence from onshore exposures in south Cornwall. *Journal of the Geological Society*, 154, 163–168.

Shail, R.K. and Leveridge, B.E. (2009). The Rhenohercynian passive margin of SW England: Development, inversion and extensional reactivation. *Comptes rendus géoscience*, 341(2), 140–155.

Shail, R.K. and Wilkinson, J.J. (1994). Late- to post-Variscan extensional tectonics in south Cornwall. *Proceedings of the Ussher Society*, 8, 162–270.

Shail, R.K., Stuart, F.M., Wilkinson, J.J., Boyce, A.J. (2003). The role of post-Variscan extensional tectonics and mantle melting in the generation of the Lower Permian granites and the giant W-As-Sn-Cu-Zn-Pb orefield of SW England (extended abstract). *Applied Earth Science*, 112, 127–129.

el-Sharkawi, M.A.H. and Dearman, W.R. (1966). Tin-bearing skarns from the north-west border of the Dartmoor granite, Devonshire, England. *Economic Geology*, 61(2), 362–369.

Shepherd, T.J. and Darbyshire, D.P.F. (1986). Fluid inclusion Rb-Sr geochronology of mineral deposits. In *Geology in the Real World; The Kingsley Dunham Volume*, Nesbitt, R.W. and Nichol, I. (eds). Institution of Mining and Metallurgy, London.

Shepherd, T.J., Miller, M.F., Scrivener, R.C., Darbyshire, D.P.F. (1985). Hydrothermal fluid evolution in relation to mineralization in southwest England with special reference to the Dartmoor-Bodmin area. In *High Heat Production (HHP) Granites, Hydrothermal Circulation and Ore Genesis*, Darnley, G.A. (ed.). Institution of Mining and Metallurgy, London.

Sheppard, S.M.F. (1977). The Cornubian Batholith, SW England: D/H and 18O/16O studies of kaolinite and other alteration minerals. *Journal of the Geological Society*, 133, 573–591.

Simons, B. (2015). Processes controlling critical metal (Li, Be, Ga, Ge, Nb, Ta, In, Sn, Sb, W and Bi) distribution in the peraluminous granites of the Cornubian Batholith. PhD Thesis, University of Exeter.

Simons, B., Pirrie, D., Rollinson, G.K., Shail, R.K. (2011). Geochemical and mineralogical record of the impact of mining on the Teign estuary, Devon, UK. *Geoscience in South-West England*, 12, 339–350.

Simons, B., Shail, R.K., Andersen, J.C.Ø. (2016). The petrogenesis of the early Permian Variscan granites of the Cornubian Batholith: Lower plate post-collisional peraluminous magmatism in the Rhenohercynian zone of SW England. *Lithos*, 260, 76–94.

Simons, B., Andersen, J.C.Ø., Shail, R.K., Jenner, F.E. (2017). Fractionation of Li, Be, Ga, Nb, Ta, In, Sn, Sb, W and Bi in the peraluminous Early Permian Variscan granites of the Cornubian Batholith: Precursor processes to magmatic-hydrothermal mineralisation. *Lithos*, 278–281, 491–512.

Smith, M., Bank, D.A., Yardley, B.W.D., Boyce, A. (1996). Fluid inclusion and stable isotope constraints on the genesis of the Cligga Head Sn-W deposit, SW England. *European Journal of Mineralogy*, 8, 961–974.

Smith, W.D., Darling, J.R., Bullen, D.S., Lasalle, S., Pereira, I., Moreira, H., Allen, C.J., Tapster, S. (2019). Zircon perspectives on the age and origin of evolved S-type granites from the Cornubian Batholith, Southwest England. *Lithos*, 336, 14–26.

Stanier, P. (1999). *South West Granite: A History of the Granite Industry in Cornwall and Devon*. Cornish Hillside Publications, St Austell.

Štemprok, M. and Seifert, T. (2011). An overview of the association between lamprophyric intrusions and rare-metal mineralization. *Mineralogia*, 42(2–3), 121–162.

Stimac, J.A., Clark, A.H., Chen, Y., Garcia, S. (1995). Enclaves and their bearing on the origin of the Cornubian Batholith, Southwest England. *Mineralogical Magazine*, 59, 273–296.

Stone, M. (1968). A study of the Praa Sands elvan and its bearing on the origin of elvans. *Proceedings of the Ussher Society*, 2, 37–42.

Stone, M. (1971). Form and structure of the granites of South-West England. Symposium on structure of South-West England. *Proceedings of the Ussher Society*, 2, 253–263.

Stone, M. (1992). The Tregonning granite: Petrogenesis of Li-mica granites in the Cornubian Batholith. *Mineralogical Magazine*, 56(383), 141–155.

Stone, M. (1997). A geochemical dichotomy in the Cornubian Batholith. *Proceedings of the Ussher Society*, 9, 206–210.

Stone, M. and Exley, C. (1989). Geochemistry of the Isles of Scilly pluton. *Proceedings of the Ussher Society*, 7, 152–157.

Stone, M. and George, M.C. (1983). Some phosphate minerals at the Megiliggar Rocks, Cornwall. *Proceedings of the Ussher Society*, 5, 428–431.

Stone, M., Exley, C.S., George, M.C. (1988). Compositions of trioctahedral micas in the Cornubian batholith. *Mineralogical Magazine*, 52(365), 175–192.

Tapster, S. and Bright, J.W. (2020). High-precision ID-TIMS cassiterite U–Pb systematics using a low-contamination hydrothermal decomposition: Implications for LA-ICP-MS and ore deposit geo-chronology. *Geochronology*, 2(2), 425–441.

Taylor, G.K. (2007). Pluton shapes in the Cornubian Batholith: New perspectives from gravity modelling. *Journal of the Geological Society*, 164(3), 525–528.

Thorpe, R.S., Cosgrove, M.E., Calsteren, P.W.C. (1986). Rare earth element, Sr- and Nd-isotope evidence for petrogenesis of Permian basaltic and K-rich volcanic rocks from south-west England. *Mineralogical Magazine*, 50(357), 481–489.

Tischendorf, G., Gottesmann, B., Förster, H.J., Trumbull, R.B. (1997). On Li-bearing micas: Estimating Li from electron microprobe analyses and an improved diagram for graphical representation. *Mineralogical Magazine*, 61(409), 809–834.

Turpin, L., Velde, D., Pinte, G. (1988). Geochemical comparison between minettes and kersantites from the Western European Hercynian orogen: Trace element and PbSrNd isotope constraints on their origin. *Earth and Planetary Science Letters*, 87(1), 73–86.

Vernon, R.H. (1984). Microgranitoid enclaves in granites – Globules of hybrid magma quenched in a plutonic environment. *Nature*, 309(5967), 438–439.

Warner, M., Spaargaren, B., Jones, R., Watts, D., Doody, J. (1994). High resolution images of the lower crust: Deep seismic reflections from 15 to 180 Hz. *Tectonophysics*, 232, 235–237.

Warr, L.N., Primmer, T.J., Robinson, D. (1991). Variscan very low-grade metamorphism in south-west England: A diastathermal and thrust-related origin. *Journal of Metamorphic Geology*, 9(6), 751–764.

Wilkinson, J.J., Rankin, A.H., Mulshaw, S.C., Nolan, J., Ramsey, M.H. (1994). Laser ablation-ICP-AES for the determination of metals in fluid inclusions: An application to the study of magmatic ore fluids. *Geochimica et Cosmochimica Acta*, 58, 1133–1146.

Williamson, B.J., Spratt, T.J., Adams, J.T., Tindle, A.G., Stanley, C.J. (2000). Geochemical constraints from zoned hydrothermal tourmalines on fluid evolution and Sn mineralization: An example from Fault Breccias at Roche, SW England. *Journal of Petrology*, 41(9), 1439–1453.

Willis-Richards, J. and Jackson, N.J. (1989). Evolution of the Cornubian ore field, Southwest England. Part I: Batholith modeling and ore distribution. *Economic Geology*, 84(5), 1078–1100.

Wilson, M., Neumann, E.-R., Davies, G.R., Timmerman, M.J., Heeremans, M., Larsen, B.T. (eds) (2004). Permo-Carboniferous magmatism and rifting in Europe. *Geological Society Special Publications*, 223.

Yeomans, C.M., Shail, R.K., Grebby, S., Nykänen, V., Middleton, M., Lusty, P.A.J. (2020). A machine learning approach to tungsten prospectivity modelling using knowledge-driven feature extraction and model confidence. *Geoscience Frontiers*, 11(6), 2067–2081.

Ziegler, P.A. and Dèzes, P. (2006). Crustal evolution of Western and Central Europe. In *European Lithosphere Dynamics*, Gee, D.G. and Stephenson, R.A. (eds). Geological Society Memoir No. 32, London.

3

The W Deposit at Panasqueira (Portugal): A Critical Bibliographical Review

Christian MARIGNAC and Michel CATHELINEAU
GeoRessources, CREGU, University of Lorraine, Nancy, France

3.1. Introduction

Variscan Western Europe is one of the world's great tin-tungsten provinces, within an initial endowment (past production + reserves + resources) that can be estimated to have been on the order of a million tonnes of tungsten (Appendix A3.1). With greisens and skarns, the peribatholic Sn-W mineralized veins play a major role. The Portuguese Panasqueira deposit is emblematic of the latter. This world-class deposit, situated in the Central Iberian Zone (CIZ) of the Variscan orogeny, contained 147 kt W when it was discovered at the end of the 19th century, that is, more than one-third of the total endowment of the CIZ, estimated at 350 kt W (section A3.1).

Exploited for a century and a half, this deposit has been the subject of many studies since the 1950s (Thadeu 1951; Bloot and Wolf 1953). The final two decades of the 20th century saw a succession of works by Kelly and Rye (1979), Bussink (1984), Bussink et al. (1984), Snee et al. (1988), Polya (1988, 1989), Noronha et al. (1992), Foxford et al. (2000), Polya et al. (2000) and Lourenço (2002). Then, after a relative lull, the most recent period saw a renewal of interest, motivated (at least in part) by a desire to help the mining industry renew reserves, which are becoming

For a color version of all figures in this chapter, see www.iste.co.uk/decree/metallic1.zip.

exhausted, in particular with publications by Pinto et al. (2015), Lecumberri-Sanchez et al. (2017), Codeço et al. (2017, 2019, 2020), Jaques and Pascal (2017), Martins (2017), Jacques et al. (2018), Launay (2018), Launay et al. (2018, 2019), Carocci (2019), Carocci et al. (2021), Cathelineau et al. (2018, 2020), Marignac et al. (2020), Martins et al. (2020) and Mateus et al. (2020).

Many aspects have been tackled through these publications, including the geometry of the mineralized system, the details of how it is filled and the composition of the mineralized phases, the alterations associated with them, the nature of the fluids and relationships with the granite present at depth, and the age of magmatic and hydrothermal phenomena.

It may therefore be thought that the Panasqueira deposit is particularly well understood; however, this is not the case. As Gonçalves et al. (2018) emphasize, the very accumulation of so much structural, geochemical and mineralogical data demonstrates the need to rethink the genetic models resulting from these works, most of which are based on the idea of a simple magmatic-hydrothermal system centered on the Panasqueira granite.

Many of these publications present contradictions between them, which, for some, may seem irreducible. For example, for Kelly and Rye (1979), followed by a large number of authors (Bussink et al. 1984; Polya 1989; Polya et al. 2000), the deposit was formed at less than 360°C, indeed 320°C for the wolframite itself, and at very low depth (from 600 m to 2 km), albeit with overpressure excursions (up to 100–150 MPa). On the contrary, according to Jaques and Pascal (2017), Cathelineau et al. (2020) and Mateus et al. (2020), this formation took place at a depth of more than 10 km and at temperatures well above 400°C.

In the same way, the role played by the small greisenized cupola, which mining has shown to exist at depth, is the subject of much controversy. It pre-dates the mineralization for some (Kelly and Rye 1979; Polya 1989; Lourenço 2002; Marignac et al. 2020) and is contemporary and genetically linked (magmatic fluids) for others (Bussink et al. 1984; Codeço et al. 2017; Launay et al. 2018). Alternatively, Lecumberri-Sanchez et al. (2017) believe that the hydrothermal alterations at the vein walls (tourmalinization, muscovitization) are synchronous, which leads to significant consequences in terms of generation of the deposit; but this synchronicity is much contested by Carocci et al. (2021). Finally, the tectonic model for the opening of the mineralized veins invoked by Jacques et al. (2018) contradicts the structural observations reported by Launay (2018) in his PhD. In such a context, the present chapter proposes discussing a series of these controversial subjects by presenting the opposing theses in a contradictory fashion and trying to reach a conclusion wherever possible.

After presenting the deposit's architecture (which is the subject of broad consensus), we will address successively the chronology of the alteration and mineralization phases, the mechanism for opening the veins, the P and T mineralization conditions, the nature and origin of the fluids, and the Panasqueira granite relationships to mineralization.

3.2. Geological context

3.2.1. *Regional context*

The polymetallic deposit at Panasqueira (W-Sn-Cu) is situated in the Beira Baixa province (Central Portugal). It lies in the CIZ, which represents the axial zone of the Iberian Variscan belt (Figure 3.1(a)). The CIZ essentially includes a multikilometer-thick sequence of schists and graywackes dating from the end of the Ediacaran and the Cambrian, the Portuguese "schisto-graywacke complex", also called the Beira group, which is overlaid discordantly by Armorican quartzite from the Upper Ordovician (Diez Fernandez et al. 2013).

Tectonic and thermal evolution in the context of collision occurred over the whole Carboniferous, ending at the start of the Permian (Martinez Catalan et al. 2009; Diez Fernandez et al. 2016; Pereira et al. 2018). Giving way to initial thickening of the crust (episode D1, start of the Carboniferous; recumbent P1 folds), the start of evolution from the Upper Carboniferous was characterized by an extensive episode (D2 episode, recumbent P2 folds) at 320–315 Ma and by crustal melting events on a large scale (migmatites), resulting from the combination of post-thickening warming of the middle crust and the input of heat derived from the mantle at the base of the continental crust (Dias et al. 1998). Since this heat input is more or less cyclical, another consequence is the location of syn- to post-kinematic granitic plutons in the interval of 315–295 Ma (Mateus and Noronha 2010; Sant'Ovaia et al. 2010; Fernandez-Suarez et al. 2011; Diez Fernandez and Pereira 2017).

In parallel, deformations have successively taken place in the context of crustal-scale shear zones. At approximately 315–310 Ma (episode D3), upright P3 folds of multi-kilometer amplitude (1–10 km) developed in a NW–SE to N–S direction, with S3 schistosity in the axial plane. They precede and/or accompany large shear zones, first sinistral (approximately 309–305 Ma) and then dextral (approximately 304–295 Ma) (Diez Fernandez and Pereira 2017 (and included references)). Many mineralized systems developed in the CIZ in direct association with the tectonics from the Upper Carboniferous and the emplacement of granitoids. More than 150 deposits are known, corresponding to peri- and intragranitic rare metal systems (Nb-Ta, Li, W, Sn) and to shear zones (Au-Ag-As-Sb-Pb–Zn-Cu) (Tornos et al. 2000; Neiva 2002; Mateus and Noronha 2010; Noronha 2017; Roda-Robles et al. 2018).

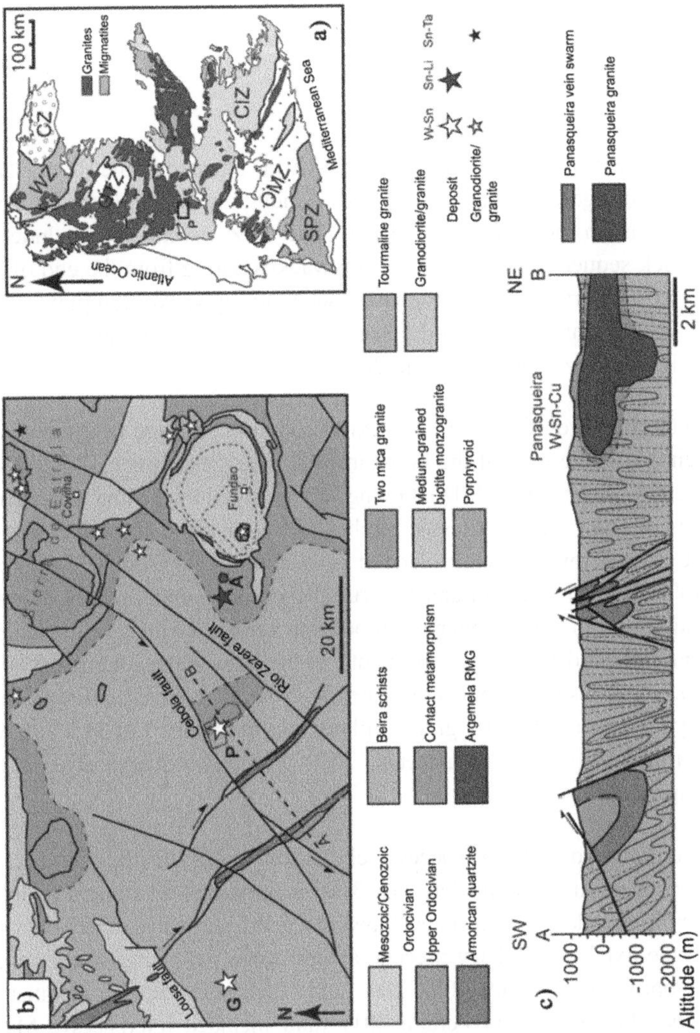

Figure 3.1. *The regional context at Panasqueira (redrawn from Marignac et al. 2020)*

COMMENT ON FIGURE 3.1.– *(a) The great Variscan units and the Panasqueira location (simplified from Diez Fernandez et al. 2016). Allochthonous units: CZ (Cantabrian Zone), GTZ (Galicia–Trás-os-Montes Zone), OMZ (Ossa Morena Zone), SPZ (Southern Portuguese Zone); autochthonous units: CIZ (Central Iberian Zone), WZ (Western Asturian–Leon Zone); P: location of figure (b). (b) Geological map of the Panasqueira region (located in (a)) (adapted from Bussink (1984), Lourenço (2002) and Launay (2018)); A: Argemela, G: Gois, P: Panasqueira, line A–B section of figure (c). (c) Geological section A–B (simplified from Launay 2018).*

3.2.2. *Local geological setting*

The Panasqueira deposit is found at the heart of an upright anticlinorium in the NW–SE direction of Upper Carboniferous age (D3) and formed by the Beira schists, framed by narrow synclines of Ordovician Armorican quartzite, to the southwest of the Serra da Estrela granitic complex (Launay 2018; Figures 3.1(a) and (b)). The Beira schists have been metamorphosed in the lower greenschist facies (epizone). The veins of synmetamorphic quartz (exudation quartz) are for the most part very rare in these schists, where decametric to metric veins of late kinematic milky quartz are encountered only sporadically. However, these veins are much more abundant in the area of the deposit, where they form a swarm; miners there call them *Seixo Bravo* (SB), and they are generally sterile. According to observations in the horizontal plane, the SB systems seem to result from the opening under fluid pressure of joints corresponding to vertical shears, for the most part conjugate, without displacement (virtual shears) and resulting from a state of compressive constraints, with an overall NE–SW shortening direction. These joints have been filled with quartz, and the resulting veins have been deformed plastically (boudinage, micro-folding, mullions) in relation to regional schistosity S3 (Figure 3.2(a)). The SBs are therefore clearly late kinematic in the D3 event. Although NNW–SSE and ~E–W conjugated directions coexist on the edge of the swarm, the N120-130°E direction broadly predominates at the heart of the deposit (Lourenço 2002). Ten kilometers to the east of the deposit are the Fundão granodiorite (Garcia 2004) and the small Argemela cupola, a rare metal granite (RMG) associated with tin mineralization and with minor veins of wolframite (Lima et al. 2019; Michaud et al. 2020; Figure 3.1(b)).

At Panasqueira itself, the presence of a deep granitic intrusion is indicated by the presence of a large contact metamorphism aureole (Figure 3.3), represented by mottled schists (biotite-cordierite), with the appearance of andalusite in proximity to the granitic contact (Clark 1964). These mineral assemblages constrain pressure to approximately 300 MPa at the time of emplacement of the Panasqueira granite

(Pattison 2001; Kuhn 2018). The Serra de Estrella granitic massif would have been emplaced at a comparable depth (300–350 MPa: Neiva et al. 1987). The presence of this granitic massif is probably associated with the aplitic dikes oriented NW–SE found in the Cabeço do Pião region (where they are intersected by wolframite quartz veins) but are also encountered sporadically along the eastern boundary of the metamorphic aureole (Launay 2018). Mafic dikes are also encountered, generally oriented in the same direction (Kelly and Rye 1979) (Figure 3.3). According to a recent geophysical model (Ribeiro 2017), the main body of the Panasqueira granitic massif has the shape of a small granitic laccolith, approximately only 1 km thick, 6.5 km in length and 4.5 km in width, with a voluminous keel descending to approximately 2 km and elongated NNW–SSE, and with a rather flat roof from which there emerges a cupola, also elongated NNW–SSE, which is offset from the keel (Figures 3.1(c) and 3.4) and encountered during mining work (section 3.3). The volume of the intrusion can be estimated at 32 km^3.

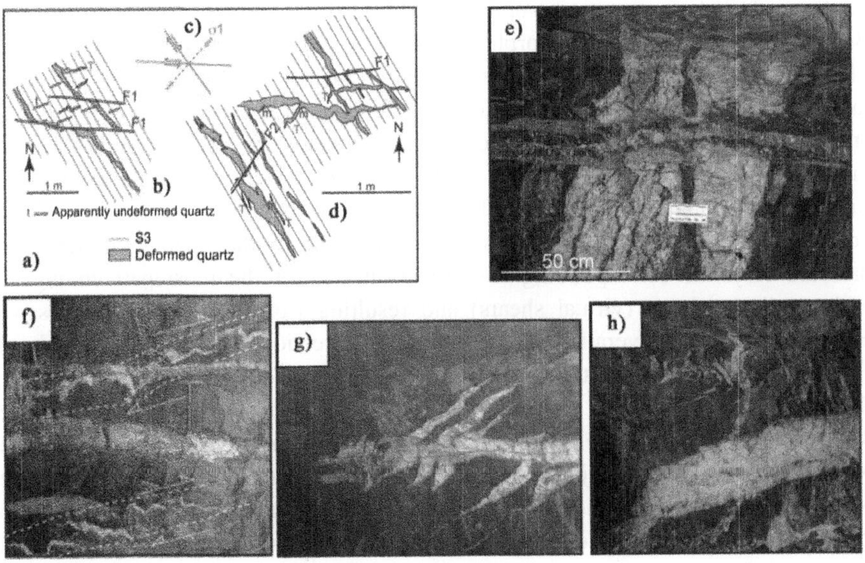

Figure 3.2. *Late kinematic D3 quartz veins ("Seixo Bravo" SB) at Panasqueira*

COMMENT ON FIGURE 3.2.– *(a) Geometry of the subvertical SBs observed in the horizontal plane (banks of the Rio Zezere). (b). (d) Set of veins interpreted in conjugate strike-slips, with subordinate T tension joints (m: mullions). (c) Kinematic interpretation (F1, F2: fragile strike-slip faults). (e)–(h) Observations in mining work: (e) subvertical SB intersected by mineralized veins (note the boudinage*

(L2-D29-R3-AW20)); (f) SBs organized into sets of overlapping joints interpreted in dextral relays of a flat virtual shear (L1-D13-R13-AW11); (g) SB vein occupying a flat shear, with the associated tension joints (L2-D23-R6A-AW22); (h) idem, the flat system seems superposed to a subvertical SB (L3-D9-R1-AW13).

3.2.3. *Mineralized vein system*

Although mineralized veins exist on the surface, where they are still visible and enabled the discovery of the deposit at the end of the 19th century, most of the vein system lies at depth, over a surface of approximately 10 km^2, with a vertical extension of more than 500 m (Kelly and Rye 1979; Foxford et al. 2000).

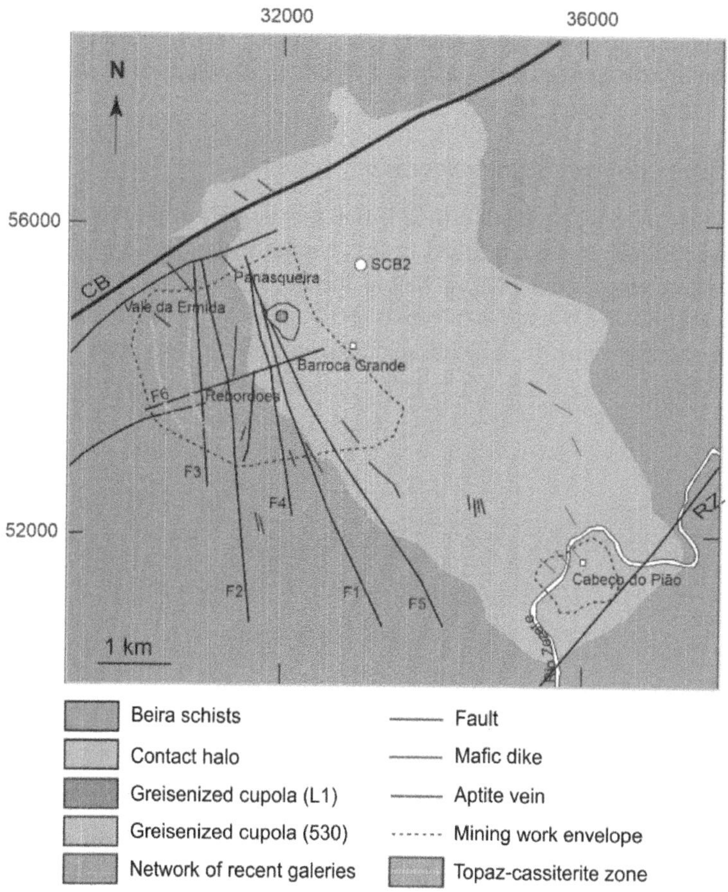

Figure 3.3. *Map of the Panasqueira region*

COMMENT ON FIGURE 3.3.– *This map shows the thermal aureole on the surface and the mining work below (adapted from Kelly and Rye 1979; Lourenço 2002; Gonçalves et al. 2018). SCB2 location of the SCB2 borehole. Faults: CB Cebola fault; F1 main fault; F2 Lameiras fault; F3 Valle de Freiras fault; F4 1W fault; F5 Masso fault; RZ Rio Zezere fault.*

The exploitation method, used for more than 50 years, uses unblasted chambers and pillars and provides excellent conditions for observation in three dimensions and over large distances. Several sectors (Panasqueira, Valle de Ermida) have been or are still (Barroca Grande, Rebordoes) being exploited actively; offsetting the SE, a subordinate set was also formerly exploited at Cabeço do Piao (Figure 3.3). The deposit is exploited over four levels (L0 at 680 m; L1 at 620 m; L2 at 560 m; L3 at 470 m). In the Barroca Grande sector, at level L0, a set of particularly thick and rich veins was found 50 years ago ("Bank"). The "alpine" set of Cebola and Rio Zezere faults would have caused the Panasqueira block to collapse, which would also have protected it from erosion (Launay 2018).

3.2.3.1. *The geometry of the mineralized system*

Economic W-Sn-Cu mineralization is found in a series of shallow dipping quartz veins, most of which are contained in the thermal metamorphism envelope. They intersect the greisenized cupola as well as the Seixo Bravo veins. A system of late subvertical ENE–WSW and "N–S" faults (N10°E and N170°E) intersect the granitic intrusion and the veins; it is better identified in mining work than on the surface (Figure 3.3).

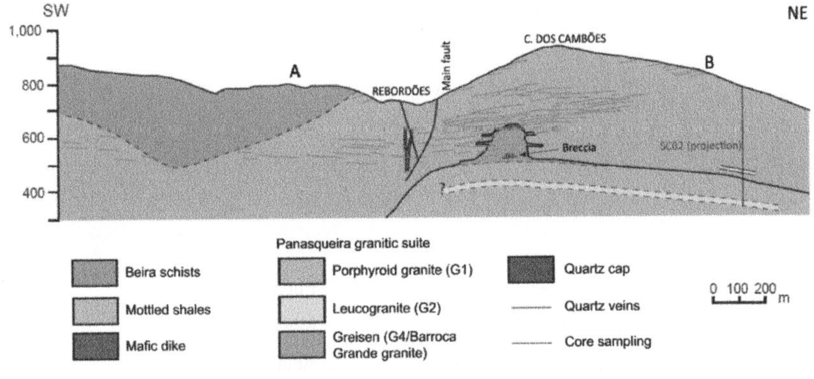

Figure 3.4. *Section of the Panasqueira deposit (Kelly and Rye 1979; Bussink 1984; Martins 2017; Marignac et al. 2020). The lateral extension of the leucogranite G2 sheet is hypothetical, such as the prolongation toward the bottom of the G1 facies out of the influence of SCB2 core drilling (located in Figure 3.3)*

The abundance of mineralized veins is greatest above the greisenized cupola, but the system in its entirety is largely offset from this apex (Figure 3.4). The deposit is divided into two parts, approximately at the level of the main fault: to the east, the veins have an average dip of 10–12° to the SW with a NW–SE orientation, whereas to the west, the dips of 7–8° are generally oriented to the SE, but this change is not localized by the cupola. A 3D mapping of the dilation of the mineralized veins (calculated as the ratio of the cumulated thickness of the veins to the height of a vertical reference cross-section) was created by Lourenço (2002) from the description of numerous surveys by Hebblewaith and Antão (1982). From this, this author concluded that there was still no favored relationship with the greisenized cupola because instead of observing a maximum dilation above the cupola, he obtained a quasi-isotropic distribution of the dilation over the whole deposit. However, thanks to the mapping of the four sections corresponding to levels L0, L1, L2 and L3, a trend toward a linear organization became apparent, especially sensitive at the deepest levels (L2, L3) where the NW–SE direction dominated (Lourenço 2002, Figure 69, p. 138).

At the scale of the pillars left by mining, the elementary veins, the thickness of which can vary very quickly, are presented as a series of sections, which dip alternately between NW and SW. The passage from one vein to another is made by a relay with the characteristic "eel's tail" (*rabdos de anguilhas*) morphology, sometimes with an oblique connection (Figures 3.5(a) and (c)). Despite this appearance, however, the elementary veins are interconnected in space, following a pattern termed "helicoidal" by Kelly and Rye (1979) (Figure 3.5(b)). Their dimensions are highly variable: from some centimeters to more than 1 m in thickness and from some decimeters to more than 1 km in length. Nevertheless, according to Foxford et al. (2000), the average elemental vein can be represented as an elliptical disk (100 × 300 m), with an average thickness of 20 cm (Figure 3.5(d)). The veins are affected by alterations at their walls, notably with systematic tourmalinization, which is moreover asymmetrical and more developed at the roof (30 cm on average) than at the wall (15 cm on average) (Carocci et al. 2021) (section 3.3.3).

In addition to the flat veins interconnecting them, we find straighter veins, also mineralized, called "galo veins" by miners, which can reach 1–5 m in vertical development. It might be thought that these veins represent an exacerbation of the bridges that connect the "eels' tails". The "galo veins" are thought by some (Ribeiro and Pereira 1982; Launay 2018) to be more common in the region of the greisenized cupola, around which they would form a concentric system, analogous to a ring-dike. However, measurements by Lourenço (2002) show that the "galos" are systematically oriented in the NE–SW direction, and this is the case in the entire Barroca Grande sector, without any particular check on their location.

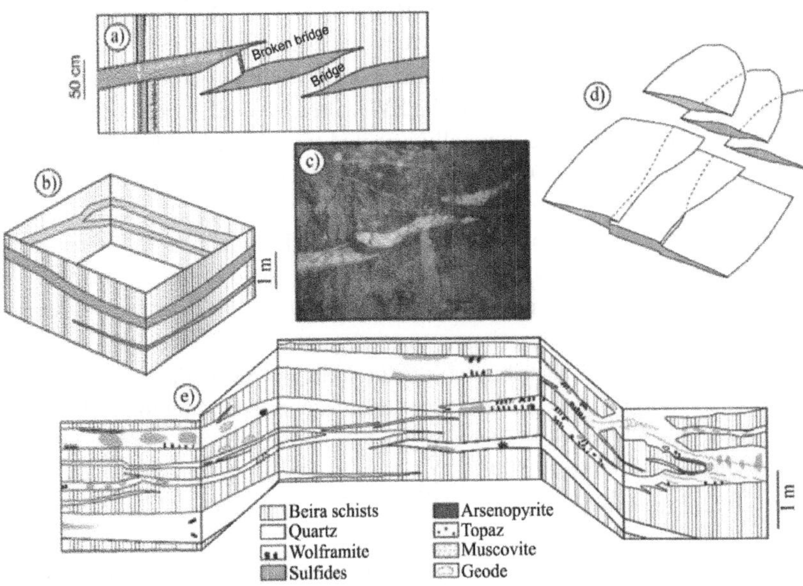

Figure 3.5. *General aspect of the Panasqueira veins*

COMMENT ON FIGURE 3.5.– *(a) Schematic 2D diagram of the "eel's tail" relay; note the opening in mode I shown by the offset of the* Seixo Bravo marker *(redrawn from Foxford et al. (2000, figure 1)). (b) "Helicoidal" behavior of veins in 3D (redrawn from Kelly and Rye (1979, figure 12)). (c) Example of an "eel's tail" structure (L1D5R7AW15). (d) 3D extension of figure (a); the elementary veins take broadly ellipsoidal forms and are interconnected; note the overlappings that cause the "eels' tails" and the cracking of the bridges (redrawn from Foxford et al. (2000, figure 1)). (e) Example of the development of veins at the scale of the pillars; note the strong nugget effect of the distributions of both the wolframite and the sulfides, and that these minerals do not coexist systematically; note also the oblique slits of muscovite (L0D15R15AW10 "Bank", drawn in situ [CM, 1970]).*

3.2.3.2. *Mineralogy and zonings*

Observation of veins in the mining sites shows that they present essentially as networks of quartz veins, where the economic minerals (wolframite, cassiterite, chalcopyrite) are randomly distributed (Figure 3.5(e)). Many other minerals are associated with them: silicates (tourmaline, muscovite, topaz, chlorite), sulfides (pyrrhotite, sphalerite, stannite, galena, pyrite, marcasite, arsenopyrite, many phases of Bi-Ag), phosphates (apatite, wagnerite, isokaite) and carbonates (siderite, dolomite, calcite). In total, approximately 50 phases have been described by most authors, the most notable being Kelly and Rye (1979), Polya (1989), Martins (2017), Martins et al. (2020) and Mateus et al. (2020). A particular mention is "muscovite", which is rich in fluorine, lithium and sometimes iron (Li- and Fe-Li-muscovite) (Mateus et al. 2020). The Panasqueira deposit is renowned moreover for its spectacular collection of minerals found in open vugs within the veins (Figure 3.6). The primary minerals can themselves be large, with quartzes or wolframites that can reach 50 cm in the richest sites (the "Bank", for example).

Figure 3.6. *Examples of automorphic minerals from the Panasqueira vugs*

COMMENT ON FIGURE 3.6.– *In this figure and in all the following ones, the abbreviations for the minerals conform to the IMA's recommendations (Whitney and Evans 2010). (a and b) Quartz-wolframite assemblages; note that the quartz covers the wolframite; (c) wolframite-arsenopyrite-muscovite assemblage, in the order of crystallization in the vug; (d) layer of muscovite and arsenopyrite on a quartz wall; (e) euhedral chalcopyrite aggregate (Ccp2, see section 3.4.2.7); (f) sphalerite and chalcopyrite growing (in this order) on euhedral quartz.*

The wolframite is practically the only tungsten mineral: scheelite is in fact extremely rare at Panasqueira, where it was actually found only in a vein intersecting a mafic dike, a clear manifestation of a fluid–rock interaction, the calcium evidently comes from the dike and the tungsten from the fluid. This wolframite is a ferberite in reality, with a hubnerite content between 12 and 20 mole %, and it is often rich in niobium (up to 0.74% Nb_2O_5) (Polya 1988a). This author believed it would be possible to link the size of the Nb content of the wolframites to their greater or lesser proximity to the NW–SE faults. Like other authors before or after him (Lourenço 2002), he thus envisaged an active role for these faults in generating the mineralization. However, as Foxford et al. (1991) have clearly shown, the brittle faults encountered during mining work are clearly later than the mineralized system, the veins of which they intersect and displace. It is probable that Polya (1988a) was the victim of a sampling bias, similar to those before him who thought it could be shown that the manganese content of the wolframites increased closer to the greisenized cupola, which his own work led him to deny.

On the scale of the deposit, some minerals show a zonal distribution. This is how Pinto (2014) and Pinto et al. (2015) observed an increase in the quantity of topaz and sphalerite from the NE to the SW of the mine at all levels. Mateus et al. (2020) similarly described enrichment in topaz in the western sectors, with a concomitant increase in the quantity of cassiterite (Figure 3.3). These observations are in agreement with the geochemical data presented in Pinto et al. (2014), who have analyzed the average content of the veins and their host rock for a range of elements. They thus conclude that, on the one hand, there are no relationships at level L1 between the Sn content and the proximity of the greisenized cupola, and on the other hand, an enrichment is emerging in the western part of the deposit. Moreover, the surrounding rock near the mineralized veins shows higher levels of F, Li and Sn in this same western sector. Mateus et al. (2020) also observe that the muscovite selvages seem more developed to the NE and W, which they think could be linked (similar to the data on the topaz) to the distance from the cupola.

3.2.3.3. *Quantitative aspects*

The number of elementary veins can be estimated using the published values for the system's total volume, estimated between 3 and 6×10^{-3} km^3 (Polya 1989), and assuming an average thickness for the veins of 0.2 m (Foxford et al. 2000). A number of elementary veins between 750 and 1,500 (Carocci et al. 2021) are therefore found. Polya (1989) is the only author to have sought to quantify the filling of mineralized veins beyond economic metals (W, Sn, Cu). With ~90% in volume, quartz is by far the most abundant mineral (Figure 3.5(e)), while with an estimated tonnage of 250 kt, iron is by far the most abundant metal, present in wolframite, chalcopyrite, sphalerite (~8% Fe) (Martins 2017), arsenopyrite, siderite and above all pyrrhotite (and associated pyrite; section 3.5.2.6), which is the most abundant sulfide. Tungsten is unsurprisingly the second (~150 kt), followed by copper (70 kt). Aluminum (60 kt), concentrated in topaz and muscovite, follows, far ahead of tin (25 kt), essentially in cassiterite but also in stannite and zinc (20 kt). The estimated abundance of sulfur (200 kt) underlines the importance of sulfides in the Panasqueira polymetallic deposit. Silver and bismuth are present fairly systematically, but gold is absent; selenium, tellurium and antimony are very rare, as well as molybdenum (section 3.3.2.6.2).

3.2.3.4. *The anomalous halo*

The deposit is also characterized by a halo of geochemical anomalies, which has been studied repeatedly using surface schist sampling. Oosterom et al. (1984) are interested in the near halo, in a radius on the order of one kilometer from the greisenized cupola, taken to be the center of the deposit. As shown in Figures 3.7(a) and (c), significant anomalies (up to 60 ppm in tungsten and 110 ppm in tin) follow axes in a broadly NW–SE direction centered on the mineralized zones. These directions are approximately those of the late faults known in mining work (F1–F2 in Figure 3.7). A neat correlation with fluorine (Figure 3.7(c)), lithium and rubidium is also observed. A comparable mapping, but over a larger surface, was carried out by Gonçalves et al. (2018); significant anomalies in W and Sn were again observed, still centered on the deposit, however with the difference that a very large tungsten anomaly (up to 409 ppm) appears more to the west (without any mineralization being known at present). The Sn and W anomalies are clearly superposed and organized in a NW–SE direction (Figures 3.7(d) and (e)). The authors note a strong correlation between the Sn, F, Li and As anomalies, whereas the tungsten shows a strong correlation with Nb, Ta and Ti. The authors link this latter characteristic to the existence of early rutiles rich in rare metals, the abundance of which in the rock surrounding the veins has been demonstrated by Carocci et al. (2021). Moreover, it is interesting to note that Pinto (2014), studying the geochemistry of the rock surrounding the veins, observed the same correlation between Sn, F and Li. The Cebola fault does not seem to play the lesser part in the distribution of the anomalies, except to some extent for fluorine, so long as the western anomaly is not attributed to it (as Gonçalves et al. (2018) do).

Polya (1988b) characterized the imprint of mineralization in the distant environment, and he produced the profiles in Figures 3.5(f) and (g), which show that the anomaly halo extends more than 5 km from the center of the deposit. From this,

Polya (1988b) deduced an estimate of the tonnages of metals that the hydrothermal system was able to export in the rock surrounding the deposit, and the values that he found are higher than those of the exploitable tonnages, often very much so: 300 kt W (compared to 150 kt in the deposit), 900 kt Sn (compared to 25 kt in the deposit), 800 kt Cu (compared to 70 kt in the deposit), 4 Mt As (compared to 90 kt in the deposit) and 100 Mt K (compared to 30 kt in the deposit).

3.2.4. *The RMG granitic suite at Panasqueira*

3.2.4.1. *History*

The existence of a granitic complex under the deposit, suspected since the contact halo was identified, was confirmed when miners encountered the greisenized cupola. At the top of this cupola, Kelly and Rye (1979) described a lenticular quartz cap 15 m thick, emitting veins of quartz into the nearby surrounding rock, veins taken up moreover in a later quartz-wolframite episode. These observations are no longer possible today; the corresponding mining works were sunk a long time ago. Toward its base at level L2, Kelly and Rye (1979) were able to observe a breccia in the cupola where large angular disjoint hornfelse clasts (up to 2 m) are cemented by the greisenized granite. The observation is no longer possible today, but a new gallery at level L2 has intersected a similar breccia, the elements of which are fragments of plastically deformed quartz veins, perhaps Seixo Bravos (Marignac et al. 2020).

Old, exploratory drill cores (now lost) were summarized by Bussink (1984), who was able to construct, under the cupola, a vault of porphyric biotite granite, recognized over a hundred meters of depth, affected at its roof by a greisenization that always exceeds that of the cupola. This author draws on his section vertical greisen roots that sink into the Panasqueira granite but are purely speculative, as they are not crossed by any of the drill cores shown on this same section.

The greisenized cupola emits sill-shaped apophyses (Bloot and De Wolf 1953; D'Orey 1967), which overprint the Seixo Bravos (Launay 2018). An old survey, external to the cupola and described by Bloot and De Wolf (1953), encountered a shale panel 50 m below the summit of the granite. A recent forage (SCB2) encountered thin sills of a strongly weathered granite in the first tens of meters above the roof of the pluton (Marignac et al. 2020). It can therefore be envisaged that the Panasqueira granite (in any case, its apical part) was emplaced in the form of successive intrusive sheets, in agreement with recent thinking on the emplacement of the granitic plutons, according to which they were built by the accretion of thin successive sills, ≤100 m thick, fed by the dikes (Michel et al. 2008; Leuthold et al. 2012).

Bussink's reconstruction (1984) served as a reference for most later works, including the most recent ones. Only Lourenço (2002, 2006) was a disclaimer, describing the greisenized cupola as the product of the alteration of a granite different from the porphyroid biotite granite: for Lourenço, this granite ("the Barroca Grande granite") was an RMG comparable to that at Argemela.

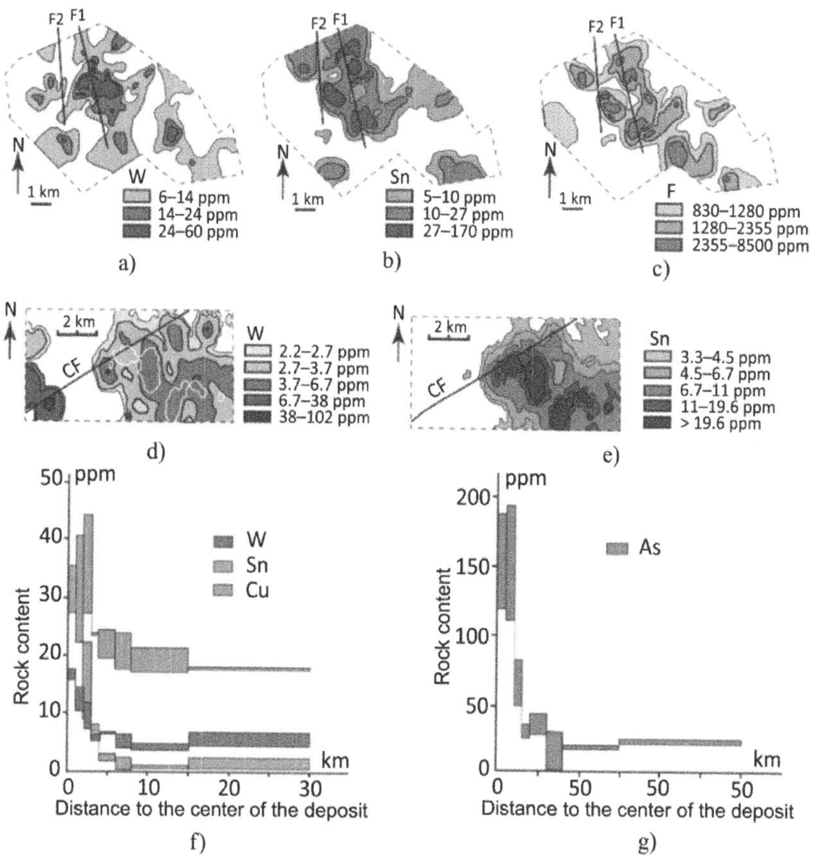

Figure 3.7. *The anomaly halo associated with the Panasqueira deposit*

COMMENT ON FIGURE 3.7.– *(a–c) Mapping of the proximal anomalies in W (a), Sn (b) and F (c), redrawn and simplified from Oosterom et al. (1984); F1 ("main fault") and F2: late faults. (d and e) Mapping of anomalies at kilometric scale, for W (d) and Sn (e), redrawn and simplified from Gonçalves et al. (2018); CB: "Cebola fault". In (d), the most pronounced Sn anomalies are superposed on the tungsten map (white lines). (f and g) Profile of anomalies at large scale for W, Sn, Cu (f) and As (g), redrawn from Polya (1988b).*

3.2.4.2. *The G1–G4 suite at the pluton roof*

The petrographic study detailed on samples from the SCB2 borehole, combined with a geochemical database created by Pinto (2015) from the same drill core and from a systematic sample of the greisens, allowed Marignac et al. (2020) to show that to the two-mica porphyroid leucogranite (G1) identified by Bussink (1984), which forms the apical part of the pluton modeled by Ribeiro (2017), and possibly the greater part of its mass, are added several more evolved facies of leucogranites, G2–G3 in sheets intercalated in the apex, and G4 that forms the greisenized cupola (Figures 3.4 and 3.8(a)). By disentangling the hydrothermal alterations (section 3.8), Marignac et al. (2020) demonstrate that sets G1–G4 represent a magmatic suite of peraluminous leucogranites belonging to an RMG series of the Beauvoir type rich in phosphorous (Linnen and Cuney 2005) (Figure 3.8(b)). The G1 granite, a protolithionite granite, is the least evolved. The G2 and G3 granites, which are Li-muscovite and columbotantalite granites, represent an intermediate stage. The G4 granite, which is the most evolved, resembles the Argemela granite, as already suggested by Lourenço (2002, 2006), whose "Barroca Grande" granite is clearly identified in G4.

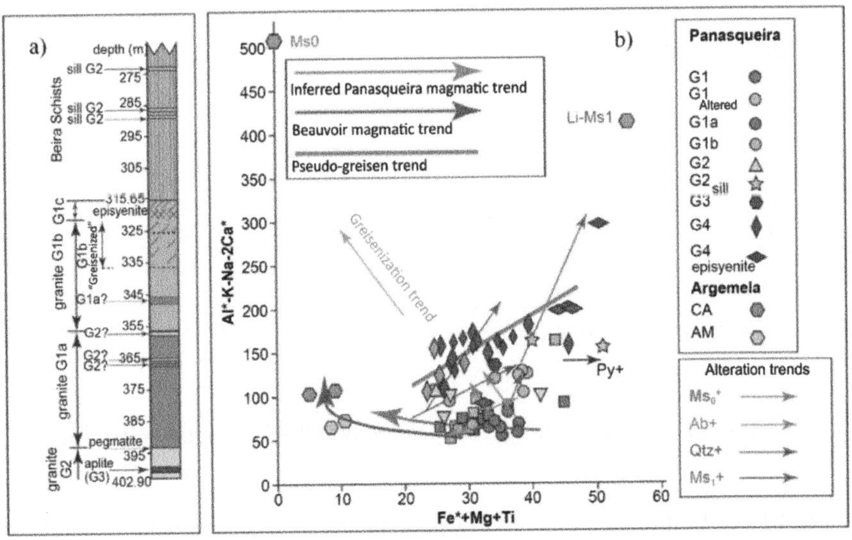

Figure 3.8. *The Panasqueira granitic suite*

COMMENT ON FIGURE 3.8.– *(a) Interpretation of the SCB2 drill core (located in figures 3 and 4 from Marignac et al. (2020)). (b) Representation of analyses of the G1–G4 suite on diagram A = Al-Na-K-2Ca versus B = Fe+Ti+Mg (in millications) by Debon and Le Fort (1988) (modified from Marignac et al. (2020)). In this diagram, where the quartz and the feldspars are projected back to the origin of the axes, parameter A is defined to represent the peraluminous character, whereas parameter B represents the mafic components. This diagram allows easy evaluation of the effects of muscovite on alteration trends. In conformity with the petrographic data, it thus appears that in the G1 suite, it is the effects of early muscovitization (Ms0) that predominate, whereas in the G2–G4 suite, the succession of quartzification and muscovitization (Ms1) leads to a mass alteration trend, nearly orthogonal to the actual greisenization trends (the "pseudo-greisen" trend). By reversing the different alteration trends, a magmatic evolution can be reconstructed for the Panasqueira granitic suite, which proves very similar to that of the Beauvoir-type peraluminous RMG, to which the Argemela granites also seem to belong.*

3.3. Relative chronology of the alteration and mineralization stages

3.3.1. *Historical overview and methodological considerations*

The establishment of a reliable and robust paragenetic succession (i.e. one resistant to the examination of new samples and/or new mining sites) is an obvious prerequisite for all geochemical studies when they aim to reconstruct the evolution of the P–T-t-X conditions governing the formation of the deposit and from this, to deduce how it was created.

Since Kelly and Rye (1979), numerous paragenetic successions have been proposed at Panasqueira (Polya 1989; Polya et al. 2000; Lourenço 2002; Pinto et al. 2015; Martins 2017; Carocci 2019; Cathelineau et al. 2020; Mateus et al. 2020). Some consensus has been established over time distinguishing an early stage of oxides (wolframite, cassiterite) and silicates (quartz, muscovite, tourmaline, topaz), the "Oxide Silicate Stage (OSS)", followed by a sulfur stage (pyrrhotite, sphalerite, chalcopyrite, pyrite), the Main Sulfide Stage (MSS) and a Late Carbonate Stage (siderite, dolomite, calcite), but there have been numerous variations on this basic theme over time, without calling into question the basic statement that all the silicates participated in the early wolframite deposition (Figure A.3.1(a)). In particular, Lourenço (2002) identified two distinct oxide–silicate stages, one wolframite–cassiterite and the other cassiterite, separated by the sulfur stage (he takes up the viewpoint presented in Snee et al. (1987)). Recently, Martins (2017) and Martins et al. (2020) proposed paragenesis characterized by many recurrences, with as many oxide as sulfur stages. However, in other recent publications, Carocci (2019) and

Cathelineau et al. (2020) questioned this schema: as seen in Figure 3.9, there are very many differences; they hinge in particular on the early stages, where only quartz and wolframite coexist, and on the distribution of topaz and muscovite over time.

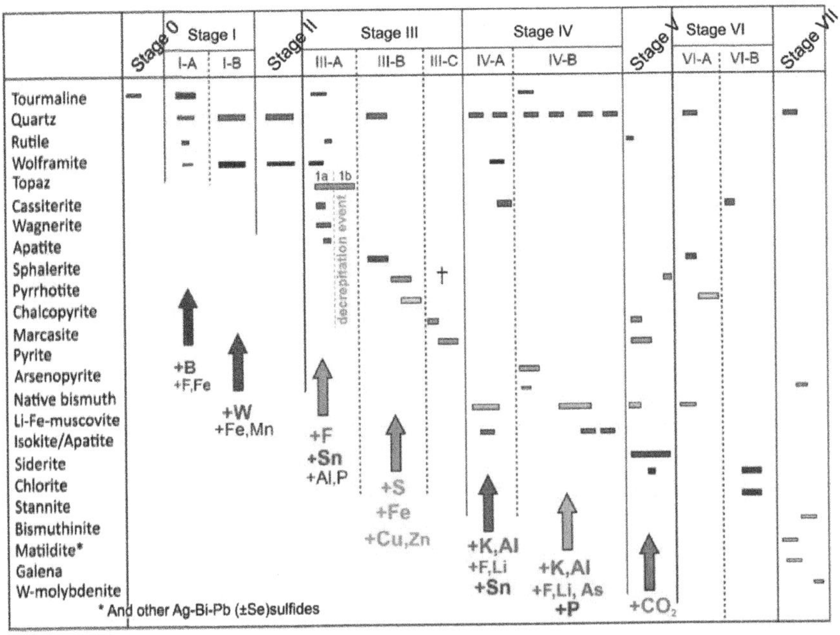

Figure 3.9. *The alternative paragenesis by Carocci (2019) and Cathelineau et al. (2020). This paragenesis differs radically from the previous ones, mainly because of the consideration of robust paragenetic criteria for the mineral phases (explanations in the text)*

The primary cause of these divergences about the paragenetic succession could be methodological in nature. All of the authors, without exception, share macroscopic observations made at the scale of the mine's pillars, although the oldest publications remain at this level; this is how Polya (1989, p. 1139) writes that "these stages can be readily discerned in a typical vein, which consists of a muscovite selvage overgrown by wolframite and then quartz, and a central filling of massively concentrated sulfide". It is therefore because it is commonly observed that the wolframites are separated from the surrounding by often palmar muscovite walls (Figure A3.2), arsenopyrite (Figure A3.3), topaz, or a combination of these phases that the idea of an OSS including all these phases may have seemed natural at first glance, although some macroscopic observations may be of a nature to bring more nuance to, or indeed to contradict, this idea (Figure A3.2). The problem arises because microscopic observations are not, most often, used to validate (or refute) the apparent macroscopic chronology or because, when they are, the most commonly used criteria are corrosion or inclusion criteria (the latter at the scale of the mineral or using the identification "container = contained" at the scale of the veins), which leads above all to the addition of phases at different stages (Martins 2017; Mateus et al. 2020). When intersection criteria are used in these conditions, they can complicate rather than clarify the paragenetic succession. This is how, for example, the triple distribution of the Sp-Ccp-Stn association in Martins' design (2017) is an artifact, due to the multiplication of topaz–muscovite associations on a macroscopic base, and then consideration only of the intersection relationships between the sulfides and these supposedly different generations of silicates.

Indeed, these criteria are dangerous because they are often ambiguous: Carocci (2019) and Cathelineau et al. (2020), who use microscopic observations systematically, state that they rely above all on two criteria:

– an intersection criterion, which is the main criterion, based particularly on the existence of fissure manifestations (veins and veinlets) of the different phases (oxides, silicates, sulfides);

– a filling order criterion in cavities that were known to be open.

In this, they add the principle of parsimony (also called "Ockham's razor"), which stipulates that the number of hypotheses should be kept to a minimum consistent with factual observations, that is, in a search for a robust paragenetic schema, reducing the number of stages to the minimum consistent with the set of relationships observed. As we will see below, this methodological rigor seems to guarantee the veracity of the paragenesis obtained.

It should be recognized that the reality of the deposit will complicate the use of these criteria. On the one hand, the superimposition phenomena (which often involve processes of dissolution and replacement) tend to obscure temporal relationships: microscopic observation often reveals that apparent macroscopic succession results from a succession of preferential reopenings (especially on the walls), which combine with dissolution-replacement phenomena, especially of quartz (Carocci 2019; Cathelineau et al. 2020). On the other hand, the scale of observation required to study the samples can often be greater than the dimensions of the thin blade, a difficulty clearly linked to the wealth of the deposit (with, for example, wolframite crystals that can exceed 10 cm).

One solution to this latter difficulty can be to use scanning systems at the scale of the sample (micro-XRF) before moving to microscopic scale proper (photonic and electronic tools: SEM, micro-analysis). This remains unavoidable and is sometimes decisive, as we will see in the following examples (see, for example, Figures 3.12 and 3.14).

Moreover, Cathelineau et al. (2020) have shown that the quartz most intimately associated with the growth of wolframite in large quartz-wolframite veins contains only decrepitated fluid inclusions (FIs) but that they may present recrystallization indices in the form of patches characterized by the disappearance of these early inclusions, replaced by more recent, non-decrepitated inclusions. This observation gives these authors an additional criterion, which sometimes seems decisive, to evaluate, at the level of the walls, the contemporaneity or not of the silicate phases with the early wolframite, depending on whether these phases are associated with recrystallization of the early quartz.

3.3.2. *The paragenetic succession at Panasqueira*

Given the decisive importance of establishing a reliable chronology and of the magnitude of the divergences concerning it, this section will be particularly well developed. There is a consensus that the late carbonate and base metal assemblages (sphalerite, galena) belong to Mesozoic and Tertiary events and are linked to the set of faults (this is the LCS stage: Figure A3.1).

They will not be discussed here as the description is limited to the mineralization that is generally thought to be associated with Variscan events. However, Pinto et al. (2015) observed cassiterite in the carbonate filling of a fault parallel to fault 8E, demonstrating the late remobilization of tin.

3.3.2.1. *Division into stages: the main filling events*

The succession of a quartz-wolframite stage and a quartz-sulfur stage is evident everywhere in mining works and is the logical basis of Kelly and Rye's OSS–MSS distinction (1979) taken up by the authors who followed them. Nevertheless, even though Polya et al. (2000) and Foxford et al. (2000) recognize that this sulfur stage corresponds to a resumption of the opening of the early veins, most authors believe all the fillings to be the result of a single general logic of opening (even though, since the work of Snee et al. (1987), it is assumed that this process extended over several million years). However, Carocci (2019) and Cathelineau et al. (2020) emphasize that sulfur reopenings, which present as completely sealed, giant vugs (Figure 3.10) cut at an angle from previous mineralizations and thus correspond to new logics of opening, oblique to the previous ones (Figure A3.4).

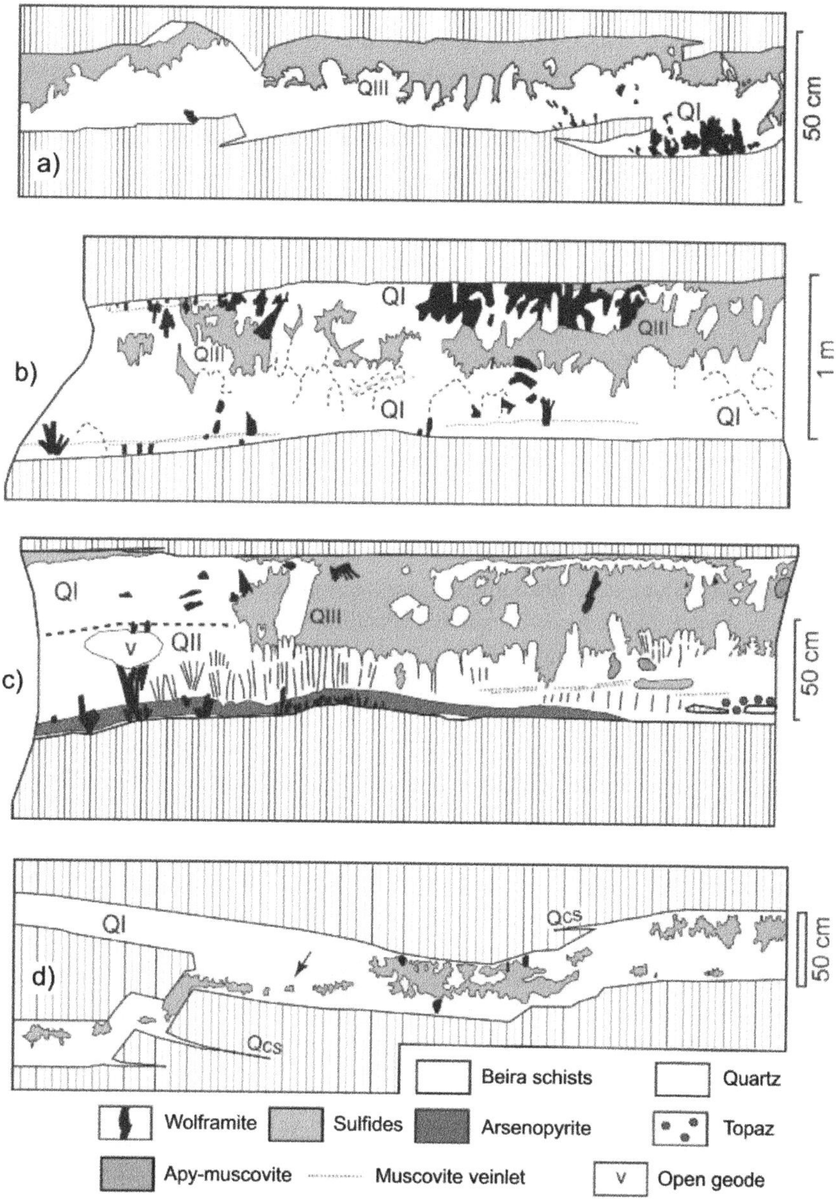

Figure 3.10. *Characterization of the stage III vugs sealed by the sulfides (III-B)*

COMMENT ON FIGURE 3.10.– *(a) A plurimetric vug develops at the roof of the vein (drawing from photo, Figure 3-A from Foxford et al. (1991); L2D23R3AW25). (b) Superposition of stage III on a stage I vein; the morphology of the "alpine" style QI quartz is underlined by the growth zones; note that the logic of opening III is not in continuity with the QI quartz: the early vein was sealed completely (blue dashed line) before the reopening; note also that the muscovite veinlets are parallel with the walls (drawing in mine workings [CM, 1970] L0D15R15AW10 "Bank"). (c) The plurimentric sulfife-filled vug III obliterates the contact between a QI and a QII quartz vein; note that the morphology of the QIII quartz largely reflects that of the early quartz; note also that the arsenopyrite selvage and the muscovite veinlets are parallel with the walls (drawing in mine workings [CM, 1970], L0D15R15AW10 "Bank"). (d) The development of sulfide-filled vugs occurs obliquely in the direction of the walls (arrow); note the prolongation of the sections of QI vein with alternating dips by veinlets of early palissadic quartz (Qcs) (drawing from photo, Foxford et al. (2000), Figure 5-a, unspecified mine work).*

The distinction in the early fillings of two quartz styles, on the one hand, of large "alpine" style prismatic crystals and, on the other hand, of more acicular prisms associated with crack-seal phenomena, was recognized by Foxford et al. (1991, 2000), but these authors have considered whether this involves two more or less contemporary modes of opening, albeit at different positions in the vein system, or a continuous evolution of the deposition conditions.

On the contrary, Cathelineau et al. (2020) show that it involves two generations of early veins characterized, one (stage I, QI) by "alpine" quartz and the other (stage II, QII) by "acicular" quartz (Figure 3.10). Observation of the pillars in fact unambiguously demonstrates that stage II corresponds to reopenings of the stage I veins and not to a prolongation of the crystallization of the quartz in a continuous filling process (Figures 3.10 and A3.4).

Wolframite is present in both generations of veins, where it is in fact the only mineral to accompany quartz (Cathelineau et al. 2020) (Figure 3.11); the other commonly associated phases, especially in selvages, have clearly proven to be later (Figure 3.12).

The reopening and sulfide precipitation stage, which equally affects the stage I or II veins, is therefore stage III in the paragenetic scheme in Cathelineau et al. (2020) A new stage, stage IV, can also be defined as the one when the vugs were filled, from which come the spectacular crystals that make Panasqueira's reputation among collectors (Carocci 2019). Identification of this stage results from the observation that the envelope of the vugs obliterates all the previous stages (the vugs can go as far as to occupy the entire height of a vein; section 3.3.2.4).

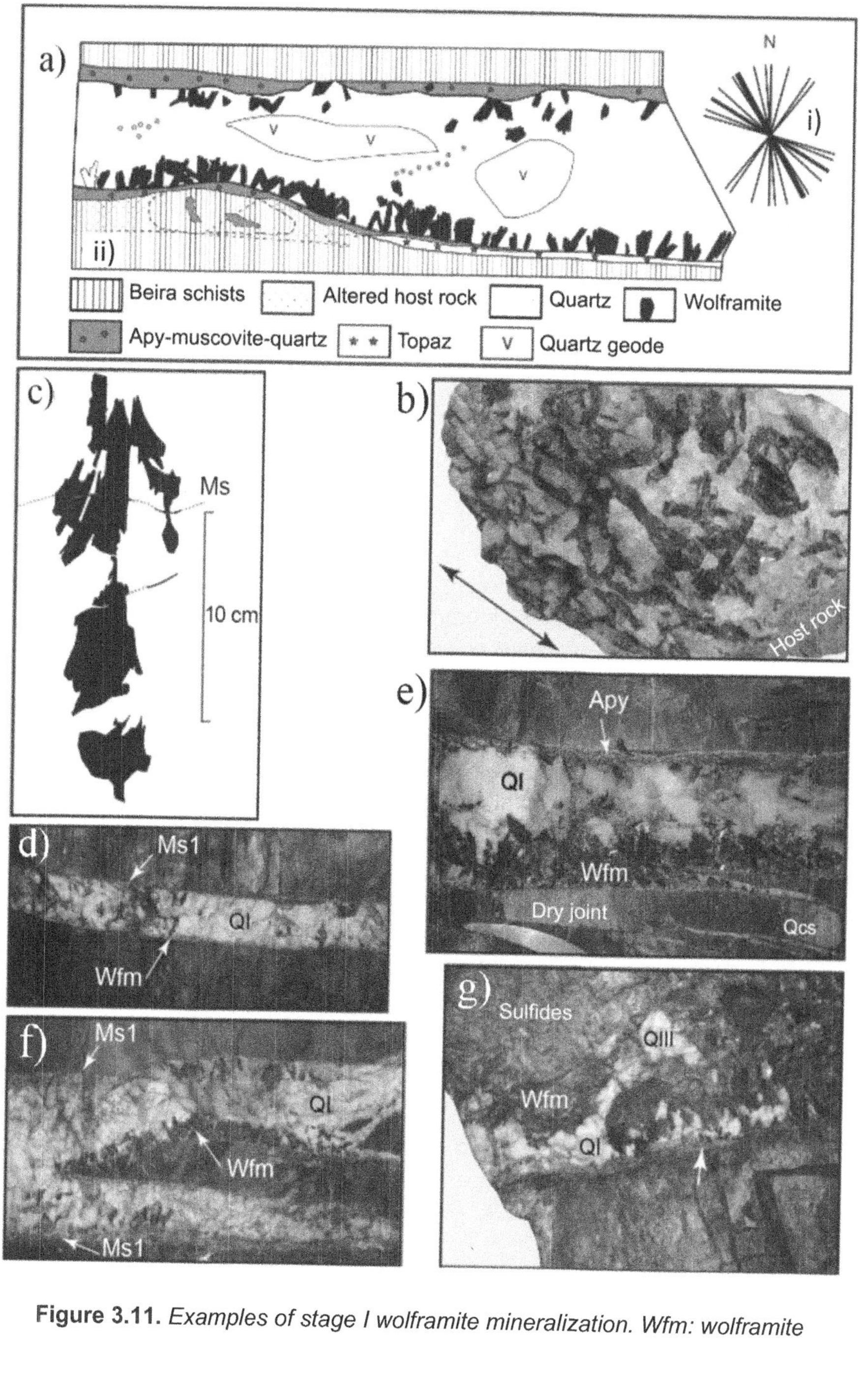

Figure 3.11. *Examples of stage I wolframite mineralization. Wfm: wolframite*

COMMENT ON FIGURE 3.11.– *(a) A case of stage I wolframite hyperaccumulation: i) note the selvages and the importance of the muscovitization of the wall (drawing in mine workings [CM, 1970]; L0D15R15AW10 "Bank"); ii) orientation statistic of the wolframite prisms in a) (15 face measurements), showing a general preferential orientation ~N130°–140°E.*

(b) Quartz and wolframite in a tangential section; a statistical orientation is perceptible (double arrow) (L0D15R15AW10 "Bank").

(c) Detail of the development of a stage I wolframite; note the secant muscovite veinlets (drawing in mine workings [CM, 1970]; L0D15R15AW10I2 "Bank").

(d) Quartz vein with small, dispersed wolframites; note the weak development of the muscovite selvage (L1D13R10AW17).

(e) Detail of a vein showing direct contact of the stage I wolframite with the wall rock; note the arsenopyrite selvage at the roof (L1D29R7AW14).

(f) Hyperaccumulation of wolframite, with prisms in direct contact with the wall; note the muscovite selvage (Ms1).

(g) Detail of the wall of a vein where stage III is superimposed on stage I; note the direct contact of QI quartz or the wolframite with the schist wall, as well as the early generation of small wolframite crystals (arrow) (L1D29R7AW14).

Some of these stages can be subdivided. This is how, for example, Polya et al. (2000) and Foxford et al. (2000) introduce a very early stage characterized by crack-seal quartz veinlets (their QT stage), some of which locate the opening of the earliest generations of mineralized veins (section 3.4.2). According to Carocci et al. (2021), the veinlets at this stage (renamed Qcs) are closely linked to tourmalinization of the walls, a precursor event to the main wolframite mineralization, with which it is intimately linked over time. These authors therefore define a stage I-A, corresponding to these early phenomena, which moreover also sees the very beginning of the crystallization of the wolframite (Figure 3.10). As we will see below, it is also possible to subdivide stages III and IV.

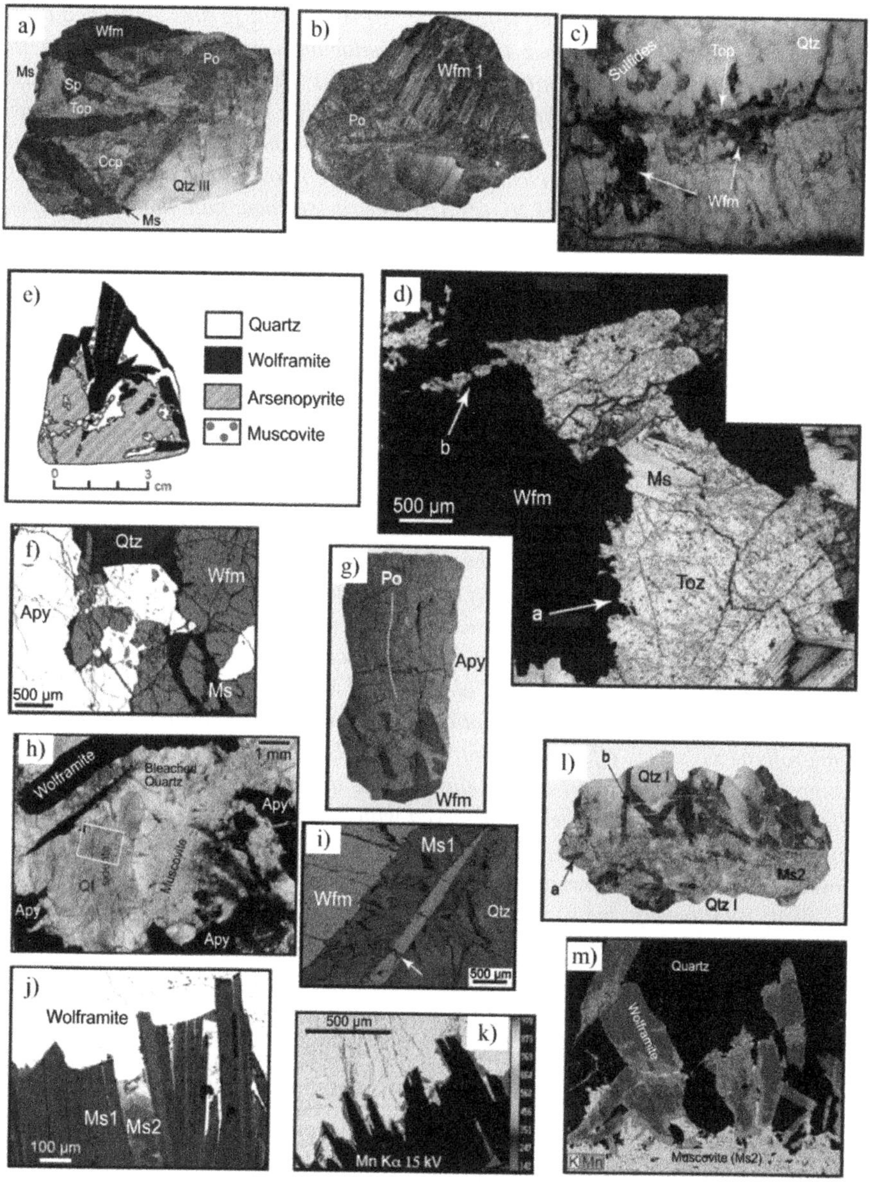

Figure 3.12. *Examples of relationships between wolframite 1 and quartz I and the other mineral phases*

COMMENT ON FIGURE 3.12.– *Wfm: wolframite. (a) Wolframite prisms embedded in the sulfides; note the euhedral topaz (in fact, pseudomorphoses by muscovite) at the interface between the wolframite and the sulfides) (E1). (b) Wolframite prism encased/intersected by the pyrrhotite (E2). (c) Wolframite intersected by an intravein topaz veinlet (L1D29R13AW14). (d) Detail of the relationships between the wolframite and a topaz selvage; it seems clear that the topaz corrodes (arrow a) and cracks (arrow b) the wolframite (LMP, LN). (e) Arsenopyrite selvage; note the clearly secant character of the muscovite (Ms2) (clearroom drawing, E3). (f) Detail of the relationships between the arsenopyrite of one selvage and the wolframite; the wolframite is clearly brecciated and cemented by the arsenopyrite (LMP, reflected light, E4). (g) Wolframite prisms embedded/dislocated in the massive pyrrhotite; the pyrrhotite forms lamellar prisms (dashed line) (E5). (h) Relationships between the muscovite (Ms1) and the wolframite; QI, directly associated with the wolframite, loses its early FIs ("bleached" quartz) in contact with the muscovite, which is therefore later; note that the arsenopyrite seems to obliterate this muscovite (LMP, natural light, E6). (i) Detail of H, showing that the palmar muscovite pushes on the wolframite (and not vice versa); the arrow emphasizes a cracking of the wolframite blade, sealed by the muscovite; this blade could have been torn from the large wolframite prism (note its rough surface) (reflected light). (j and k) Control of the Mn alteration of the wolframite (Wfm1) by the later muscovite (E6). (j) BSE image showing the two generations of muscovite in contact with the wolframite; the Ms2 is more ferriferous and more finely zoned than Ms1. (k) False colored X-ray (EPMA) map in of the same area; the development of the Mn alteration, at the muscovite–wolframite interface and in microfissures, seems effectively controlled by Ms2. (l) Development of a band of fine-grained muscovite (Ms2) cutting through the quartz I-wolframite 1 association; relics of quartz and wolframite are observed within the muscovite (arrow a) and Ms2 veinlets intersect quartz and wolframite (arrow b) (E7). (m) Detail of the contact between Ms2 and the wolframite in the previous sample: false colored μ-XRF map; the clear relationship between the muscovite (K Kα map) and the manganese alteration of the wolframite (Mn Kα map) is very much apparent (cross-section). An inventory of samples can be found in section A3.3.*

3.3.2.2. *Stage III subdivisions*

3.3.2.2.1. Relationships between the topaz and the sulfides

At the macroscopic scale, Cathelineau et al. (2020) observed that topaz, alone or linked to sulfides, cassiterite, or, more rarely, wolframite, can form veins independent of large quartz veins while remaining parallel to them (Figure A3.5). Inside the large quartz veins, aside from the topaz at the walls, topaz veins clearly intersect the wolframite, or masses of topazite replace the quartz (Figures 3.12(c) and A3.5). Such observations suggest that the topaz does represent an early phase, and they confirm microscopic observations, notably the observation of topaz infiltrating the wolframites at the walls (Cathelineau et al. 2020) (Figure 3.12(d)). Moreover, when topaz is associated with sphalerite or pyrrhotite, it appears clearly earlier (Figures 3.5 and 3.13). The logical conclusion of this collection of arguments is that the topaz, later than wolframite 1 and 2 and immediately earlier than the stage III sulfides, defines a III-A episode of this stage, the deposition of the massive sulfides characterizing stage III-B (Cathelineau et al. 2020) (Figure 3.10). Moreover, Cathelineau et al. (2020) demonstrated that the growth of the topaz occurred systematically in two stages separated at many points by a crack-seal episode, thus defining two generations of topaz (Top1a and Top1b) (Figures A3.5 and A3.8). During deposition of the topaz, an episode of decrepitation of the early FIs occurred (Cathelineau et al. 2020, section 3.5.1.2.2).

3.3.2.2.2. Existence of a stage III wolframite

There are wolframite crystals intimately linked to the topaz, which can also be shown to precede Top1b (Figure A3.6). There therefore exists a generation of wolframite linked to the start of stage III-A. It is difficult to evaluate the quantitative importance of this Wfm3-III-A, but it seems limited. However, there are quartz and wolframite veins whose walls are not marked by stage I tourmalinization: this is especially the case with the small veins (5–10 cm in thickness) that can be observed on the surface (Figure 3.25(b)). Without being certain of this, we could believe that such veins are also representative of stage III. It makes sense nevertheless to remark that the quartz of these veins contained numerous decrepitated inclusions.

3.3.2.2.3. Stage III sulfides

At stage III-B, chalcopyrite and pyrrhotite show contradictory relationships (Figure 3.13), which suggest their syngrowth, with chalcopyrite occurring slightly later. On the other hand, the stage III sphalerite systematically and in abundance contains inclusions of chalcopyrite, typical of chalcopyrite disease (Figure 3.13). This phenomenon is well known for demonstrating that chalcopyrite is later than sphalerite (Barton et al. 1987), even though some erroneously still speak of "exsolution" (Jaques and Pascal 2017; Martins 2017). The sphalerite also systematically contains inclusions of pyrrhotite, depicting a real "pyrrhotite disease" (Figure 3.13). It can therefore be concluded that the crystallization of sphalerite occurred earlier than that of the pyrrhotite-chalcopyrite association. The pyrrhotite is almost always affected, often almost entirely replaced, by an episode of pyritization, which defines stage III-C and will be considered further because of its relationship with later stages (section 3.3.2.6).

Aside from the filling of the large, open stage III cavities, the sulfides of this stage can develop in more restricted patches obliterating quartz I and II. In particular, a privileged development of such patches can be observed along the joints of the quartz QII prisms or at the junction between the QI and QII veins (thereby confirming that the stage III reopenings are in no instance in continuity with the previous openings) (Figure A3.7).

3.3.2.2.4. Stage III phosphates

There are two generations of phosphates at Panasqueira. The most evident is the late F-apatite, clearly exposed in the stage IV vugs (section 3.3.2.4). However, another phosphate is encountered sporadically in direct macroscopic association with the topaz, which can clearly be recognized from its pink tints (Figure 3.14). This is a fluorophosphate, a member of the triplite family, with Mg ≥ Fe + Mn, and hence called wagnerite here. Microscopic observations and μ-XRF imaging confirm that this wagnerite is intimately associated with topaz and pyrrhotite, thus forming part of stage III-A (Figure 3.14). Rare occurrences of alumino–phosphates from the goyazite family (nevertheless with a significant F content), constantly associated with the stage III tourmaline (Tur2) (section 3.3.2.5) and the topaz, also form part of this stage.

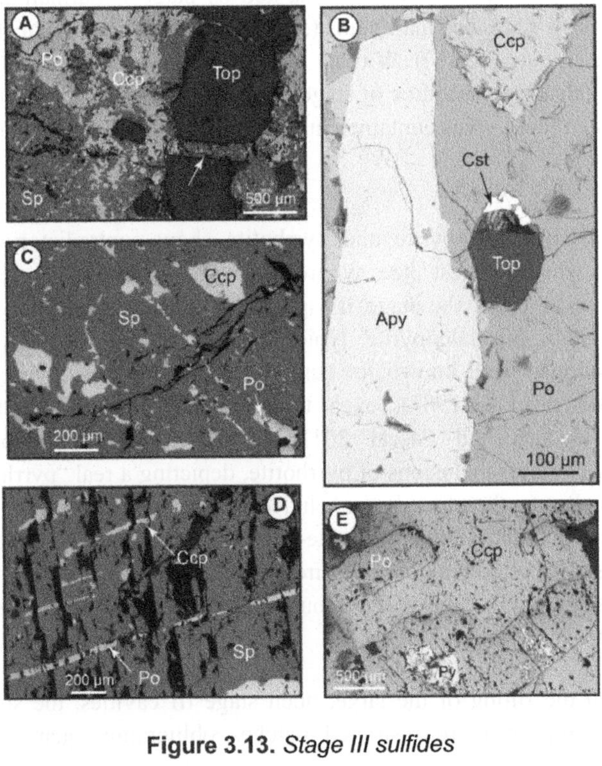

Figure 3.13. *Stage III sulfides*

COMMENT ON FIGURE 3.13.– *(a) Relationships with the topaz: the euhedral topaz is cracked and cemented (arrow) by sphalerite; chalcopyrite here seems to intersect pyrrhotite, which itself seems to intersect sphalerite (LMP, reflected light, E13). (b) Topaz-cassiterite association included in pyrrhotite (image BSE, E14). (c and d) "chalcopyrite disease" and "pyrrhotite disease" in sphalerite (LMP, reflected light, E13). (e) Chalcopyrite-pyrrhotite relationships: these two sulfides here seem to be penecontemporary; note the beginning of pyrrhotite pyritization (LMP, reflected light E13).*

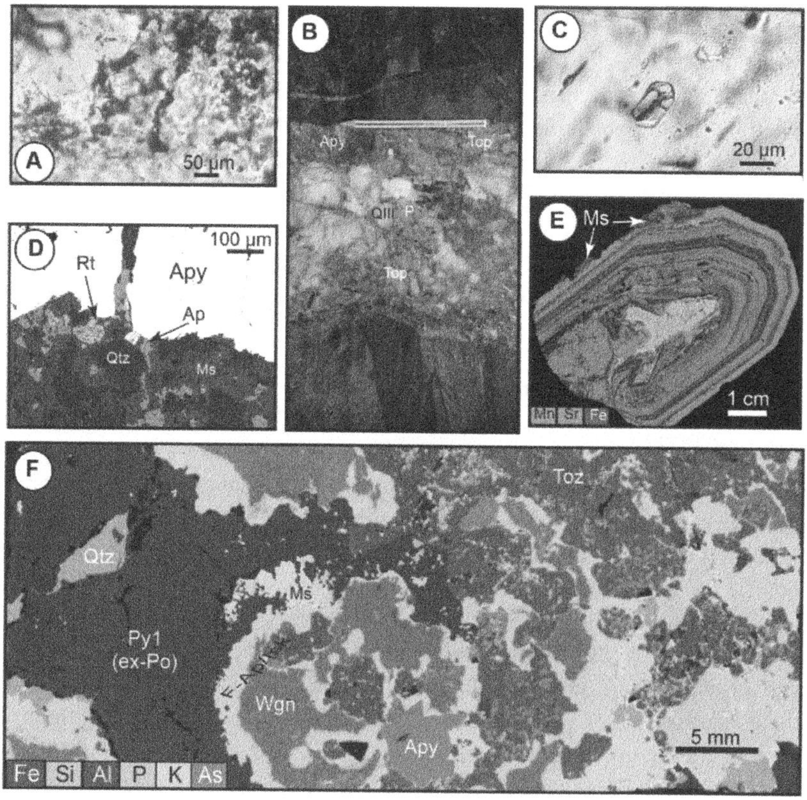

Figure 3.14. *The two generations of Panasqueira phosphates*

COMMENT ON FIGURE 3.14.– *(a) Decrepitated fluid inclusions in a stage III apatite (thick blade, natural light, sample). (b) Association with topaz of phosphates from the triplite (P) family (L1D29R13AW14E). (c) Plastically deformed fluid inclusions in a stage III apatite (thick blade, natural light, same sample as in (a)). (d) Stage IV apatite in microcracks affecting a quartz-muscovite-rutile-arsenopyrite assemblage (BSE image; sample PAN-IV-1a, L0D9XR17AW13). (e) Zoning of a stage IV apatite (false-colored μ-XRF map, vug sample L0D15R15AW10 "bank"). (f) Association of wagnerite with topaz and replacement by an isokite-apatite succession (μ-XRF, E18).*

The wagnerite is systematically replaced by an F-apatite in a majority of occurrences with the intermediate formation of isokite (although they are defined at Panasqueira, panasqueraite (Isaacs and Peacor 1981) and thadeuite (Isaacs et al. 1979) could not be found in the present work) (Figure 3.14(a)). The formation of this apatite may have occurred later than the pyritization of the pyrrhotite. This is suggested by the presence of sulfur in significant quantities in its structure, where it is in contact with pyrite 1 (it would therefore be a stage IV apatite). However, apatite veins have been observed, intersecting the greisenized cupola, albeit earlier than the muscovite of these greisens. These apatite veins have primary FIs that are typically "decrepitated" (Figure 3.14(a)), which suggests that they belong to stage III-A (section 3.5.1.2.2). It therefore seems that there are two generations of F-apatite at Panasqueira.

3.3.2.3. *The "vug" stage (stage IV)*

3.3.2.3.1. Overview

As we have seen above (Figure 3.6), this stage is characterized by incomplete fillings of cavities that remained open, where it is possible to observe the sequential crystallization of nice euhedral crystals (including minerals belonging to later stages; sections 3.3.2.5 and 3.3.2.6). The walls of the vugs may result either from rupture or from dissolution of the host minerals (most often quartz). Their opening is therefore without doubt partly of tectonic origin, especially since it is not uncommon to observe, on their floor, quartz (or shards of quartz) that have fallen from the walls and the crystallization of which had continued: bipyramid quartz, one of the prisms of which is clearly growing from the crack; lamellar quartz developed from shards and on which stage IV minerals may grow (a phenomenon well described by Kelly and Rye (1979), who link this to the abundance of FI planes observable toward the base of these vuggy crystals). The opening of the vugs often affects the whole thickness of the vein (Figure 3.15(a); Figure 3.22(d)); their morphology is tubular overall, with a dominant direction to which secondary diverticula may connect. A statistical survey carried out on vugs from the "Bank" site from the Barroca Grande sector has shown that despite a high dispersion, the axes of these tubes define a preferred, broadly E–W direction (Figure 3.15(b)). These features suggest that the geodes may appear on E–W tension joints affecting the veins, from which they would develop by dissolution and subsequent precipitation. This geodic phenomenon is highly developed in the upper levels of the Barroca Grande sector, where vugs reaching several meters in width and depth in the pillars were clearly visible in the 1969–1970s. It is on the whole much more discreet (although with comparable characteristics) in the deeper and more southern levels toward which recent exploitation had moved (Figure 3.15).

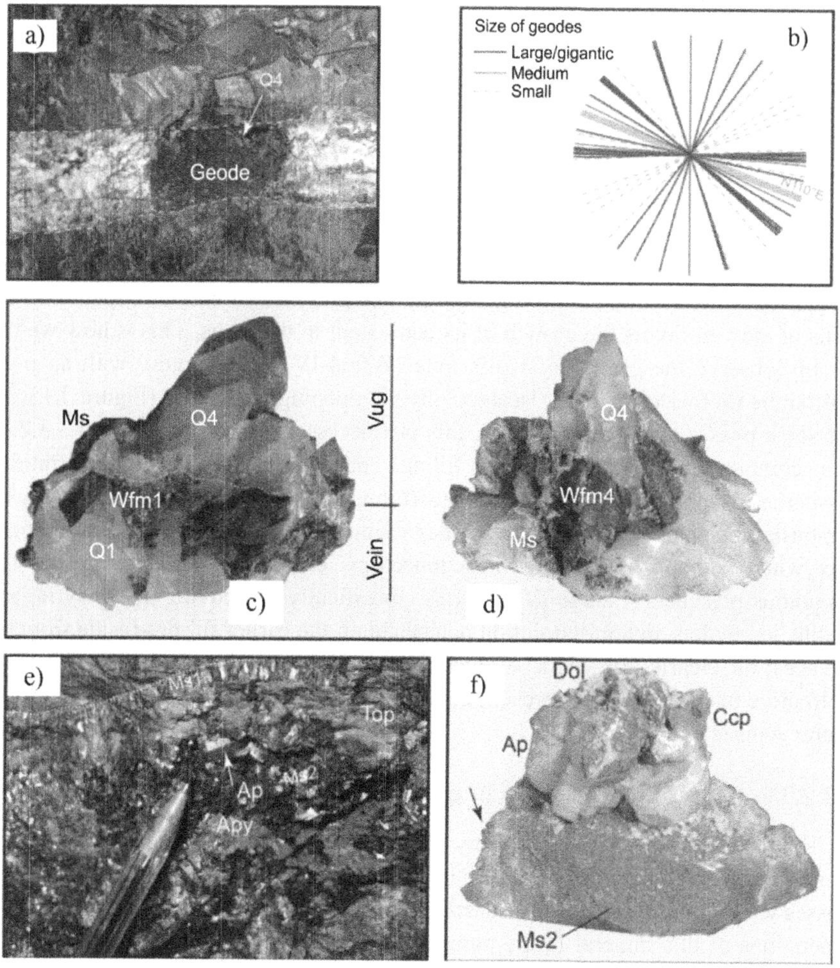

Figure 3.15. *Illustration of the geodic phenomenon*

COMMENT ON FIGURE 3.15.– *(a) An (emptied) vug open in a Q1 quartz vein (L0-D11-R15-AW17); the vug occupies the whole thickness of the vein; some Q4 quartz that has not been extracted is still visible. (b) Statistics of the orientation of the tubular stage IV vugs in the veins of the "Bank" workings (observations CM 1970). (c and d) Wall of a quartz-wolframite-muscovite vug in a stage I vein (L0D15R15AW10 "Bank"). (c) Back view: the Q4 quartz grows from a Q1 quartz wall; this wall is a crack, which has "scalped" the wolframite 1 (dashed). (d) Front view: the automorphic wolframite 4, covered by Q4 quartz and (Ms2 muscovite), grew from the wolframite 1 visible on the back of the sample. (e) Example of an arsenopyrite-muscovite (Ms2)-apatite microvug (L3D9R1AW36). (f) Q4 quartz and apatite of a stage IV vug (L0D15R15AW10 "Bank"): this quartz covered by apatite was found laid in the geode, detached from the wall along a healed crack by the growth of small crystals (arrow).*

Where it is present, wolframite (Wfm4-IV) is always the first phase to appear, followed by quartz. The growth of the quartz (which often germinates on earlier quartz intersected by the geodic opening) is, in the vugs, practically continuous. It accompanies the succession of vug fillings. The presence of an earlier phase on the walls of the vug favors the growth of its equivalent in the veins. This is how we can notably observe the growth of wolframite (Wfm4-IV) in continuity with an older wolframite (Wfm1 or Wfm2), "scalped" by the opening of the vug (Figure 3.15(c)). For some phases (tourmaline, topaz), this is a necessary condition (section 3.3.2.4). The complete succession of the fillings defining stage IV is wolframite-cassiterite-arsenopyrite-muscovite-apatite (Figure 3.9). This succession is regular, demonstrating that the network of geodes is interconnected, but it is not complete everywhere, demonstrating that the connections can have been interrupted. The ubiquitous presence of stage IV minerals, specifically muscovite, arsenopyrite and apatite, in veinlets and/or dissolution patches in all the earlier fillings (aside from the selvages) is clearly the mark of these connections. It does not seem that the wolframite of this episode is present outside the geodes, with the exception of some quartz veins at deeper levels.

3.3.2.3.2. Arsenopyrite, a stage IV phase

The arsenopyrite can occupy many positions in the vein system: euhedral in the tourmalinized walls, or in the late vugs (Figure 3.16), in more or less anhedral masses within the sulfides or the quartz, which has led many authors to multiply the generations of this mineral in the paragenetic succession (Figure A3.1). Indeed, as Carocci (2019) has shown, all arsenopyrites, regardless of their position in the veins or their apparent relationships with other mineral phases, have the same set of characteristics: they are zoned, with a core and two series of envelopes (associated with the phenomena of corrosion and patchy zoning); they contain native bismuth in inclusions, quite systematically in the central zones (Figure 3.16); some crystals also contain loellingite inclusions, which are invariably found in the central zones as well. For Martins et al. (2017), this native bismuth is an "exsolution resulting from the löllingite breakdown", but Carocci (2019) found no textural evidence for such a phenomenon: quite the contrary, löllingite, in euhedral inclusions, shows no sign of decomposition, and the native bismuth, which is much more common than löllingite, shows no correlation with this latter. In addition, each zone is characterized by As/As+S ratios within a specific range, the same regardless of the context, with an evolution marked by a decrease in this ratio from the central zone toward the outer rim (Figure 3.17). It therefore seems improbable (principle of parsimony) that all of these characteristics could have been produced more than once, and it is therefore quite rightly that Carocci (2019) concludes that there is only one generation of arsenopyrite at Panasqueira. The presence of this phase in the vugs is therefore sufficient to demonstrate that the arsenopyrite belongs to stage IV (Figure 3.9). This viewpoint is reinforced by observations showing, for example, that the arsenopyrite is:

– later than the wolframite when it is in the selvage position (Figure 3.12);

– later than the pyrrhotite, which it replaces (Figure 3.18);

– or later than the topaz, which it obliterates (Figure A3.5).

Within the veins, arsenopyrite is associated with the "bleaching" of QI or QII quartz (Figure 3.18(a)) and is often accompanied by a micromosaic QIV quartz (Figure 3.18(b)). At the deepest levels and toward the SW, arsenopyrite is greatly developed and is accompanied by larger quartz (Figure A3.3).

3.3.2.3.3. The two generations of stage IV muscovite

At Panasqueira, there are multiple generations of muscovite, a fact well documented by Snee et al. (1988), Lourenço (2002), Martins (2017) and Mateus et al. (2000), among others. The muscovite of the selvages is unanimously considered the earliest (Ms1) and is attributed to the first stages by most authors (Figure A3.1). Without calling into question its status as the first muscovite formed, Carocci (2019) and Cathelineau et al. (2020) contest the early character of the palmar muscovite of the walls: they in fact observe that the decrepitated inclusions are absent from the quartz of the veins close to these muscovites, which proves the posteriority of these muscovites to the stage I quartz and wolframite. Favorable observations make it possible to confirm this posteriority (Figure 3.19(b)).

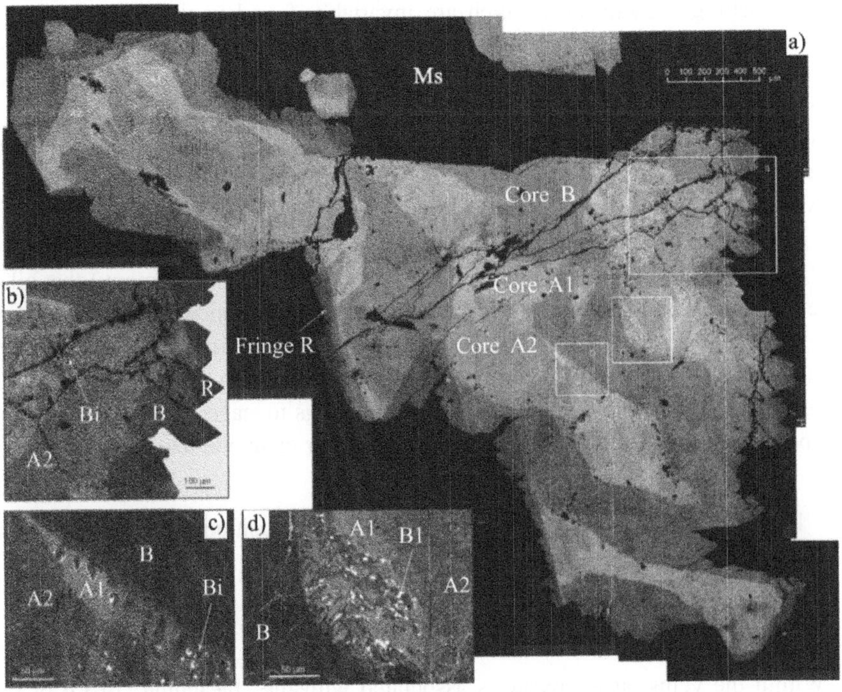

Figure 3.16. *The zoning of the arsenopyrites: a representative example demonstrating all the characteristics typical of the arsenopyrites in the deposit (images BSE, E19, see Figure 3.10(a))*

COMMENT ON FIGURE 3.16.– *(a) Mosaic image illustrating the complex zoning: three zones, with decreasing As content (core A, core B, fringe R); distinction between two generations (A1, A2) in core A; mixed structure: both the concentric (growth) zoning and the "patchy zoning" (corrosions). (b to d) Details showing that the native bismuth (Bi°) is confined in A cores; and more particularly here, cores A1. In B and C, note the complex interface between cores A and B.*

It can be shown that the arsenopyrite obliterates the palmed muscovite of the walls (Ms1) (Figure 3.19(b)), whereas in the vugs, one generation of muscovite is later than the arsenopyrite, which it covers systematically; this latter (Ms2) therefore belongs to stage IV. It is moreover associated with spectacular alterations of the primary wolframite, characterized by an increase in the manganese content, which is remarkably well illustrated by μ-XRF mapping (Figures 3.12(k) and (m)). The muscovite of the selvages (Ms1), which often forms the wall of the stage IV geodes, can be considered a sign of the start of this stage (section 3.3.2.4).

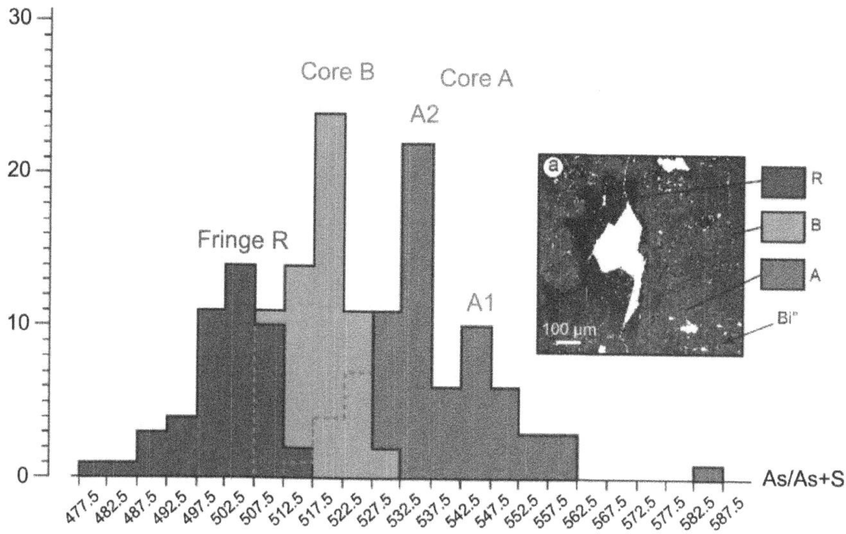

Figure 3.17. *Statistics of the sulfur and arsenic values of the Panasqueira arsenopyrites*

COMMENT ON FIGURE 3.17.– *The zoning of the arsenopyrites is well marked in the statistical distribution of the As/As+S ratios: in spite of a limited overlapping, the histograms for each of the zones identified in Figure 3.16 and in (a) are well separated (a: BSE image, sample PAN-VI-8b, site L0D15R15AW10 "Bank").*

The frequent association of the muscovite (Ms1) and arsenopyrite in the selvages and the recurrence of the muscovite (Ms2) crystallization later than arsenopyrite suggest the subdivision of stage IV into two substages IV-A and IV-B, the first associated with crystallization of wolframite (Wfm4-I-A) and cassiterite (Cst2-IV-A) and the second with crystallization of the main generation of apatite (Ap2-IV-B), which follows Ms2 (Figure 3.9).

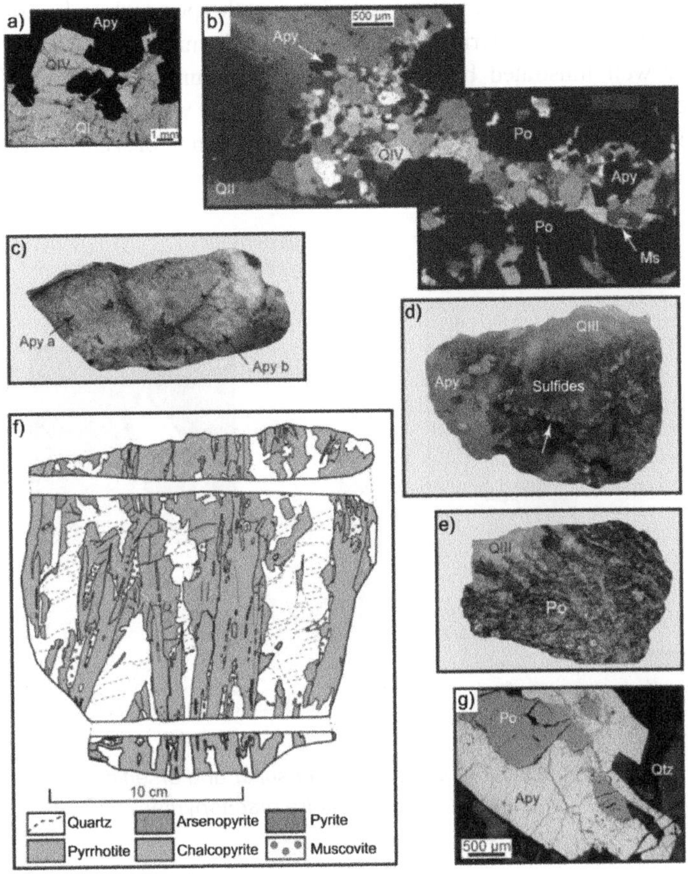

Figure 3.18. *Examples of relationships between arsenopyrite and stages I to III*

COMMENT ON FIGURE 3.18.– *(a) "Bleaching" of the QI quartz (transformed into QIV quartz) in contact with the arsenopyrite (LEP, sample PAN-III-3b). (b) Partitioned "crack-seal" affecting the pyrrhotite in a QII quartz; the crack is filled by an arsenopyrite–micro-mosaic QIV quartz association (LMP, polarized light, sample PAN-I-12-2, LOD15R15AW10 "Bank"). (c) Massive pyrrhotite sample, with arsenopyrite: either in "rods" in apparent syngrowth with the pyrrhotite (a), or in bands with a more secant aspect (b) (sample PAN-VI-14 "Bank"). (d) Sample of stage III sulfides associated with the arsenopyrite; note the string of small euhedral arsenopyrites (arrow) obviously of fissural origin (sample PAN-V-8 L0D15R17).*

(e) Blades of pyrrhotite intercalated with QIII quartz (sample PAN-VI-1 "Bank"). (f) Clearroom drawing of a section in a sample of lamellar pyrrhotite intercalated with QIII quartz (the upper and lower faces of the section are also represented); note the abundant arsenopyrite capping the pyrrhotite (sample PAN-IV-11 "Bank"). (g) Detail of a pyrrhotite cap corroding the arsenopyrite (LMP, reflected light, sample PAN-VI-4 "Bank").

3.3.2.3.4. Stage IV apatite

The stage IV apatites are famous in collectors' circles (De Ascenção Guedes 2002) (Figure 3.6). They are very common and are observed in most cavities. Their color varies from green to colorless, and they are characterized by highly developed growth zoning, marked in the first order by the variation in strontium and manganese content levels (Figure 3.14(e)). Knutson et al. (1985) demonstrate, however, by studying the zoning of luminescence responses, governed by rare earth elements, and fission tracks governed by uranium, that these elements also participate in zoning. These authors conclude from their study that the "stratigraphy" defined by the intensity of light emissions and the fission track density is common to all of the apatites from a given vug but believe that their data invalidate any correlation from one vug to the other over distances of more than ten meters. A reexamination of their data (Figure A3.9), however, shows that this is not the case but that on the contrary, the zoning of the apatites can serve to define a microstratigraphy, which correlates with the scale of the deposit, thus demonstrating the interconnection between these cavities. However, the frequency of gaps in this stratigraphy similarly shows that the connections can be interrupted in time and space.

3.3.2.4. *Several generations of tourmaline*

According to Carocci et al. (2021), the tourmaline of the tourmalinized walls is truncated by the large opening of the stage I veins. Indeed, there are mineralized veins where long acicular tourmalines have grown on the walls (Launay et al. 2018), which may seem contradictory. These prisms take root on the edges of the tourmalinite (which differ from the tourmalinized walls through the absence of quartz, muscovite, etc.), which can be seen as locally obliterating the stage I-A Qcs veinlets and overlying the early tourmalinization (Figure 3.20). The quartz associated with these prisms is characterized by the absence of decrepitated FIs; therefore, it is no older than QIII quartz. Because, moreover, the acicular tourmaline prisms can be enveloped by topaz, it can be concluded that this tourmaline is not contemporary with tourmaline 1 of the wallrock tourmalinization: these are tourmaline 2 from stage III (and more precisely III-A) (Figure 3.20(e)).

Figure 3.19. *Example of a muscovite selvage developed later than the stage I quartz-wolframite assemblage (E24)*

COMMENT ON FIGURE 3.19.– *(a) LMP, natural light: the muscovite selvage is interpreted as being the result of a filling following a "scissor" logic (arrows); note the large blade of muscovite penetrating the wolframite. (b) LMP, polarized light: the selvage is clearly developed in several stages corresponding to the successive crack-seals. (c) Interpretation: growth of the selvage is broken down into three steps (Z1 to Z3); the first two correspond to syntactic fillings, which have been eroded/dislocated by the third episode, which is single-oriented: small laths detached from zones Z1 and Z2 (a) are found dispersed in zone Z3; the growth of some crystals is continuous however (b); a more or less marked halt in growth (c) defines two subzones in Z3. On the wall of the selvage, the remainders of a Qcs veinlet (initially, at the wall of the stage I vein) are isolated in a border formed of an aggregate of microcrystals of muscovite: this replacement predates the opening of the selvage, as the erosion of Z0 by Z1 in several places demonstrates (d). Note the corrosion of wolframite (e) and quartz (f) by muscovite. Note also that arsenopyrite is superimposed in a "cookie-cutter" fashion on muscovite (g). The chalcopyrite here is a late phase (Ccp2).*

The prisms of these two generations of tourmaline can show overgrowths (reaching as far as the development of new prisms) when they are covered by Ms1 muscovite (Figure 3.20). These overgrowths and these third-generation tourmaline prisms (Tur3) are well characterized chemically (especially by their particular ferriferous character). When the walls of the stage IV vugs are formed by the tourmalinites (Tur2), they are covered with needles of a very ferriferous tourmaline, identified as a Tur3 tourmaline, which is covered systematically by the earliest minerals of the vug fillings (cassiterite and arsenopyrite). This is a strong argument for making the Ms1 muscovite the earliest phase of stage IV.

3.3.2.5. *The pyritization episode (the PAS of Kelly and Rye (1979)) and its relationships with siderite: definition of stage V*

All the authors recognize that an episode of penetrative pyritization affects the stage III-B pyrrhotite. Marcasite (which precedes pyrite) is associated with this episode. Siderite is fairly systematically present in this assemblage, and observation of its apparently symplectic relationships with pyrite (Figure 3.21) has led all of the authors to associate siderite with the pyritization stage. This nevertheless poses a problem. While pyrite and marcasite are clearly obliterated by all the stage IV phases, justifying attribution of the pyritization episode to stage III-C (Carocci 2019) (Figure 3.9), it can be observed just as clearly that in some stage IV vugs, siderite crystallizes later than apatite. Despite appearances, however, there is no need to consider the existence of two generations of siderite. The systematic observation of the pyrite-siderite associations does in fact show that the emplacement of the latter is linked to microcracks and is accompanied by dissolution–recrystallization of the pyrite (Py1), giving rise to a network of vugs and micro-vugs with euhedral pyrite (Py2) on their walls, which takes the appearance of a symplectite at the peak of its development (Figure 3.21).

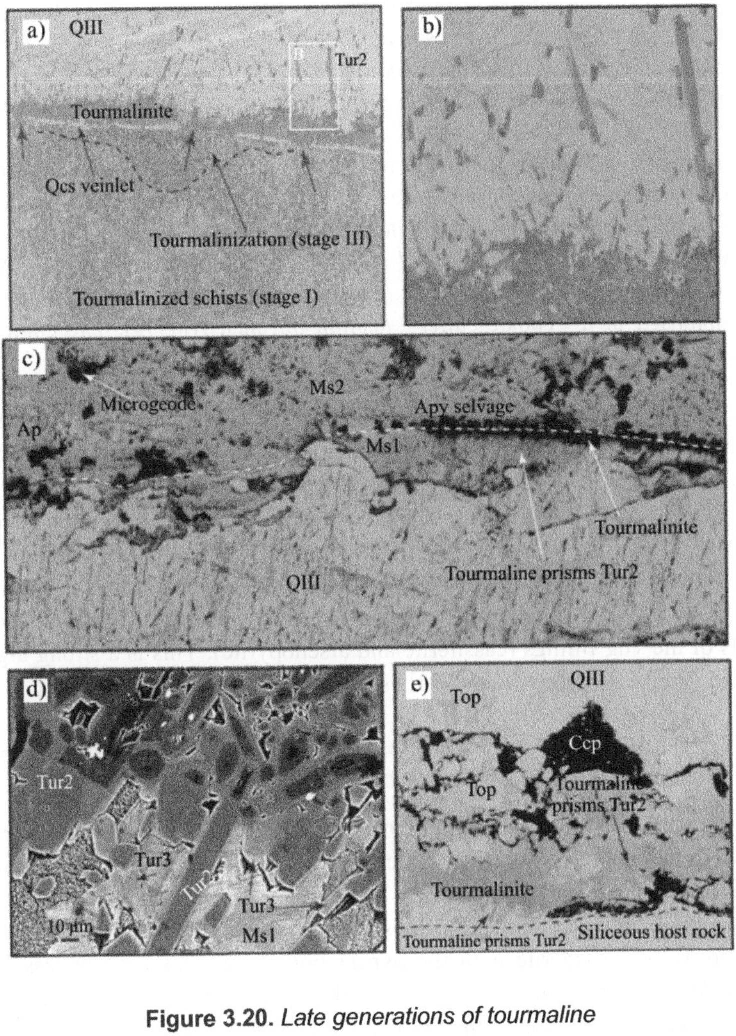

Figure 3.20. *Late generations of tourmaline*

COMMENT ON FIGURE 3.20.– *(a) Tourmalinite fringe and acicular tourmaline prisms (Tur2) salient in the quartz vein (QIII); the tourmalinized host rock is irregularly affected by a return of tourmalinization (overgrowths on the Tur1 prisms); note the local obliteration of a quartz Qcs veinlet (green arrows) by the tourmalinite (LMP, natural light, sample Pan-4b, Barroca Grande road). (b) Detail of A; note the limpid QIII quartz (without decrepitated FI) (LMP, natural light). (c and d) Muscovite reworking of a tourmalinite fringe analogous to that in figure (a) (wall of a stage IV vug). (c) Overview: muscovite 1 developed at the expense of the QIII quartz, preserving the acicular tourmaline prisms (Tur2) (LMP, natural light). (d) Detail of (c), showing the tourmaline overgrowths (Tur3) on the tourmaline 2 (BSE image). (e) Tourmalinite fringe at the wall of a topaz-quartz (stage III) vein; large acicular tourmaline prisms (Tur2) are wrapped by the topaz (Top1b); note the late chalcopyrite (LMP, natural light, sample PAN-XIX-16-a2-b1, D29R13AW14-E).*

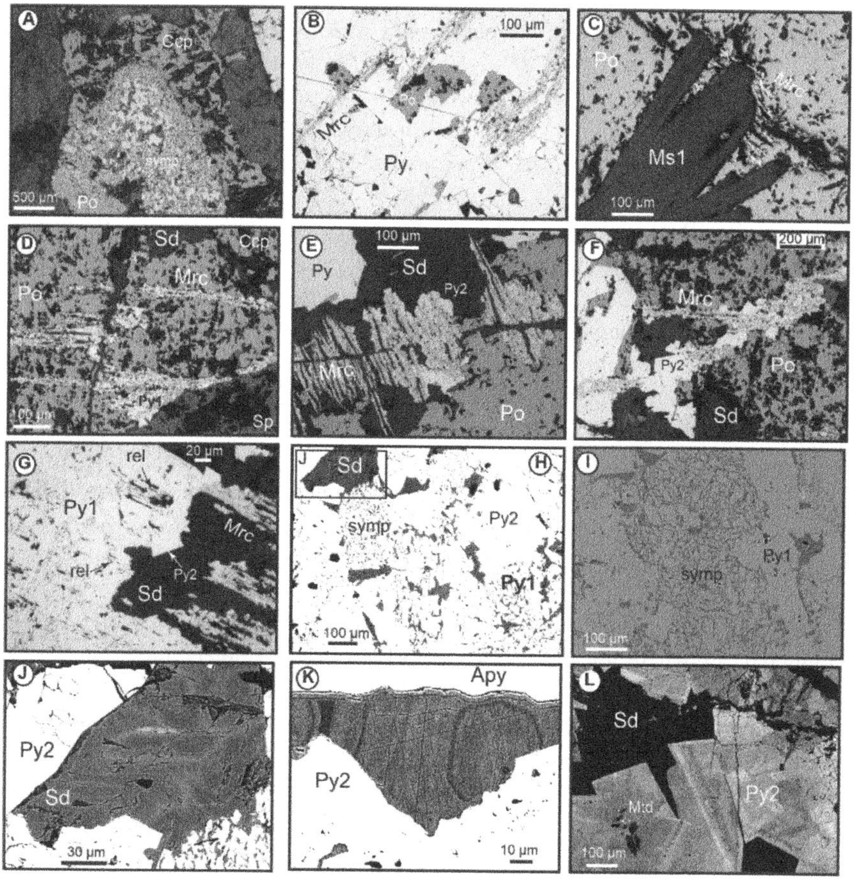

Figure 3.21. *Pyrrhotite pyritization stage (stage III-C) and its relationships with siderite (stage V)*

COMMENT ON FIGURE 3.21.– *(a) Symplectic association of siderite and pyrite (symp) replacing the pyrrhotite; note the more granular patches of a pyrite-siderite assemblage at the core of the symplectite (LMP, reflected light, sample PAN-V-6-c1, L0D15R15I2). (b) Pyrite and marcasite replacing the pyrrhotite, of which some relics remain (LMP, reflected light, sample PAN-II-a2, "Bank"). (c) Muscovite (Ms1) lamella overprinting marcasite in the pyrrhotite; (LMP, reflected light, sample PAN-XV-1c, L1D15R17AW15). (d) Siderite crack intersecting marcassite and pyrrhotite; note that the marcassite is beginning to be replaced by pyrite (Py1) (LMP, reflected light, sample PAN-II-a2, "Bank"). (e) The siderite dissolves/ replaces both pyrite and pyrrhotite, but does not, or barely, affects the marcassite; note the overgrowths over the marcassite of small euhedral pyrite crystals (Py2) (LMP, reflected light, sample PAN-II-a2, "Bank"). (f) As in E, but with a greater development of pyrite 2 (LMP, reflected light). (g) Selective dissolution of the pyrrhotite and pyrite 1 by siderite; the marcassite is preserved; note the porous remains of marcasite (rel) and the development of the pyrite (Py2) in contact with the siderite (LMP, reflected light, sample PAN-II-a2, "Bank"). (h) Dissolution of the primary pyrite (Py1) yielding interconnected microvugs with siderite; the siderite forms symplectic (symp) assemblages with a new pyrite (Py2) (image BSE, sample PAN-V-6-c1, L0D15R15I2). (i) Detail of a symplectic association of siderite and pyrite 2 (image BSE, sample PAN-II-5-2a,L0D9R17AW13). (j) Detail of (h) showing the zoning of the siderite (variations of their magnesium levels) (BSE image). (k) Detail of a siderite microvug in contact with arsenopyrite and pyrite, (images BSE, sample PAN-VI-14, "Bank"). (l) Euhedral pyrite (Py2) bordering a siderite vug (zoning revealed by the film of oxidation on the surface); the matildite (Mtd) is found here in micro-inclusions in the pyrite; elsewhere in the sample, it forms microcracks: thus, it is later than Py2, demonstrating the "inclusion paragenetic criterion" is unreliable (BSE image, sample PAN-VI-11, L0D3).*

Independent of the reworking of the pyrite (Py2-V), this siderite can be encountered anywhere in the veins, in the form of veinlets (sometimes with structures typical of a shear deformation: Figure 3.22(b)), or of microvugs, which, when in contact with the earlier muscovite (Ms1 or Ms2), are associated with overgrowths or neoformations of ferriferous muscovite (Ms3) (Figures 3.22(e) and (f)); such microvugs in the muscovite selvages were wrongly taken by Martins (2017) as "evidence" of the existence of an early generation of siderite ("interstitial in the muscovite"). Real quartz-siderite vugs, sometimes sealed (Figure 3.22(d)), may also develop. A relatively magnesian chlorite (Mg-Chl1) is locally associated with the siderite. For Carocci (2019), this carbonation episode has therefore been characterized sufficiently to define a stage V.

3.3.2.6. *Late stages*

Numerous mineral phases of the Panasqueira assemblages were deposited later than the stage V siderite. These phases, which are generally very minor and dispersed in the deposit, are not observed, with rare exceptions, in macroscopic form in the geodic fill, which complicates the establishment of chronological relationships. Nevertheless, it is possible to distinguish at first order a set dominated by the association of Cu-Zn-Sn (stage VI) and a later set characterized by the association of Ag-Bi-Pb (stage VII). It seems that these two sets are themselves earlier than the carbonated assemblages (siderite, dolomite, calcite), summarized by various authors under the label LCS (Late Carbonate Stage) and generally considered to be linked to the (post-Variscan) "alpine" activity of the deposit's fault systems (Polya et al. 2000; Lourenço 2002; Mateus et al. 2020).

3.3.2.6.1. Stage VI

Stage VI begins with the arrival of a second generation of sulfides, ferriferous sphalerite and chalcopyrite. These sulfides are observed, in this order of appearance, in some vugs, where they clearly succeed the siderite, while still being earlier than the late carbonates (see Figures 3.6 and 3.23). It is difficult to appreciate the quantitative significance of this chalcopyrite 2 in the copper budget on the scale of the deposit. It is not exceptional to encounter quartz-sphalerite-chalcopyrite microgvugs at random in the veins or to observe chalcopyrite microcracks at the microscopic scale (more rarely sphalerite microcracks), indifferently intersecting the stage IV arsenopyrite and muscovite (Figure 3.23(a)). However, at the scale of the veins, it is rare to observe large patches of these sulfides obliterating the arsenopyrite selvages, if they are observed at all. It therefore seems probable that most of the deposit's copper stock was deposited at stage III. Detailed observations show that chalcopyrite (Ccp2) is associated with a new generation of muscovite (Ms4), which is only observed in microcracks (Figure 3.23(e)).

A late generation of ferriferous chlorite is expressed in the topaz, in the form of networks of microcracks and dissolution patches, sometimes preceded by kaolinite, which intersect the muscovite (Ms1 and Ms2), which may itself have affected the topaz. Relationships with the stage VI sulfides are difficult to observe; some microscopic observations show convincingly that the chlorite intersects chalcopyrite 2 (Figure 3.23(e)). The stannite, which is directly associated with the chalcopyrite, is clearly synchronous with this chlorite (Figure 3.23). Moreover, there is no reason to consider the existence of several generations of stannite: it is because he assigns without clear evidence the chalcopyrite 2 of Carocci (2019) to two distinct episodes of crystallization, that Martins (2017), who clearly sees the direct link between stannite and chalcopyrite, deduces that there are as many generations of stannite as of chalcopyrite.

The cassiterite (Cst3-VI) also belongs to stage VI: in microvug fillings, in association with the chlorite (Figure 3.23(b)), in cracks, affecting the arsenopyrite and later than the siderite (Pinto et al. 2015) (Figure 3.23(c)). The hypothesis in Pinto et al. (2015), according to which this cassiterite would be homologous to the cassiterite they observe in a fault carbonated fault box, is not the most likely.

3.3.2.6.2. Stage VII

The minerals of the Bi-Pb-Ag association are still later; Figure 3.23(g) shows, for example, that matildite intersects stannite. The most common mineral assemblage is matildite-galena-native bismuth–bismuthinite, which appears noticeably in this order. Matildite and galena generally contain minor quantities of selenium (up to 5% weight Se in galena, which is otherwise often rich in Ag and Bi). The bismuthinite replaces the native bismuth (sulfur contribution).

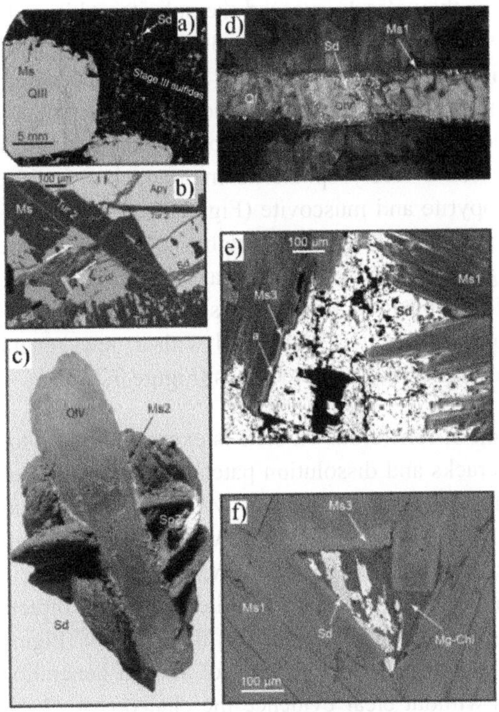

Figure 3.22. *Expression of siderite when not associated with the recrystallization of pyrite 1*

COMMENT ON FIGURE 3.22.– *(a) Network of siderite veinlets in the stage III sulfides; note the muscovite inserted at the QIII-sulfur interface (LEP, natural light, samplePAN-XV-1c, L1D15R7AW15). (b) Siderite veinlets overprinting a complex selvage (see Figure 3.20(a)); note that the Sd veinlets correspond to a shear deformation (C–S structure, white arrows); note too that the tourmaline 1 prisms of the wall tourmalinization are truncated in contact with the vein (BSE image, sample PAN-II-8b, L0D15R15I2). (c) Euhedral siderite wrapping a QIV quartz and its muscovite coating (Ms2) in a vug (L0D15R15AW10 "Bank"). (d) Example of a vug sealed by the siderite, occupying the whole thickness of the vein and locally obliterating the muscovite border on the roof (arrow) (L2D1P4I15). (e) Siderite microvug in a muscovite selvage; a Li-Fe-muscovite (Ms3) fringe develops in contact (and only in contact) with the siderite, with a final rim of Mg-Li–muscovite (a) (BSE image, sample PAN-XX-3-1b, L3D9R1A). (f) Siderite microvug in a muscovite selvage; a Li-Fe-muscovite (Ms3) fringe develops in contact with the siderite; a magnesian chlorite accompanies the siderite (BSE image, sample PAN-XI-196-1, L1D15R17AW15).*

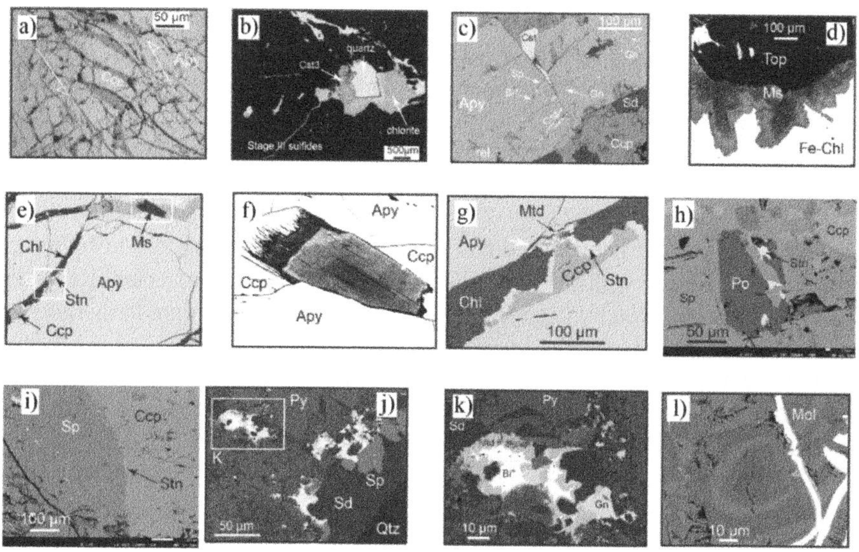

Figure 3.23. *Some aspects of the late stages*

COMMENT ON FIGURE 3.23.– *Mtd matildite, Bi° native bismuth. (a) Microfissural chalcopyrite in an arsenopyrite ($Ccp_{2\text{-}VI}$); note the "sinistral" relays (LMP, reflected light, sample PAN-VI-8a, "Bank"). (b) Cassiterite (Cst3)-quartz-chlorite microvug in the massive sulfides of stage III; note the microcrack feeders to the microvug (LMP, natural light, sample PAN-X-10, L0D15R15AW10I2). (c) Fissural cassiterite* *in the arsenopyrite (Cst3); other cracks are filled with sphalerite, native bismuth or galena, without a clear relative chronology here (BSE image, sample PAN-V-6-c1, L0D15R15I2). (d) Muscovite and chlorite in a micro-cavity formed by the topaz dissolution (BSE image, sample PAN-XIII-8-b2, L1D29R7AW14). (e) Network of chalcopyrite-muscovite-chlorite-stannite micro-cracks in an arsenopyrite (image BSE, sample PAN-XIII-8-b2, L1D29R7AW14). (f) Detail of (e); note the absence of stannite in contact with the muscovite (BSE image). (g) Other detail of (e) showing the direct association of the chalcoyrite and the stannite, at the chlorite-chalcopyrite interface; the chlorite, which is later, inserts itself between the chalcopyrite and the arsenopyrite (white arrow; the matildite granules are clearly later than the chlorite (image BSE). (h and i) In the stage III pyrrhotite-sphalerite-chalcopyrite assemblages, the stannite is an overprinting phase (images BSE, sample PAN-II-5-2a, L0D9R17AW13): (h) the stannite patches obliterate the pyrrhotite/sphalerite contacts. (i) Strings of micro-inclusions of stannite in either chalcopyrite or spahalerite, note that one string transects the chalcopyrite to sphalerite boundary, showing that is a sealed microcrack (another example of the unreliability of the "inclusion criterion"). (j and k) The late Bi-Ag-Pb assemblages (BSE images, sample PAN-II-5-2a, L0D9R17AW13): (j) overview, the stage V symplectites are obliterated; (k) detail of (j) showing the native bismuth later than the matildite and the galena. (i and l) The molybdenite obliterates the zoned siderite of a microvug in a muscovite selvage (image BSE, sample PAN-II-6).*

Other minerals appear at this stage: Pirard (1988) indicates pavonite. Tellurium is very rarely expressed in the form of Te-canfieldite (Wimers 1985); Martins (2017) moreover encountered benjaminite and ikunolite. Antimony, reputed to be absent from the assemblages with the exception of the Vale da Ermida sector (Kelly and Rye 1979), was observed by Carocci (2019) in the form of a very rare freibergite.

Molybdenite is very rare at Panasqueira, where it is among the latest phases, encountered exceptionally in the muscovite selvages, or intersecting the wolframite (Clark 1965). It is rich in tungsten (up to 5.2% weight W).

3.3.3. *Wallrock alterations*

The wallrocks of the mineralized veins are systematically affected in the mine by varied hydrothermal alterations: to the ubiquitous tourmalinization (stage I-A) can be added silicification, muscovitization, topazification, veins of apatite, arsenopyrite, sulfur veins and siderite veins (Figure A3.10). These alterations obviously reflect the corresponding phases that participated in filling the veins: their chronology is therefore reasonably parallel to that of these phases, as numerous observations have demonstrated. Since silicification is, on the one hand, later than stage I tourmalinization and, on the other hand, affected by all the other alterations, silicification is therefore very likely also of stage I in a probable relationship with QI quartz (Carocci et al. 2021; Figure A3.10). The muscovitization, in bands of a few centimeters at the walls, is by far the most frequent alteration after the tourmalinization; as has been seen (section 3.3.2.5), it is associated with the Ms1 muscovite. Topazification, which is not common, is obliterated by muscovitization, which is coherent with its attribution to stage III (Figure 3.23(f)). Arsenopyrite is common in the altered walls in the form of small euhedral crystals. In particular, it lines slits that are parallel to the schistosity (Carocci et al. 2021). Furthermore, these slits in diaclase form, which use schistosity and guide the crystallization of the arsenopyrite, clearly intersect the muscovitized fringes, thus indirectly confirming that the arsenopyrite is later than the Ms1 muscovite.

3.3.4. *Consequences and implications of revising the paragenesis*

3.3.4.1. *Questioning some results*

As the paragenesis revised by Carocci (2019) and Cathelineau et al. (2020) seems robust, its consideration will lead some results to be questioned, which illustrates the decisive importance of establishing a paragenetic succession when exploiting geochemical data. This is how the tourmaline studied by Launay et al. (2018) to characterize the fluid circulation dynamic using the asymmetries of the growth forms (application of Curie's law) is described by these authors as forming "fine acicular crystals (1–3 mm) with < c > axes vertical and perpendicular to the vein edge" (Launay et al. 2018, p. 4). It is therefore clearly tourmaline 2, not tourmaline 1, the prisms of which are always truncated by opening the veins (Carocci et al. 2021; section 3.3.2.5; Figure 3.c in Launay et al. 2018), which is compared directly with Figure 3.23(a), and confirms this affirmation. It follows that the fluid circulations conditioned by the greisenized cupola that these authors reconstructed relate to the start of stage III and therefore teach us nothing about the fluid circulations at the main stages of mineralization (stages I and II). Similarly, by relying on the paragenesis established by Kelly and Rye (1979), who argue that the muscovite on the walls is an early phase, Lecumberri-Sanchez et al. (2017) treat the alterations where a proximal fringe of muscovitization and a more distal tourmalinization are observed as a unitary phenomenon. They deduce from this a model according to which precipitation of the wolframite is conditioned by the interaction of mineralizing fluid (tungsten carrier) and the surrounding rocks, the alteration of which would free the iron needed to precipitate the ferberite. As tourmalinization and muscovitization are disjoint phenomena, the first earlier and the second largely later than the precipitation of the wolframite, the foundations of this attractive theory have been removed.

In the same way, the temperature estimates proposed by Codeço et al. (2019), using a boron isotopic geothermometer applied to the tourmaline-muscovite equilibrium, are biased by the fact that since the tourmaline is of stage I or stage III-A (the authors do not clearly differentiate), it cannot be in isotropic equilibrium with the stage IV muscovite that seems to be associated with it, or if we consider that the tourmaline is isotopically reequilibrated with this muscovite, it cannot nevertheless be used to estimate the temperature of the tourmalinization, contrary to the authors' intentions. Equally biased are the temperature estimates made using the thermometer based on the exchange of F and OH between topaz and several generations of muscovite (Mateus et al. 2020), since these phases are not contemporary, and the topaz is systematically corroded by the muscovite.

At best, these estimates could therefore reflect the temperature of this alteration: since the Ms1 is associated with a pseudomorph of topaz 1a (section 3.3.2.2.1), this would therefore in all likelihood affect only the Ms2. However, the dispersion of these values obtained by Mateus et al. (2020), between 475°C and 393°C, does not make us optimistic about this interpretation: disequilibrium is more probable. Still biased are the estimates by Jaques and Pascal (2017) or Martins et al. (2020) of the temperature at which the arsenopyrite is formed (several generations in reality, in their ideas; section 3.3.2) using data from Kretschmar and Scott (1976), but for arsenopyrite-pyrite equilibriums that need not be considered either, since the arsenopyrite is later than the pyritization of the pyrrhotite or later than the late generation of pyrite (Figure 3.9).

3.3.4.2. *Implications for the geochemical evolution of the deposit*

The paragenetic evolution presented in Figure 3.9 shows that the history of the deposit is also one of the successive introductions of a series of elements, with some interruptions but also cyclicities or at the very least repetitions (an aspect to which Martins (2017) and Mateus et al. (2020) have also drawn attention).

The B (tourmaline) and W (wolframite) contributions at stages I, III-A, and IV-A are clearly cyclical. The F, P, Sn, and Li contributions are repetitive but in different mineral and temporal combinations: topaz, phosphates and cassiterite in III-A; cassiterite then Li-muscovite and F-apatite in IV; cassiterite and stannite in VI; Li-muscovite at stages IV-A and IV-B. The contribution of iron is permanent, since the ferriferous phases crystallize at all stages, which involves the permanence of more or less reductive conditions throughout the hydrothermal history, since iron can only be transported in Fe^{2+} form ($FeCl_2°$ complex in solution (Simon et al. 2004)) and is also found in this reduced form in ferriferous minerals of the different stages. However, from stage III-B to stage VI, we observe repetition of Fe, Cu and Zn contributions in association with a concomitant sulfur contribution.

Another way of seeing this evolution would be to contrast an episode of tungsten alone (stages I and II) with an episode of tin (stages III-VI), with the latter marked by an alternation between rare metal deposition (W, Sn, Li, associated with F and P) and sulfide deposition (Fe, Cu, Zn).

In the course of this complex evolution, the pyritization of the pyrrhotite at stage III-C (increase in fS_2), the massif contribution of arsenic at stage IV-A, and the influx of CO_2 that characterize stage V are moreover singular events, markers of a rupture in the hydrothermal evolution.

3.3.5. *Absolute chronology*

A consensus has long been established, attributing the genesis of the Panasqueira deposit to the start of the Permian. Nevertheless, in the recent past, it has been possible to consider that this opinion, based both on the ages obtained on the Panasqueira granite in the 1970s and on the muscovite datings by Snee et al. (1988), has been confirmed by Launay's datings (2018) of the apatites (those in the veins just as much as those in the granite) and those of Mateus et al. (2020) on the uraninites (Figure 3.24).

However, in recent years, other data have come to undermine these certainties. This is how, for example, Marignac et al. (2020) show that the dating of the Panasqueira granite to ca. 294 Ma is in fact the dating of a hydrothermal imprint that can be attributed to stage IV. Moreover, Zhang et al. (2019) dated a Panasqueira cassiterite to ca. 303 Ma; considering these associated uncertainties, Mateus et al. (2020) believe that this dating confirms a mineralizing event at approximately 300 Ma. However, Carocci et al. (2021) obtained a U–Pb (LA-ICP–MS) age of 305.2 ± 5.7 Ma on early rutile (stage I-A), which therefore dates the start of wolframite mineralization. Although Zhang et al. (2019) unfortunately do not specify the cassiterite they have dated, it is plausible, considering its relative abundance, that it is a stage III cassiterite; this is in any case the opinion of Carocci (2019).

In total, it seems possible to conclude that there is a hydrothermal history of long duration, beginning at the very end of the Pennsylvanian (Kasimovian–Gzhelian boundary, approximately 305–306 Ma) and extending into the Lower Permian (Cisouralian) (at least up to the Artinskian, at ca. 292 Ma, or even the Kungurien), over nearly 15 Ma.

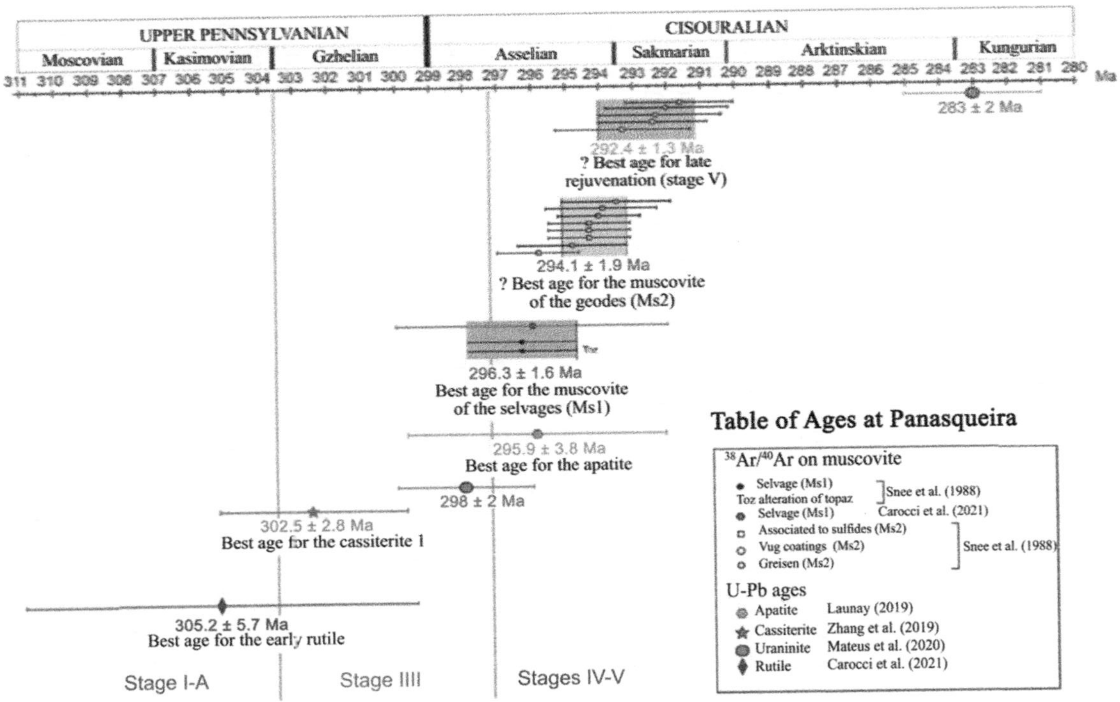

Figure 3.24. *Table of the available ages for the Panasqueira deposit*

3.4. The opening of the veins

3.4.1. *The opening conditions*

In a general analysis of the mechanism by which the veins can open, Nüchter and Stöckhert (2008) conclude from considerations on the effective pressure that, to keep the crack open, the fluid pressure (Pf) should remain on the order of size of the $\sigma 3$ constraint and that sealing occurs as soon as $Pf < \sigma 3$. They note that the textures of the filling depend on the opening speed: fibrous quartz and crack-seal when the fill rate of the crystals (the sealing) is faster than the opening rate, free cavity growth in the opposite case. These considerations account for the differences observed between stages I and II in Cathelineau et al. (2020).

On the one hand, there is a consensus in publications on Panasqueira, which considers that the opening of the veins is conditioned by fluid overpressure. At the outset, Kelly and Rye (1979) envisaged a simultaneous opening of all the veins, which would immediately have reached their current opening and would have been sealed progressively, leaving residual cavities (the current vugs). This view is contradicted by the ^{39}Ar-^{40}Ar datings in Snee et al. (1988), which have demonstrated a minimum duration of hydrothermal activity of almost 5 Ma, wholly incompatible with a joint system remaining open, as envisaged initially, as well as from study of the vein structures (section 3.3.2.1). Studying these structures led Foxford et al. (1991, 2000) to accept that there were episodes of sealing and reopening, according to the crack-seal model. Nevertheless, these authors, as well as Polya et al. (2000), continued to believe that their OSS and MSS stages correspond to a single tectonic event, the successive repetitions of opening occurring in continuity with the previous ones, and that, moreover, the veins did not undergo any reactivation after the MSS stage, since later episodes of deposition took place in a network of residual cavities. These views clearly contradict those of Carocci (2019) and Cathelineau et al. (2020), who highlight the independence of the successions of sealing and reopening from stage I veins to stage III, each with its own logic, and show the independence of the development of stage IV vugs, which are not therefore residual (see above, section 3.3.2.1). Nevertheless, at least at stages I and II, the logic of the veins' opening generally conforms to a mode I dilatation (Figure 3.5). More precisely, as Foxford et al. (1991) have shown, each crack at the origin of an elementary vein begins in mode I, similar to a tensile crack; however, when neighboring veins begin to overlap, they enter into mechanical interaction and are converted into mode III shears, as a consequence of which the schists within the overlap area undergo torsion, which leads to the formation of "eel's tail" structures. The repetitive crack-seal opening of these subhorizontal veins indicates for these authors that a compressive stress state (i.e. σv = σ3) prevailed during the period corresponding to the three main mineralization stages.

3.4.2. *The origin of the flat joints*

Nüchter and Stöckhert (2008) emphasize that since veins originate in fissures, "the stages of fragile rupture and crack should be distinguished from the opening and sealing stages". The origin of the flat joints of the Panasqueira system has been the subject of contradictory propositions. Kelly and Rye (1979) envisaged an effect by which the lithostatic pressure decreased in connection with erosion, combined "with the effects of residual tectonic deformation", without further precision. This point of view, which could be admissible in the low-pressure context acknowledged by these authors, can no longer be admissible, now that it is accepted that a depth of 10–12 km presided over the first stages of the deposit's history (Cathelineau et al. 2020) (section 3.5.1.2.2). For Ribeiro and Pereira (1982), the veins result from the reopening, governed by the intrusion of the granitic cupola, of tensile slits perpendicular to S3, created at the end of the D3 deformation. The first point faces the lack of any correlations between the vein system and the cupola (see above, section 3.2.3.1), and the second does not explain the fact that the veins do not form a parallel and regular system. Jacques et al. (2018) take up this tensile joint model but on the basis of a new analysis of local tectonic structures. These authors in fact observe a subvertical intersection lineation (formerly described by Marignac (1973)) and conclude from this the existence of kilometer-scale isoclinal folds (their "F1"), with subvertical axes, and an axial-plane schistosity ("S1") subparallel to S0; these folds are folded again by a second generation of mesoscopic folds ("F2"), also with vertical axes and an axial-plane schistosity ("S2") NW–SE. The asymmetry of these folds is interpreted as the result of transpressive deformation characterized by the simultaneity of the horizontal shortening and of a right strike-slip displacement generally NW–SE to N–S.

Jacques et al. (2018) also observed that flat joints are very widespread in the Panasqueira region. Noting the lack of any trace of movement on the surface of these joints and the fact that they cut through the "S1" and "S2" schistosities without torsion, Jacques et al. (2018) take up a suggestion by Marignac (1973) and interpret these joints "open in mode I and perpendicular to S0-1" as < ac > joints of "F2" folds. The existence of different families of flat joints is attributed to the non-cylindricity of these folds. It might be objected that this explanation does not take proper account of the coexistence of two conjugate families in a single place or of variations in the dip of the veins over short distances (Figures 3.25(a) and

3.25(b)). However, above all, this interpretation comes up against a major difficulty: it is incompatible with the structures at the regional scale.

The whole Panasqueira region is in fact characterized by upright folds of decakilometric size and NW–SE orientation (regional phase D3), with a structure in which tight synclines (marked by bars of Armorican quartzite that form a relief) are separated by vast anticlines, which Figure 3.1(c), taken from Launay (2018), clearly shows: this author moreover observed D3 structures on a metric scale in the Rio Zezere, not far from Panasqueira. Indeed, there is no evidence in the regional map to support an episode of ductile NW–SE to N–S strike-slip of a size compatible with the structures described by Jacques et al. (2018) (Figure 3.1(b)). The only faults of this direction, observable in mining work (Figure 3.3), show only very small displacements, which are not ductile, and their activity is moreover thought to be alpine. It is highly probable, in these conditions, that the structures described by Jacques et al. (2017) are truly older structures inherited from regional phases D1 and D2 (responsible for the formation of more or less recumbent folds; section 3.2.3.1), redeformed and verticalized by the D3 folds; this would explain the generality of a vertical intersection lineation between S0 and the schistosity: this would in fact be a verticalized, subhorizontal early lineation.

Flat joints in fact exist on a regional scale, far beyond the mineralized system. They have been described to the south of the Sierra de Estrela (to the north and east of the deposit) by Derré et al. (1986). These authors in fact observe ubiquitous conjugate joint systems, extending largely well beyond the thermal aureole, and characterized by dips less than 40° and often close to horizontal. In a given location, it is common for one of the two families to be better represented than the other; indeed, one may even be absent, and this behavior is observed particularly when the dip of the joints is lower than 20°.

In most cases, the S3 schistosity is intersected by these flat joints, but sigmoidal torsions are observed locally, evidence that they occupy the position of conjugate shears in the context of horizontal shortening. Their association with "kink folds" adds to the evidence. The authors unfortunately do not specify the direction of shortening, but it could be thought that it is compatible with the one associated with the F3 folds. The observation by Derré et al. (1986) of aplitic or pegmatitic sills or indeed of biotite leucogranites injected into some of these joints or in a bisector plane confirms that this system is indeed Late Carboniferous, and the observation of tourmalinization or muscovitization phenomena governed by these joints is evidence of their capacity to drain hydrothermal fluids. This flat joint system develops widely toward the west since the joints can be found up to the vicinity of the bars of *Daedalus* quartzite from the Lower Ordovician at Cabeço Mosqueiro (personal observations). They are, however, particularly abundant in the environment close to the Panasqueira deposit: these are in fact the joints mentioned by Jacques et al. (2018).

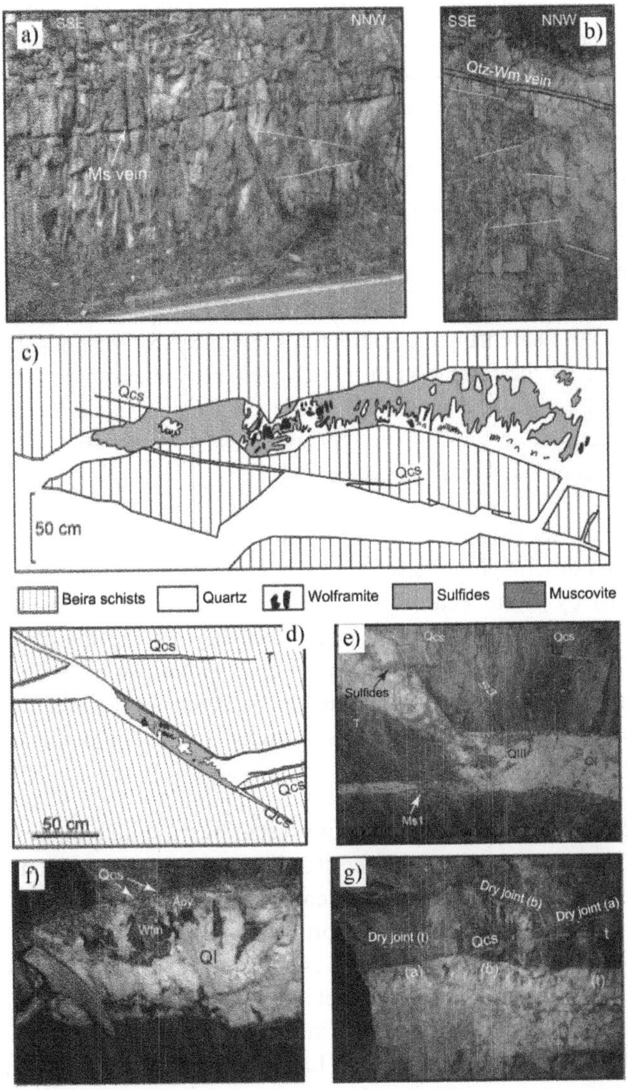

Figure 3.25. *Flat joints and their relationships with vein systems*

COMMENT ON FIGURE 3.25.– *(a) Outcrop on the road between Barroca Grande and Panasqueira; the conjugate flat joints are highlighted in white; note the use of one of the families by the muscovite vein (= muscovite Ms1 of the selvages in mining workings). (b) Outcrop on the road from Rio Zezere to Barroca grande: a quartz-wolframite vein of the Sao Francisco de V. sector; the vein uses one of the two directions of the conjugate joints (highlighted in white). (c) The mineralized veins use the same directions as the early Qcs veinlets (drawing in situ [CM, 1970], L0D9R17AW13). (d) The quartz veins use both conjugate directions, expressed by the Qcs veinlets; additionally, there is a Qcs veinlet in a position bisecting both conjugate directions: expression of a T joint? (Drawing from photograph, L3D17R03AW39). (e) Conjugate Qcs joints parallel to the "dry" joints (white dotted lines); the mineralized vein uses one of these directions, as well as a bisecting direction also expressed by the "dry" joints (T) (L2D1P14I1S). (f) Qcs veinlets underlining the hangingwall of a stage I vein (with also an arsenopyrite selvage). (g) The quartz vein can be broken down into three segments in directions (a), (b) and (t), parallel to flat joints with the same direction in the wallrock: two conjugate sets (a) and (b), and a bisector joint (t) indicating a tension joint; Qcs veinlets also use these joints; the t veinlets seem to be organized according to a "dextral" relay (overthrust) (L1D15R7AW15).*

However, Foxford et al. (2000) have shown that the early millimetric QT veins (Qcs veins in Carocci et al. (2021)) form two conjugate sets with low dips (10°–30°) (Figures 3.25(d) and (e)). "Low but real shear displacements of the QT walls" would indicate that these veins developed in a mixed mode, that is, including an oblique opening component (in lithostatic conditions and with a strong stress contrast). These authors recognize the continuity of this QT/Qcs network with the systems of flat joints described by Derré et al. (1986), which therefore amounts to saying that these veins occupy pre-existing joints. For Foxford et al. (2000, p. 1072), the main veins have used the earlier QT/Qcs veins: "mode I reopening of a selection of QT veins under fluid overpressure". In fact, the remains of Qcs veins are often present at the walls or in the continuation of the veins (Figures 3.13, 3.15 and 3.25). At the scale of the veins, there is evidence (also infrequent) of tangential movements associated either with the formation of flat joints or equally with the opening of veins as such. It is in fact possible to observe in places the non-orthogonal opening of active veins (Figure 3.26(c)), attested by the mismatch, on either side of the veins, of markers such as subvertical SBs (Figure 3.26(b)). More spectacular, but unique to our knowledge, is the observation of twisted wallrock panels involved in the "eel's tail" structures (Figure 3.26(a)). At the microscopic scale, it has been possible locally to observe bends affecting the mimetic tourmalines of the main schistosity (Tur 1) and that the bends are sealed by the Qcs veinlets that mark the start of stage I

(Carocci et al. 2021). It is also not exceptional to observe a twisting of the schistosity over ≤100 μm at the hanging or footwall of the veins, indicating shear on the joint defining the wall (Figure 3.26(d)). Most commonly, the muscovite selvages are not affected, thus indicating deformation prior to stage IV. However, Launay (2018) indicates an occurrence where the Ms1 muscovite selvage is itself sheared. Finally, many observations attest to the existence of plastic deformation of the quartz of the veins, without it being possible, unfortunately, to determine the associated kinematic. It is in any case certain that these deformations are early and that they predate stage IV muscovite (Ms1) (observations by Foxford et al. (1991)). It may therefore seem certain that the opening of the flat veins occurred in tectonic continuity with the formation of the flat joints that gave way to them.

3.4.3. *The place of flat joints in the regional history*

From the above, it follows that the system of flat conjugate joints was necessarily formed in P and T conditions in the same order as the conditions that prevailed at the start of the hydrothermal circulations, between 450°C and 500°C, that is, in a generally ductile context for quartz (section 3.5.1). It might therefore be thought that this system is a response to a transitional episode in which the deformation accelerated, allowing temporary transition to a fragile regime. These are the same conditions that allowed the formation of Seixo Bravos: it is therefore relevant to enquire about the relationships that may exist between these SBs and the flat joints.

Surface observations suggest that the development of subvertical SBs is compatible with the regional context of NE–SW shortening (Figure A3.2): indeed, “the appearance of compressional and transpressional structures, which are formed in close collaboration over the course of the same regional shortening event, suggests that the tectonic stresses along the X and Y axes were similar and could be reversed in response to the fluid pressure conditions linked to the hydrothermal activity” (Launay 2018). In mining work, flat SBs are encountered, which intersect the subvertical SBs (Noronha et al. 1992), and late-kinematic slit systems that can be interpreted as tension joints in relay associated with virtual thrusting (Figure 3.2). Such observations tend to show that transition from a σ2 vertical state (strike-slip) to a σ3 vertical state (overthrust) occurred at the end of the D3 event in the Panasqueira region. This latter state corresponds well to the one that prevailed at the formation of the conjugate flat joints. It is therefore permissible to consider these joints as a late manifestation in the D3 event.

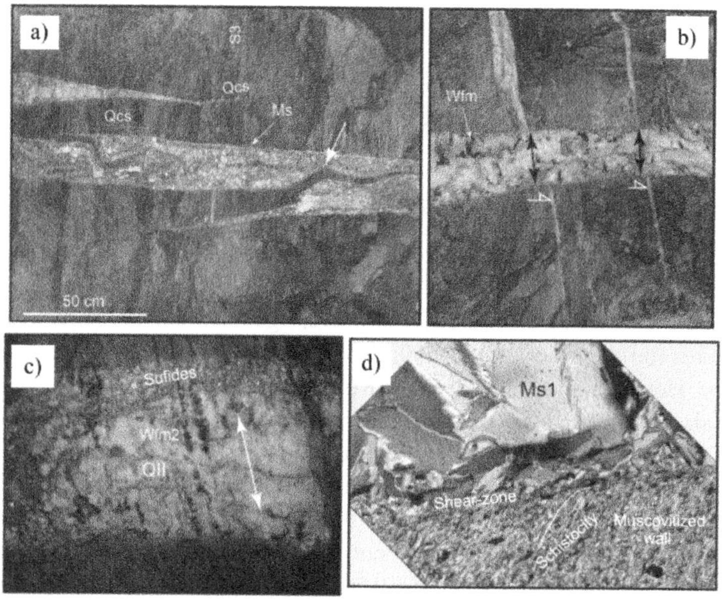

Figure 3.26. *Tangential component indices in the opening of the Panasqueira veins*

COMMENT ON FIGURE 3.26.– *(a) A unique case of the folding of "eel's tail" structures; note the slits in extrados (white arrow); both conjugate directions of the Qcs veinlets coincide with the directions of the mineralized veins (L3D5R2AAW31). (b) A mode I opening (double arrows) does not suffice to explain the offset observed on the Seixo Bravo (SB) marker; the obliquity is also attested by the average direction of small wolframite crystals (Wfm) (L1D9R15AW12). (c) An example of an oblique opening of a stage II quartz (QII) vein, well materialized (double white arrow) by the wolframite (Wfm2) prisms; note the "crack-seal". (d) Detail of a sheared contact between the muscovitized wallrock (schistosity mimetic) and the muscovite selvage (LMP, polarized light, sample PAN-XX-5, L3D19R1AW32).*

Thus, the mineralized vein system appears to be associated directly with the last manifestations of the D3 event toward the end of the Carboniferous. As this stage is thought to have begun at approximately 310 Ma (Diez-Fernandez and Pereira 2017), this would therefore fix a lower bound on the age of the deposit. Such an idea would also explain the participation of fluids of metamorphic origin in generating the mineralization (section 3.6), there again in continuity with evolution at the end of the Carboniferous, since the fluids at the origin of the Seixo Bravos quartz are themselves of metamorphic nature (Noronha et al. 1992).

3.5. P and T conditions at the early and late stages

Thermobarometric information comes essentially from studying the FIs and from the use of equilibriums between mineral phases.

3.5.1. *The early stages (stages I–III)*

3.5.1.1. *Use of mineral thermometers*

Codeço et al. (2017) used the “Ti-in-quartz“ geothermometer to evaluate high temperatures (on the order of 450–500°C) for the quartz of the tourmaline wallrock alterations. However, their confidence in Kelly and Rye’s results (1979) and those of Polya et al. (2000) was such that these authors conclude that, since the wolframite was laid down at 360°C and below, the tourmalinization therefore occurred in a decreasing temperature regime: which is clearly not the case (section 3.5.1.2). This temperature estimate is compatible with the measurements made by Cathelineau et al. (2020) (section 3.5.1.2) and does not seem to call for criticism. For the same early stage, Mateus et al. (2020) used a “Zr-in-rutile” geothermometer to obtain less realistic temperatures (between ~560°C and 640°C), probably due to inaccuracies in the measurement of Zr levels on the order of 200 ppm with the help of an electronic microprobe.

As we saw above (section 3.3.4.1), there have been many approaches based on the thermodynamic equilibrium between minerals that, since they are not truly contemporary, were not eligible for such calculations. The only plausible estimate is by Martins et al. (2020), who used the composition of pyrrhotite, and Toulmin and Barton’s (1964) data on the pyrrhotite–pyrite equilibrium. Although, contrary to what Martins et al. (2020) think, there is no pyrite contemporary with the crystallization of the pyrrhotite, it is probably admissible that this equilibrium may be evidence of conditions at the early pyritization stage (PAS, stage III-C). With an NFeS of pyrrhotite between 0.924 and 0.952 (average 0.933), this approach provided Martins et al. (2020) with a temperature of ~405°C. If we accept this, it follows that the end of stage III is characterized by a marked drop in temperature compared to the start (section 3.5.1.2).

3.5.1.2. *Thermobarometry of fluid inclusions*

3.5.1.2.1. History

Kelly and Rye (1979) were the first to describe the FIs of the Panasqueira veins. As they considered the late vugs to be the mere continuation of the mineralization processes initiated by the early phases, the inclusions contained in the large quartz of these vugs played a substantial role in their interpretations; it is true that similar inclusions are found in the earlier massive quartz. Kelly and Rye (1979) described aquo-carbonic fluids in these quartz grains and, on the basis of observations suggesting phase separation, concluded that the deposit was formed at low pressure and thus at low depth (less than 10 MPa) and at a temperature between 230°C and 360°C. These conclusions were taken up without any major changes by most later authors.

Although the sequences of opening and mineralization have been reviewed and modified since, most of the later authors described the same types of fluids and concluded that the deposit conditions were comparable, that the inclusions had been studied in the quartz (all authors), topaz, apatite or siderite (Kelly and Rye 1979; Lourenço 2002), or even in the wolframite (Campbell and Robinson-Cook 1987; Lüders 1996). The only exceptions to this consensus are the works of Jaques and Pascal (2017) and Cathelineau et al. (2020). Significantly, these latter authors are also the only ones to focus on massive milky quartz associated with the wolframite and, above all, to have plotted reliable isochores. Jacques et al. (2018) focused on early quartz veins with the aim of characterizing the conditions of their opening. They observed the coexistence of decrepitated inclusions and "healthy" inclusions in the quartz of the veins and considered that the decrepitated inclusions resulted from the destruction of the "healthy" inclusions, for which they constructed isochores. From the basis of "classical" parageneses, they used the "arsenopyrite thermometer" to extract from the isochores the barometric information they contained: with a temperature of 500°C, they found a pressure of ~300 MPa, considered to be lithostatic and therefore resulting in a depth of ~10 km. The flaw in this interpretation is clearly how the arsenopyrite is neither early nor contemporary with the wolframite quartz.

3.5.1.2.2. Thermobarometry of stages I and III-A

Cathelineau et al. (2020) were interested in the FIs of early quartz associated with wolframite and with those of topaz. These authors made the fundamental observation that the quartz most intimately associated with the growth of the wolframite in the large quartz-wolframite veins contained only decrepitated FIs (in a broad sense: IF profoundly transformed after trapping, either deformed plastically = "stretching", or truly decrepitated), some of which, however, the smallest ones, have been sufficiently spared to conserve information on their trapping (a situation well-known in metamorphic environments: Vityk and Bodnar 1995, 1998; van der Kerkhof et al. 2014; Figures 3.27(a) and (e)). These are dense aquo-carbonic inclusions rich in water, plotting the isochores of which makes it possible to broadly record the conditions in which the trapping took place by considering intersections with "reasonable" geotherms (40°C·km^{-1} and 50°C·km^{-1}). As shown in Figure 3.28, the obtained P–T domain is broad. It is possible, however, to refine the estimate by considering two conditions. The first is that, since these inclusions are without doubt partially re-equilibrated toward the lowest pressures during "decrepitation", the densest isochore is the one most representative of the trapping conditions. Considering its intersection with a geothermal gradient typical of contexts at the end of collision (40°C·km^{-1}), a pressure of 325 MPa (12 km) and a temperature of 500°C are found. This pressure is without doubt excessive, since the second condition is evidently that the depth where the vein system is found in formation could not be greater than that of the emplacement of the Panasqueira pluton, which is estimated at approximately 300 MPa, that is, approximately 11 km (section 3.3.2.2). With this last value, the trapping temperature would be established at approximately 480°C. However, this is a maximum, since it is possible, indeed probable, that the exhumation phenomena that accompanied the end of the collision also induced some system recovery. If we take into account the values estimated by Codeço et al. (2017) (section 3.6.1.1), it might be thought that 450°C is the lower bound: it would then correspond to a trapping pressure of ~285 MPa, that is, ~10.5 km depth.

As described above (see section 3.3.2.2), there are two generations of topaz, often separated by an interruption in growth, one of which (Top1a) is generally much more altered (sometimes pseudomorphosed into muscovite) than the other (Top1b). Indeed, Cathelineau et al. (2020) stated that although topaz 1a still contains decrepitated FIs, this is no longer the case with topaz 1b: the phenomenon at the root of the generalized decrepitation at Panasqueira therefore occurred at the end of stage III-A. This phenomenon explains without doubt the differential behavior of these two generations with regard to alteration, since topaz 1a was probably damaged concomitantly with decrepitation. In Top 1b crystals, Cathelineau et al. (2020) encountered aquo-carbonic FIs of three types, all primary, two predominantly aqueous, the third rich in volatiles and less dense (Figure 3.28). The isochores of the latter are fairly flat and almost parallel to the geotherms (50°C·km^{-1} and 60°C·km^{-1}), and their intersection with those of the FIs rich in water, which are denser, makes it possible to fairly precisely define the P–T conditions of stage III-A: 480°C and 235 MPa seem to be the best estimates (Figure 3.28). It might have been thought that the coexistence of "aqueous" and "volatile FIs" indicated a process of immiscibility of the hydrothermal fluid: this is not the case since, apart from the total homogenizations made over distinct ranges (250–295°C and 289–312°C, respectively), the estimated trapping conditions are all contained in the one-phase domain of the H_2O-CO_2-NaCl system (Cathelineau et al. 2020) (Figure 3.28).

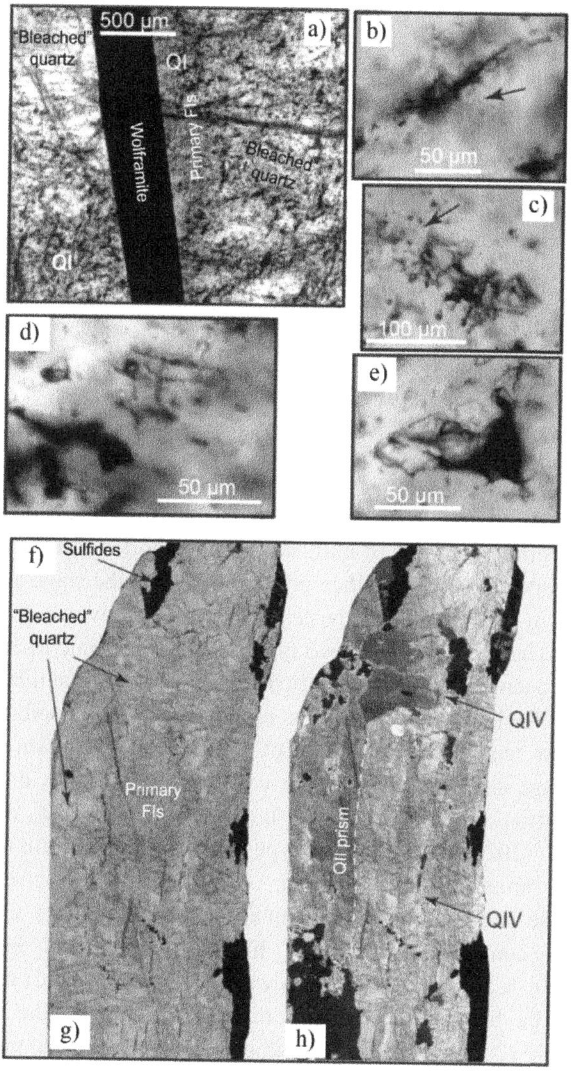

Figure 3.27. *The early fluid inclusions (FIs) and the complexity of the quartz of the veins*

COMMENT ON FIGURE 3.27.– *(a) In the wolframite synchronous quartz, "dark" patches coexist densely filled with decrepitated FIs, intimately associated with the wolframite ("primary" FIs) and more or less "bleached" patches, from which these inclusions have disappeared (LEP, natural light, sample Pan-7b, Barroca Grande road). (b–e) Examples of "decrepitated" FIs: more or less deformed inclusions, with variations in the density and volume of the volatile phase. (b and c) Decrepitated FI, with an inherited core and a halo of neoformed FIs (arrow) (LEP, natural light. (b): Sample Pan-7b, (c): sample PAN-XI-196-1, L1D15R7AW15). (d and e) Deformed FIs, which potentially carry information on the trapping conditions; note the dilation of the volatile phase in (e) (LEP, natural light, (d): sample Pan-7b, (e): sample PAN-XI-196-1). (f) Example of quartz from the stage II veins (LEP, sample PAN-XVIII-11a, L1D7R13AW14). (g) The patches of primary FIs are isolated within the quartz mass, which is penetrated by "bleached" zones, the "primary" inclusions of which have disappeared (natural light). (h) "Acicular" prism structure of the QII quartzes; the geometry of the bleached zones (see a) is independent of that of the QIIs, while the sulfides, as well as the small QIV quartzes, use the joints between the prisms (polarized light).*

3.5.2. *Late stages (stages IV–V)*

The overwhelming majority of published studies on FIs at Panasqueira relate to the late stages, especially because they are carried out in the stage IV geodic quartz, and secondarily, in the apatites or siderites of the vugs. Above all, they involve stage IV and stage V. The coexistence of two types of aquo-carbonic inclusions, one type predominantly aqueous and the others rich in volatiles, with similar Ths, has been interpreted since Kelly and Rye (1979) as an indicator of immiscibility, from which the relatively low temperatures (230–360°C) and trapping pressures (10–20 MPa) are deduced. These interpretations are not very satisfactory; they do not in fact rely on the construction of isochores, and it is very likely that, as discussed by Cathelineau et al. (2020) for these same types of inclusions, we are in reality outside the domain of immiscibility. We can certainly envisage a decrease in the temperature at the late stages: but preliminary data published by Cathelineau et al. (2018) on the contrary led to still higher temperatures being estimated, approximately 400°C, for these stages, as well as to a (lithostatic) trapping pressure of 140 ± 20 MPa being reconstructed, that is, a depth on the order of 5 km. Cathelineau et al. (2018) document a hitherto unknown IF family consisting of purely volatile liquid inclusions (rich in $CH_4 > N_2 >> CO_2$), the isochores of which intersect those of the dominating family of Lw-c type inclusions to give the estimates above. This temperature on the order of 400°C is moreover compatible with the estimate of ~405°C by Martins et al. (2020) at stage III-C (PAS) immediately earlier.

It may be noted that data on the traces of fission in the apatites, which concern evolution in the Mesozoic and Cenozoic at Panasqueira as well as regionally, also point in the direction of a significant depth toward the start of the Permian. The fission tracks in apatite indicate exhumation on the order of 3 km since the Upper Cretaceous for batholite close to the Serra de Estrela (Stapel (1999) in Cloething et al. (2002)). It makes sense, however, to note that Kelly and Rye (1979) deduce from their own data on fission tracks in the Panasqueira apatites that a clear episode of warming (≥150°C) in the Upper Jurassic (approximately 150 Ma) and perhaps another less intense episode in the Upper Cretaceous (approximately 80 Ma) affected the mineralization: they attribute these effects to episodes of subsidence. Nevertheless, studying the traces of fission in the detrital zircons of the Lusitanian Basin led Pereira et al. (1998) to the conclusion that the granites of the ancient massif had been subject to continuous erosion throughout the Jurassic and Cretaceous periods. Moreover, warming during the Lower Triassic–Jurassic is well documented in the whole ancient Iberian massif (Juez-Larré and Ter Voorde 2009), where it is generally attributed to a thermal anomaly in relation to the breakup of Pangea and the premise of the Atlantic opening. Models of the temperature–time paths made by Vasquez-Vichez et al. (2015) from measurements of fission tracks show limited warming (10–30°C) from this episode to the Lower Cretaceous, showing that if there was subsidence, it was of limited magnitude.

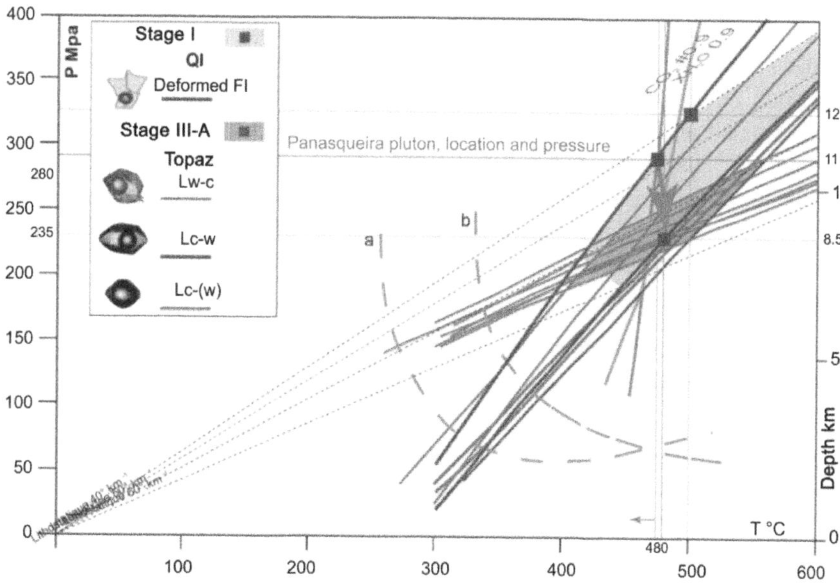

Figure 3.28. *Estimate of P–T-t conditions at Panasqueira between stages I and III-A (redrawn from Cathelineau et al. (2020))*

COMMENT ON FIGURE 3.28.– *Isochores have been drawn for the main FI families: plastically deformed inclusions from stage I in the QI quartzes; dense aqueous inclusions in the topaz: Lw-c (the volatile phase is only revealed by the formation of clathrates) and Lc-w (three-phased FI); less dense volatile inclusions (Lc-(w), with reduced aqueous phase). The colored boxes correspond to the estimates of the trapping conditions. The isopleths of the H_2O-CO_2-NaCl system (dotted gray lines) are represented for a salinity of 6% weight NaCl and for 10 (a) and 20 (b) mole % CO_2 (Weisbrod 1984). The XH_2O 0.9 and $XCO_2/(XCO_2+XCH_4)$ (CO_2#) 0.9 isopleths are taken from Huizenga (2001). Compared to stage I, there can thus be observed at stage III-A, the cumulative effects of a modest exhumation and of an increase in the temperature, which, combined with those of the plastic deformation that moreover affects the QI and QII quartzes, were probably at the origin of the generalized decrepitation of the early FIs that characterize this transition. Explanations are given in the text.*

A hydrothermal event in the Upper Jurassic is also documented in the Toledo Mountains (Barbero et al. 2005), probably linked to the Neo-Tethys, as is the case throughout Western Europe (Cathelineau et al. 2012). Alkaline magmatism from the Upper Cretaceous age is known in Central Portugal and may explain the warming inferred by Kelly and Rye (1979). In conclusion, it can legitimately be thought that the Panasqueira deposit still lay at a depth of more than ~3 km in the Upper Cretaceous. This is coherent with an observation by Groot and Wolf (1953), who stated that they found an outcrop of discordant arkoses on the Beira schists “not far from Panasqueira”. They compared this to the “Buçaco Sandstone” (of probable Eocene age): the models in Vasquez-Vichez et al. (2015) in fact suggest an acceleration of the exhumation, hitherto slow, from approximately 70 Ma; with estimated rates of 0.11–0.19 $mm.a^{-1}$, it took a mere 15–30 Ma (so covering the interval between the Upper Cretaceous and the start of the Eocene) to erode 3 km of terrain.

3.6. Characterization of the origin of fluids

It is a widespread concept that there is a favored relationship between granites (more or less specialized) and stanno-wolframiferous mineralizations, and the idea that there are genetic links between magmatic fluids, greisenization and quartz-wolframite-cassiterite veins is the basis for fairly broadly accepted models (Cerny et al. 2005). The Panasqueira deposit is no exception to this rule, and the small greisenized Barroca Grande cupola has been the subject of much speculation from this perspective. Although it has long been known that it is intersected by quartz-wolframite veins, it has been made several times, and again recently, to play a significant role in the genesis of the deposit, in agreement with general models. The most developed model was presented by Launay et al. (2018) (section 3.6.1.3).

The central question therefore involves the participation (or not) of fluids of magmatic origin in the Panasqueira hydrothermal system. This question can be approached from two angles: study of the composition of the fluids (geochemical approach) on the one hand and the quantitative examination of the availability of fluids (mass balances) on the other hand.

3.6.1. *Geochemical arguments: the early stages (I to III-A)*

3.6.1.1. *Stage I-A: analysis of the tourmalines*

At the earliest stage, FIs were found in neither the tourmaline nor the quartz of the Qcs veinlets. Nevertheless, tourmaline could serve as a tracer of tourmalinization fluids by analyzing its trace element content using LA-ICP–MS: in contrast to Codeço et al. (2020), who estimated that since the distribution of traces in tourmalines was under "stronger dependence on host rock composition", this mineral is not a good marker of fluid origin. Carocci et al. (2021) concluded that trace elements in tourmalines are effective tracers. Tourmaline 1 is in fact zoned, and variations in the concentrations of major elements and traces from one zone to another led Carocci et al. (2021) to conclude that these tourmalines crystallized from complex fluids resulting from mixing between two poles: the first (F1) was characterized by the Co-Cu-Pb-Sc-Sr-V-Cr-Nb-Ta-Sn assemblage, and the second (F2) was rich in Li-F-Fe-Mn-W. The zoning results from the relaying of F1 by F2. These two fluid poles are interpreted as being of metamorphic origin on the basis of the rare earth spectra, which do not present the characteristics expected for a fluid of magmatic origin. This does not pose any problems for the F1 pole, which is easily interpreted as a fluid of "local" origin, in equilibrium with the Beira schists. The F2 pole poses more problems because of its more "granitic" signature: Carocci et al. (2021) propose that this pole is of deeper origin, rooted in a migmatitic domain where the biotite incongruent melting took place (section 3.8). This is the F2 pole that seems to be associated with the very start of wolframite deposition (see above, section 3.3.2.1). There are tungsten-rich rutiles contemporary with the start of tungsten deposition (they contain micro-inclusions of wolframite) (Carocci et al. 2018) that also crystallize from fluid F2. In their zoning, these record short-term variations in Nb levels that are interpreted as reflecting the contribution of a third fluid pole (L3), which would be a sign of an RMG (Carocci et al. 2021).

3.6.1.2. *Analysis of the FIs: the volatile phase of stages I-B to III-A*

Most of the FIs observed in quartz and topaz by Cathelineau et al. (2020) display mainly aqueous fluids ($XH_2O > 0.9$) of medium salinity (comprised between 5 and 10% weight eq. NaCl). Data on the composition of the volatile phase of the FIs obtained using in situ Raman microspectrometry led Cathelineau et al. (2020) to envisage a majority (if not exclusive) contribution of fluids of metamorphic origin, in the broad sense of fluids resulting just as much from prograde devolatilization as from the high-temperature equilibration of any fluids with already metamorphosed rocks ("pseudo-metamorphic" fluids). Cathelineau et al. (2020) observed that the volatile fraction of the fluids that circulated in the mineralized veins permanently resulted from the mixing of a CO_2-CH_4 component dominated by CO_2 and a CH_4-N_2 component rich in nitrogen (Figure 3.29(a)). The latter is typical of the prograde devolatilization of black shales. The first is characteristic of the Fischer–Tropsch reaction between water and graphite, and Huizenga's (2001) data allow for estimation of the temperatures at which this reaction could have taken place. As seen in Figure 3.28, the reaction isopleths for the Lc-w inclusions in the topaz ($XH_2O = 0.9$ and $XCO_2 / (XCO_2 + XCH_4) = 0.9$) intersect inside the domain (P, T) estimated for stage III-A: the equilibrium (or the last equilibrium) between this fluid and graphite is therefore achieved in ambient conditions for the mineralized vein system. However, as can also be seen in Figure 3.29(a), many analyses are characterized by $XCO_2/(XCO_2 + XCH_4)$ ratios above 0.9, which, according to Huizenga (2001), means that the system is no longer buffered by graphite at temperatures above 570°C (at 300 MPa). A proportion of the fluids observed therefore come from warmer (and probably deeper) domains than the context in which the veins were formed.

3.6.1.3. *Indirect approach*

Launay et al. (2018) reconstructed the circulation dynamics of the hydrothermal fluids in the veins by studying the asymmetry between the growth zones of the tourmaline prisms, and they thus show that the vectors characterizing the flow of the fluids demonstrate a zonal arrangement around the greisenized cupola, to which they therefore attribute a central role in generating the deposit. As has been seen previously (section 3.3.4.1), these authors wrongly considered this reconstruction as relating to the early stages, whereas it only relates in reality to stage III-A. Since, according to Launay (2018), the muscovites of the vein walls are chemically identical to those of the greisens of the cupola, Launay et al. (2018) arrive at a model of the fluid transfer from the cupola in the process of greisenization ("auto-metasomatic" greisenization under the influence of the magmatic fluids leaving the cupola) toward the mineralized veins: these would therefore be magmatic fluids. Unfortunately, as Marignac et al. (2020) demonstrated, and as will be seen in more detail below (section 3.7), the muscovite of the "greisens" is in reality the result of episyenitization, since the cupola has undergone early silicification on which this unstable quartz alteration is superposed. Moreover, as it has been demonstrated (Carocci 2019) that the muscovite of the walls is stage IV (section 3.3.2.3.3), it follows that the fluids responsible for this muscovitization of the greisen are themselves of stage IV. The "demonstration" by Launay et al. (2018) of the involvement of a magmatic fluid issuing from the greisenized cupola is therefore not permissible.

3.6.2. *Geochemical arguments: the late stages (III-B, IV, V)*

The nature of the fluids that were at the origin of the mineralized veins has been the subject of contradictory considerations over the years. This debate has involved results drawn from analysis of the composition of FIs (Bussink 1984; Polya 1989; Noronha et al. 1992; Polya et al. 2000; Lourenço 2002; Lecumberri-Sanchez et al. 2017), from the study of stable isotopes (Kelly and Rye 1979; Bussink et al. 1984; Polya et al. 2000; Codeço et al. 2017), and data on rare gasses (Polya et al. 2000; Burnard and Polya 2004).

3.6.2.1. *Analysis of fluid inclusions*

All of the FI data show fluids belonging, similar to those of the previous stages, to the C-O-H-N-NaCl system. The salinity of the FIs is generally moderate, with estimated values between 5 and 10% weight eq. NaCl.

3.6.2.1.1. The volatile phase

Bussink et al. (1984) were able to use gas chromatography to measure the composition of volatile phases extracted by gas-crush from most of the major mineral phases of the deposit. Noronha et al. (1992) measured this composition directly, on the Seixo Bravos and on the quartz of the veins, using Raman in situ microspectrometry. Lourenço (2002) also used Raman microspectrometry on quartz, topaz, cassiterite and stage IV wolframite. The results obtained by these different authors are remarkably stable (Figure 3.29(b)); they find the poles of volatile constituents observed at the early stages, to which is added a pole rich in methane, which particularly characterizes the deposition of sulfides at stage III-B.

3.6.2.1.2. The aqueous phase

Data on the aqueous phase could be obtained either via crush-leach (Polya 1989; Noronha et al. 1992; Polya et al. 2000; Lourenço 2002) or, more recently, through in situ measurement of the composition of FIs by LA-ICP–MS (Lecumberri-Sanchez et al. 2017). The results can be compared to the database on magmatic fluids compiled by Audétat (2019).

Polya (1989) estimated K/Na ratios of ~0.2 (between 0.1 and 0.3 for the most part), which Noronha et al. (1992) confirmed, whereas the magmatic fluids have average K/Na ratios between 0.3 and 1.3 (mode ~0.5). Recalculated in K/Na+K, the values found by Polya (~0.17; between 0.09 and 0.23) compare to the equilibrium values at *subsolidus* temperatures of a fluid with two feldspars (one sodium, the other potassium), as measured experimentally by Orville (1963): 0.26 at 670°C and 0.16 at 400°C. However, Lourenço (2002), who found K/Na values that remained low (< 0.1) in the quartz, observed markedly higher values in the wolframites and cassiterites (also stage IV), between 0.06 and 0.44: a magmatic contribution is probable here. This is confirmed by a study of the lithium levels by the same author: while the Na/Li ratios are always very high in the QIV quartz (between 200 and 1,300), they are much lower in the wolframites and cassiterites (between 17 and 108); indeed, low values of this ratio are fairly characteristic of the magmatic fluids (Audétat 2019), in particular in the magmatic fluids that issued from the RMGs, where Na/Li < 20 (Harlaux et al. 2017). Figure 3.29(c) shows that two groups of fluids seem to be involved: one group is represented in quartz, and the other is more typical of rare metal oxides, which is coherent with Luders' observations (1996). Luders did not find the same FI types as in the "associated" quartz in the wolframites (studied in infrared).

Polya (1989) likewise estimated very low Fe/Na ratios (~0.0005), which confirm the measurements of Lecumberri-Sanchez et al. (2017). Lecumberri-Sanchez et al. found less than 2 ppm Fe in the FIs measured (remember, since these are not decrepitated, they are representative of the late phases). Lourenço (2002) obtained slightly higher values in quartz (Fe/Na < 0.05). These values are not compatible with those of magmatic fluids, which are on average significantly higher: 0.08–0.91 (mode 0.30; Audétat 2019), even in fluids from magmas as fractionated as those of Beauvoir-type RMGs (Harlaux et al. 2017).

Nevertheless, Lecumberri-Sanchez (2017), on the basis of the high Li, Rb and Cs levels that they observe in their analysis, conclude that there is a magmatic origin for the Panasqueira fluids. Polya et al. (2000) were able to measure the Br/Cl and I/Cl ratios of the FIs of their zoned quartz. These ratios are far from those of the magmatic fluids and are close to those that were observed in the orogenic gold deposits from the northern CIZ, where the fluids are of metamorphic origin (Figure 3.29(d)).

3.6.2.2. *Stable isotopes*

3.6.2.2.1. Carbon isotopes

Carbon isotope analyses carried out by Kelly and Rye (1979) on the siderites (thus at stage V) and on the FIs (thus stage IV quartz) characterize a carbon "partly derived from, or which underwent extensive exchange with, a graphitic carbon source" (Kelly and Rye 1979, p. 1796), with $\delta^{13}C$ between –13.1 and –9.3‰. This conclusion is shared by Polya et al. (2000), who calculate a $\delta^{13}C_{fluid}$ of –12 to –7‰. The latter, however, envisage (like Bussink et al. (1984) before them) the possibility of an incorporation "of organic-rich sediments in the S-type magma crystallizing to form the underlying granitoid" (Polya et al. 2000, p. 3359). These results show that interpretation using a Fischer-Tropsch reaction is valid at stage IV as well as at the previous stages. It makes sense, however, to note that Bussink et al. (1984) found markedly higher $\delta^{13}C$ values (–4 to 0‰) in the gas-crush "wolframites" (unspecified, probably stage IV).

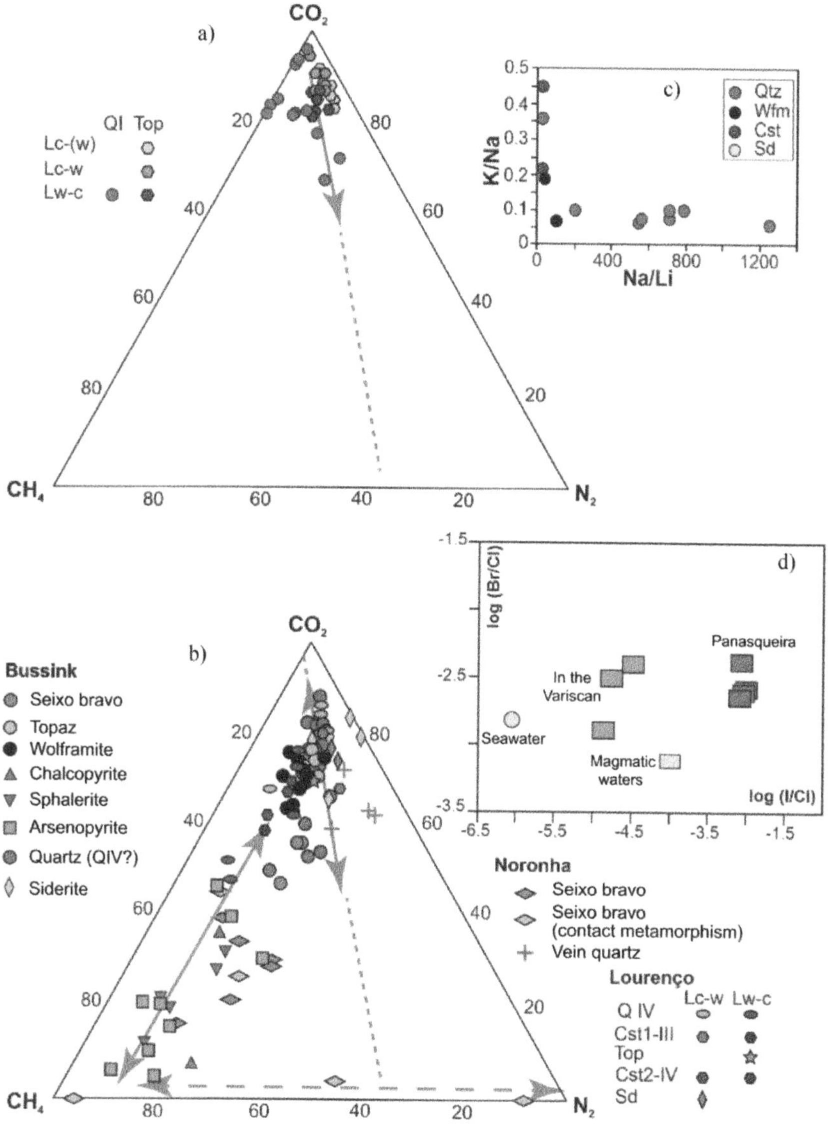

Figure 3.29. *Summary of data available on the composition of the Panasqueira fluids*

COMMENT ON FIGURE 3.29.– *(a) Composition of the volatile phase at the early stages (data from Cathelineu et al. 2020). Note the mixing trend between a* CO_2*-(*CH_4*) pole and a* N_2*-(*CH_4*) pole. (b) Composition of the volatile phase at the late stages (compilation of data from Bussink et al. (1984); Noronha et al. (1992); Lourenço (2002)). Note the two mixing trends: one is very similar to that of the early stages; the other, between a* CH_4 *pole and a pole rich in* CO_2*, is particularly well marked in the fluids extracted from the stage III-B sulfides. (c) K/Na and Na/Li ratios in the stage IV fluids (data from Lourenço (2002). Note the presence of two types of contrasting fluids, opposing the W and Sn oxides and the quartz. (d) Br/Cl and I/Cl ratios in a zoned QIV quartz (adapted from Polya et al. (2000); the data on the Variscan gold come from Boiron et al. (1996)).*

3.6.2.2.2. Oxygen and hydrogen isotopes

Stable isotopes of oxygen and hydrogen are commonly used to characterize hydrothermal fluids. At Panasqueira, they were studied by Kelly and Rye (1979), Bussink et al. (1984) and Polya et al. (2000). Since Bussink et al. (1984) measured only the $\delta^{18}O$, which are not sufficient to characterize the fluids (Sheppard 1994), their results (which are furthermore very similar to those of the other authors) cannot be used here. Kelly and Rye (1979) obtained data covering a vast range of minerals, but for the FIs, they mainly used samples coming from the vugs, and so belonging to stage IV; they also measured quartz from the "quartz cap", which can be estimated as more representative of silicification of stage I (section 3.6), as well as from Seixo Bravos; the descriptions are fairly precise for distinguishing three generations of muscovite in the lots analyzed. Polya et al. (2000) focused their efforts on a vug quartz (QIV), studying its zoning. The δD were measured by heating the FIs, or by crush-leach, without further precision, especially on the size of the mineral fragments analyzed (Kelly and Rye 1979), and using crush-leach alone by Polya et al. (2000), on samples of "several hundreds of mg". In both cases, these are therefore average values at best.

To move from isotopic values in the minerals to the values of the associated fluids, the formation temperatures must be known, on which the isotopic fractionation coefficients greatly depend, particularly for quartz. The temperatures estimated by Kelly and Rye (1979), also adopted by Polya et al. (2000), were notoriously too low (section 3.5.1); therefore, it was necessary to construct Figure 3.30 to recalculate the values with the best adapted temperatures: 400°C for stage IV minerals, 450–500°C for the "quartz cap" (which is in fact interpreted as silicification stage I; section 3.6) and 500–550°C for the Seixo Bravos, late kinematic in a tectono-metamorphic "greenschists" facies episode. The result shows a fairly weak variation in the values of $\delta^{18}O_{water}$, between 7 and 12‰, covering the boundary between "magmatic waters" and "metamorphic waters", but a very large interval of values for δD, from –40 to –140‰ (Figure 3.30). Aside from the quartz, for which the values cover the whole set of variation intervals, it can be observed that each phase is characterized by a specific box: in particular, the wolframite and cassiterite boxes are highly disjoint, the second being characterized by particularly low δD. The set of values displays a well-marked trend, which passes through the "magmatic box". These data are interpreted by Kelly and Rye (1979) (who calculated the fractionation at much lower temperatures) as conveying an origin that is either meteoric or magmatic, without being able to state which. In the same way, Polya et al. (2000) interpreted analyses that they called *nonanomalous* (relative to δD, i.e. those that are not too negative) as conveying mixtures between meteoric and magmatic waters or much exchanged meteoric waters. In fact, in Figure 3.30, it can be seen that the data define a trend that can be interpreted as the result of mixing between at least two poles, one ($\delta^{18}O$ ~11%, δD ~–40%), the other ($\delta^{18}O$ ~8%, δD ~–130‰). The first is easily interpreted as a metamorphic pole, which could be represented by the Seixo Bravos or perhaps the Qcs quartz. The second is more difficult to interpret. Campbell et al. (1984) suggested that the low values of δD could be interpreted as a sign of the participation of meteoric water in the hydrothermal system; however, as the paleomagnetic data (Scotese et al. 1979) situate Panasqueira close to the palaeoequator (in the Permian), this would imply the existence of strong relief in the domain involved, which Polya et al. (2000) do not consider to be very probable. Because of the values obtained for halogens (section 3.6.2.1.2) (Figure 3.30), these authors think it is possible to exclude a magmatic origin for these δD fluids. Their hypothesis is that it is an "organic" fluid (as defined by Sheppard and Charef (1986)), which they think could have come from a coalfield basin from the Upper Carboniferous, which would naturally have disappeared due to erosion. They exclude the contribution of a metamorphogenic methane (which would be characterized by a very negative δD: up to –250‰; Schoell 1980), on the pretext that levels of this gas in the FIs would be too low to produce the effect observed. Nevertheless, Bussink et al. (1984) measured significant molar fractions in methane, whereas Cathelineau et al. (2018) indicated the existence of FI generations with purely volatile content, which could well be the explanation for the low δD in Figure 3.30.

The quartz and wolframite are not in equilibrium: the temperatures calculated from oxygen isotopes with Zheng's equation (1992) are much too low and too variable (between 125°C and 350°C); in fact, boxes calculated taking into account the composition of the hydrogen are clearly distinct (Figure 3.30).

Polya et al. (2000) believe that the PAS marks the start of a switch in the logic of hydrothermal circulations from horizontal cracks to vertical cracks and that this change is marked by an influx of meteoric water at the origin of the alteration of pyrrhotite into pyrite.

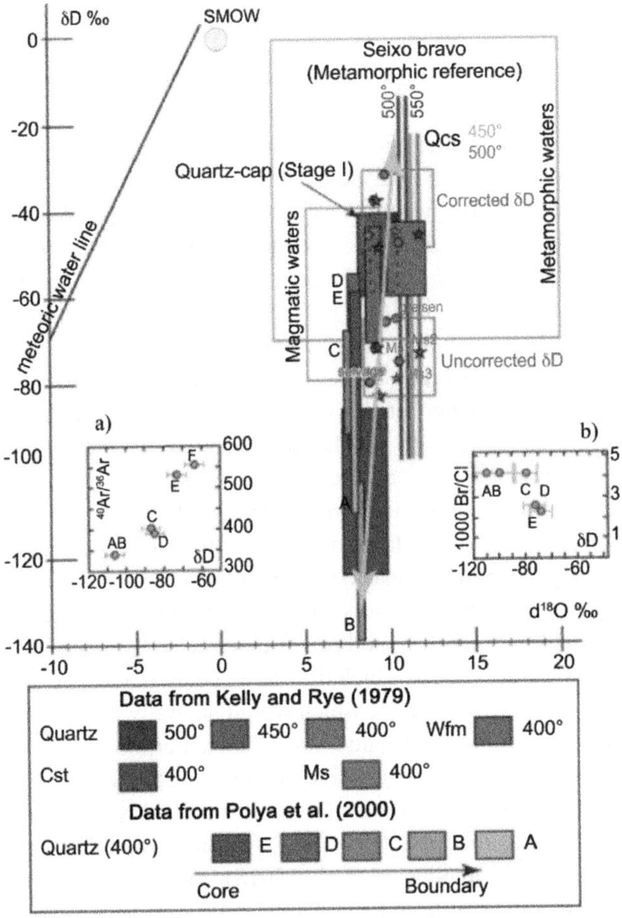

Figure 3.30. *Isotopic compositions calculated from the fluids associated with the different mineral phases of the Panasqueira deposit*

COMMENT ON FIGURE 3.30.– *The compositions have been recalculated based on data from the literature at the temperatures indicated using Zheng's equations (1992, 1993) and Richter and Hoernes' (1988) equations for the compositions of* $\Delta^{18}O_{water}$ *for quartz, wolframite and cassiterite, and muscovite. To calculate the* ΔD_{water} *associated with the muscovite, Suzuoki and Epstein's (1976) equation was calibrated for a range of temperatures (450–850°C) much higher than that considered here, and the authors suggest that the corrections are much lower at higher temperatures: two options have therefore been envisaged, and two boxes have been calculated, one with correction, the other without. Data on the Seixo Bravos come from Kelly and Rye (1979), those on the Qcs veinlets, from Polya* et al. *(2000); there are no δD data available for these quartz. In inserts a) and b): correlations between δD and the $^{40}Ar/^{36}Ar$ (a) or Br/Cl (b) ratios for the zoned quartz studied by Polya et al. (2000). Additional explanations are given in the text.*

3.6.2.2.3. Other isotopes

The sulfur isotopes have been analyzed in the stage III sulfides (MSS) by Kelly and Rye (1979), who found $\delta^{34}S$ values between –0.1 and –0.9‰; hence, they conclude that the sulfur comes either from a "deep igneous source" or a "large volume of homogenized crustal material", and the second option is thus compatible with a metamorphic origin for the fluids. Polya (1989) analyzed the isotopic sulfur composition of the pyrites from the surrounding schists, which he found to be between –10 and +3‰, so different from the sulfides in the deposit. To conclude, "this points to a granitoid sulfur source".

The nitrogen isotopes have been analyzed by Bussink et al. (1984) using gas-crush carried out on the wolframite and quartz of the veins, the "quartz cap" and the Seixo Bravos. The $\delta^{15}N$ obtained are very constant, between 3 and 5‰: they are evidence, for the authors, "of the assimilation of low grade metamorphic rocks". However, as Bebout and Fogel (1992) have shown, the isotopic composition of nitrogen resulting from metamorphic devolatilization is between –1.5 and +5.5‰, and the values obtained by Bussink et al. (1984) are probably evidence of a metamorphic origin for nitrogen.

Codeço et al. (2017) measured the $\delta^{11}B$ of the deposit's tourmaline and found values between –3.7 and –12.7‰, with a pronounced mode of – 9‰, which (at a temperature of 500°C) makes it possible to calculate a value of –7.1‰ for the fluid responsible for wallrock tourmalinization. This value is just as compatible with a magmatic origin as with a metamorphic one, but Codeço et al. (2017) opt for a magmatic origin as "the high boron concentration in the wallrocks, with >50 vol% tourmaline, along with the typical element association of Panasqueira (W, Sn, F, Nb, Ta, Rb) makes a granitic source more likely". This point will be reviewed and discussed below.

3.6.2.3. *Noble gases*

Polya et al. (2000) analyzed the $^{40}Ar/^{36}Ar$ ratios in their zoned quartz (see this section) and found that they were inversely correlated with δD (Figure 3.30). The high values of these ratios suggest a mantle origin for these authors. This is also the conclusion from the analysis of noble gases by Burnard and Polya (2004). This observation is difficult to reconcile with a classic magmatic-hydrothermal model, and discussion of this will be resumed later (see section 3.8).

3.6.3. *Mass balances and the source of the fluids*

A particularly interesting approach consists of evaluating the volumes of fluid needed to transport and precipitate the minerals involved in the mineralization, and to contrast them with potential sources. This approach was used by Kelly and Rye (1979), Polya (1989) and Carocci et al. (2021).

3.6.3.1. *A magmatic source of the fluids?*

The mass balances relating to the transport and deposition of the silica and the tungsten needed to deposit them, as well as the mass balance relating to the boron required by the tourmalinization of the walls of the veins, are presented in section A3.4. The calculations demonstrate that neither the greisenized cupola, or even the protolithionite leucogranite sheet that crowns the Panasqueira pluton (Figure 3.4), could have provided the water needed for any of these constituents.

On the other hand, the total volume of the pluton as modeled by Ribeiro (2017) would have sufficed for silica, but still not sufficed to transport the tungsten or the boron. Unfortunately, we may legitimately think that the emplacement of the pluton preceded the inception of the Panasqueira hydrothermal system: it has in fact been known for some time (Bloot and de Wolf 1953) that the iron-magnesian minerals of the contact halo are affected by the tourmalinization, which must therefore postdate them. In view of the importance of this halo, it seems that only a pluton of the dimensions estimated for the whole massif could have been responsible for it: this therefore preceded the tourmalinization and hence the tungsten mineralization.

It therefore seems to be accepted that the Panasqueira granites could not alone have been the origin (if they were the origin) of the hydrothermal fluids responsible for forming the deposit.

3.6.3.2. *Feasibility of a metamorphic source for the fluids*

Conversely, the calculations detailed in section A3.4 show unambiguously that a metamorphic source for the fluids, which is moreover strongly suggested by the geochemical data, is entirely plausible: the devolatilization of a volume of schists

$10 \times 10 \times 1$ km^3 is in fact enough to provide all the water necessary, as well as the boron from the tourmalinization.

3.6.4. *Presence of a magmatic component*

From the information and discussion above, it can be concluded that the Panasqueira hydrothermal system was dominated for the whole duration of its functioning by a single fluid type, characterized by the fluids' metamorphic origin (in a broad sense). In all the published data set, little objective data attests to the presence of fluids of unambiguous magmatic origin.

It is nevertheless highly probable that such fluids were involved, on the basis of qualitative data provided by the paragenetic study (section 3.3.2 and Figure 3.9). This is how at stage III-A, as Cathelineau et al. (2020) note, the massive contribution of fluorine (in the form of topaz), tin (in the form of cassiterite) and of phosphorous (wagnerite) is highly suggestive of the active participation of a rare metal magma. It is the same for stage IV, where the contribution of fluorine and lithium (muscovite), tin (cassiterite) and phosphorous (apatite) move in the same direction, with the nuance however that in place of being globally synchronous, as at stage III-A, these contributions are involved in sequential fashion. Indications of the presence of fluids of magmatic origin in the chemistry of the fluids associated either with the wolframite, or (and above all) with the cassiterite (see Figure 3.29(c)) evidently point in the same direction. It makes sense however to note that other geochemical indicators relating to the same minerals (the composition of volatile phases, stable isotopes) do not confirm this involvement as clearly: it therefore seems that these magmatic fluids are never involved alone, nor even, usually, dominantly.

3.7. Role of the Panasqueira granites

As has been seen repeatedly, most of the authors who have focused on the Panasqueira deposit have given the greisenized cupola a decisive role as the source of the magmatic fluids. It has just been seen that the mass balances are not compatible with purely magmatic sources of the fluids involved (section 3.6.3). It remains possible nonetheless to envisage a contribution of the Panasqueira granite in the formation of the deposit, and in particular of the G1–G4 suite of its apical part: to do so, there is a need to consider the alterations affecting this suite, and their possible temporal relationships with the stages of mineralization.

Firstly, Marignac et al. (2020) demonstrated that the alteration of the G4 cupola, far from being a classical greisenization, in reality results from the superposition of two distinct alteration processes. It makes sense to recall here that a strict definition of the "greisen" makes it the exact equivalent of the phyllitic alterations of the copper-bearing porphyry systems, that is, of an acid leaching whose emblematic reaction is one that affects the potassium feldspar:

$$3\ K[AlSi_3O_8] + 2\ H^+ = KAl_2[AlSi_3O_{10}(OH)_2 + 6\ SiO_2 + 2\ K^+$$

$$(3\ \text{orthose} + 2H^+ = \text{muscovite} + 6\ \text{quartz} + 2\ K^+)$$

The result is therefore a quartz-muscovite metasomatic rock, in which the texture of the protolith is more or less preserved, and where it is possible to observe remains of the latter's minerals. At first glance, the rocks forming the G4 cupola meet this definition: an assemblage of quartz and muscovite, with albitic remains. Nevertheless, study of the textures shows that the texture of the granite has not really been preserved. In reality, the alteration at the origin of the destruction of the albites is a silicification, which leaves behind only dentelliform ghosts of the initial plagioclase, without muscovite, and remains of phenocrysts of the magmatic quartz (but not of the quartz of the granular matrix of the protolith), as the images in cathodoluminescence show. It is therefore an ultra-acid alteration, at the origin of a "quartz-greisen", if you wish. The current state of the rocks of G4 cupola results from a muscovitization (in fact, from the development of a Fe-Li–muscovite), that is, from the dissolution–replacement of the quartz from the first alteration: this second alteration, with unstable quartz, is technically an episyenitization, which moreover leads locally to pockets or muscovite (sometimes rich in cassiterite). The final result is therefore in reality a "pseudo-greisen", which results, in an appropriate diagram (Figure 3.8) in a trend orthogonal to that of an authentic greisenization.

It can easily be shown that it is the muscovite from this final episyenitization that is equivalent to the selvage muscovites of the quartz-wolframite veins (Carocci 2019; Marignac et al. 2020): we are therefore a long way from the magmatic-hydrothermal model as conceived by Launay et al. (2018).

In addition, the G1 granite, which moderately bears the imprint of this "episyenitization", has undergone an initial phase of alteration, affecting the protolithionite, which is transformed (more or less completely) into a muscovite in the strict sense, accompanied by siderite, a rutile and sometimes by an iron chlorite. The superimposition on this early alteration of the main phase of muscovitization can be observed very clearly in the G1 granite.

The existence of high levels of Nb, Ta, W in the rutile shows that these metals, contained in the protolithionite, have been mobilized over the course of the first alteration phase. Considering the current W levels in the G1 granite (13 ppm), this makes this alteration a potential source of tungsten: a summary calculation shows that the 32 km^3 of the Panasqueira pluton represent 1 Mt W, which would be more than enough to form the deposit and its halo (Marignac et al. 2020). Naturally, the complete leaching of this volume and content level is not a realistic option. If we suppose more realistically that no more than 2–5 ppm of tungsten is extracted, the total volume of the pluton is broadly enough to provide 150 kt of tungsten accumulated in the mineralized veins, but not the additional 350 kt that Polya (1989) calculated to be stored in the deposit's halo. This statement becomes worse if we consider no more than a layer of 500 m thickness (which could reasonably be thought to be formed of G1 granite): only the vein system could therefore have been supplied (Marignac et al. 2020). It can therefore be concluded both that the Panasqueira granite represents a potential tungsten reservoir likely to have contributed significantly to the genesis of the deposit, but that a further contribution is needed in any case: this contribution may have been provided by the metamorphic fluids, whose capacity to transport and deposit the tungsten is additionally well known (Cave et al. 2016).

At this stage, it is essential to know the moment where the scavenging of this tungsten could have taken place. The G1 granite was probably representative of the granitic massif pictured by Ribeiro (2017), it therefore predates the start of the hydrothermal system's activity (section 3.2.2). Since the fluids associated with the first alteration of this granite were aquo–carbonic, as the systematic presence of siderite shows, it is highly probable that these were the same fluids as those that, in the veins, are associated with the tungsten mineralization (section 3.4.1.2; Cathelineau et al. (2020)): it thus strongly suggests that the Panasqueira granite effectively served as a tungsten reservoir for the formation of the deposit. It can be noted that this was partly the perspective of Kelly and Rye (1979), who wrote: "At Panasqueira, we believe that the fluids entered the horizontal veins in the vicinity of the cupola after deep circulation through a more extensive granite" (Kelly and Rye 1979, p. 1795), while specifying however that: "The authors prefer a more active role for the batholith which may still have been crystallizing at depth during the Sn-W mineralization" (Kelly and Rye 1979, p. 1790). However, the data available show conversely that the granite can only have played a passive role, and that it was the metamorphic fluids (in a broad sense) that probably served as vectors for the leached metals in the magmatic body to supply the mineralizing system.

3.8. Panasqueira: a magmatic and hydrothermal system of crustal scale

A major characteristic of the Panasqueira hydrothermal system is that it was dominated by the C-O-H-N-NaCl system's fluids, which were renewed for more than 10 Ma (Figure 3.24) and circulated throughout this time at high temperatures comprised between 400 and 500°C (section 3.5). The data show furthermore that it was not "one" but "some" fluids that participated in this operation: it is possible to distinguish quite clearly a H_2O-CO_2-(CH_4) pole and a H_2O-CH_4-N_2 set, where either N_2 or CH_4 dominate, and the evolution of the system is rhythmed by the mixing, in varying proportions of all these components. Some trends emerge over time: this is how the volatiles of the fluids from the initial stages (I to III-A), which see the best part of the tungsten deposition, seem to be characterized by a CO_2-N_2-(CH_4) mixture (Figure 3.29(a)), whereas from stage III-B, a CO_2-CH_4 mixture rich in methane becomes a significant component, and even the main one at stage III-B (Figure 3.29(b)).

These fluids are metamorphic by nature, in a broad sense: they are metamorphogenic and "pseudo-metamorphic" fluids, in equilibrium at depth with the metamorphic rocks. As for the early stages, Carocci et al. (2021) and Cathelineau et al. (2020) show that these fluids have a deep source, and that they are rooted in an active metamorphic zone deep in the Panasqueira granite. The participation of noble gases of mantellic origin in the composition of the stage IV fluids (Burnard and Polya 2004) demonstrates that this situation was prolonged: these gases probably in fact take their source in the mafic magmas whose interaction with the deep crust determines the development of thermal anomalies leading to the anatexis. It is sensible to recall here that Kelly and Rye (1979) already thought that the fluids mixtures are produced before their arrival in the vein system, and that we are dealing with "a very large hydrologic system that probably extended to substantial depths and [...] beyond the limits of currently known, economic veinage" (Kelly and Rye 1979, p. 1814), an opinion that fell by the wayside for half a century.

At stages I and II, the interaction of these fluids of deep origin with the already formed Panasqueira granitic pluton allowed the leaching of the magmatic tungsten (stored in the lithiniferous micas) and the partial supply of tungsten to the mineralized system, the addition potentially being provided by the metamorphic fluids (section 3.6).

The participation of magmatic fluids, little visible at the early stages, becomes perceptible however at stages III-A and IV, especially on the basis of paragenetic data. Geochemical data confirm this, while still showing that these fluids are never involved alone (section 3.6.2). It seems likely that this intervention was in particular at the origin of the introduction of tin into the system.

This observation implies that in addition to the Panasqueira pluton and the G1–G4 complex which forms its apical part, and which preceded (perhaps, by a little) initialization of the hydrothermal system, other granitic intrusions came to participate in the magmatic edifice. On the basis of the signatures observed in the mineral fillings (Al-F-Sn-P and Al-F-Li-Sn-P), these intrusions are probably related to the peraluminous RMGs as defined by Linnen and Cuney (2005). They remain unknown to this date, but it is likely that the topaz-cassiterite axis identified in the western part of the deposit (Figure 3.3) marks the location of the one associated with stage III-A (Pinto et al. 2014). Lourenço (2002) estimated that it was the Barroca Grande (G4) cupola that was responsible for the acquisition of tin at its stage EOS2 (i.e. for the stage IV cassiterite, Figure 3.9). This perspective is not acceptable, since Marignac et al. (2020) show that the G1–G4 suite was already in place at the moment when the hydrothermal system was initiated. However, an early alteration of the G4 cupola at one moment, that of the I-A tourmalinization, when the greatest part of the massif was not yet affected, may explain the Nb contribution observed by Carocci et al. (2021) in the early rutiles, and the pursuit of this alteration at stage I-B could consider the Nb enrichments described by Polya (1988a) in the wolframites.

It therefore seems clear that the Panasqueira hydrothermal system does not have the characteristics expected of a simple magmatic system. It emerges from these considerations that we are dealing with a mixed system, combining the attributes of a high temperature hydrothermal system as defined by Marignac and Cathelineau (2009) and those of several "more classic" magmatic-hydrothermal systems as associated with the RMGs.

The participation of meteoric water, which has often been mentioned to explain the composition of the fluids or to take account of the volumes of fluids involved, is not really visible in the available data. This is particularly true of the early stages and has been attributed by Carocci et al. (2021) and Cathelineau et al. (2020) to the isolating effect of the thick Ordovician quartzites cover. These are highly impermeable and the anticlinorial architecture therefore determines a vast structural trap where the fluids could accumulate and remain confined: equally the fluids produced at the structural level of the system (including the fluids resulting from the contact metamorphism), and those fluids of deeper origin; hence the excess pressures responsible for the openings of the vein system (Figure 3.31).

Carocci et al. (2021) proposed applying Connolly's model (1989) at stage I-A. According to this model, the deep fluids resulting from dehydration of the biotite gather in a collection zone, and then are concentrated and channeled into a meso-crustal fault zone, rising rapidly to a dispersion region in the cooler rocks. The half-width of the collection and dispersion zones in this model is typically in the order of 3–5 km, and the height of the system is in the order of 10–15 km. The fluid circulation causes upward thermal advection, with a time scale of around 10^4 years, leading to an increase in temperature of several tens of °C on a time scale of 100 years in the dispersion region. The model is well adapted to the Panasqueira system (Figure 3.31), its dimensions are especially compatible with the volumes required to produce the fluids involved in the system (see above, section 3.6.3) and it explains well the geotherms of more than 40°C·km^{-1} recorded by the FIs (Figure 3.28).

The model involves a structural check. Several indices converge to indicate a control in the NNW–SSE direction: elongation of the root of the Panasqueira pluton (Ribeiro 2017), the main direction of the mine's Seixo Bravos, the mafic dikes (Lourenço 2002) and the aplitic dikes (Figure 3.3) and an elongation of the proximal hydrothermal halo (Figure 3.7). Moreover, the fluid ascension will have been facilitated by the vertical permeability in the same direction (S0 and schistosity) at the core of the Panasqueira anticlinal. It is therefore logical to postulate the existence at depth of an accident crustal of the same direction, conditioning the operation of the whole of the hydrothermal system. This damaged zone probably played a role starting from the Seixo Bravos stage, and certainly conditioned the location of the Panasqueira pluton.

The model can be extended at stages I-B, II and III, at this latter stage with the interaction with the rise, along the same structure, of an RMG type granitic magma (Figure 3.31(c)), leading to an increased thermal advection, well recorded by the FIs (Figure 3.28).

At the start of the system's operation, the fluids ascending from the collection region encountered the Panasqueira granitic body at a level corresponding approximately to the base of the dispersion level as defined by Connolly (1989). According to Carocci et al. (2021), these fluids did not react with the granite: it might be thought that this is due to the fact that the granite, which was already solid, was still too hot for the fluids to be able to cross the thermal boundary layer. The ascending fluid was therefore forced around the granitic body. It was only at stage

I-B that the fluids could begin to invade the granitic mass, which coincided with the start of the main mineralization.

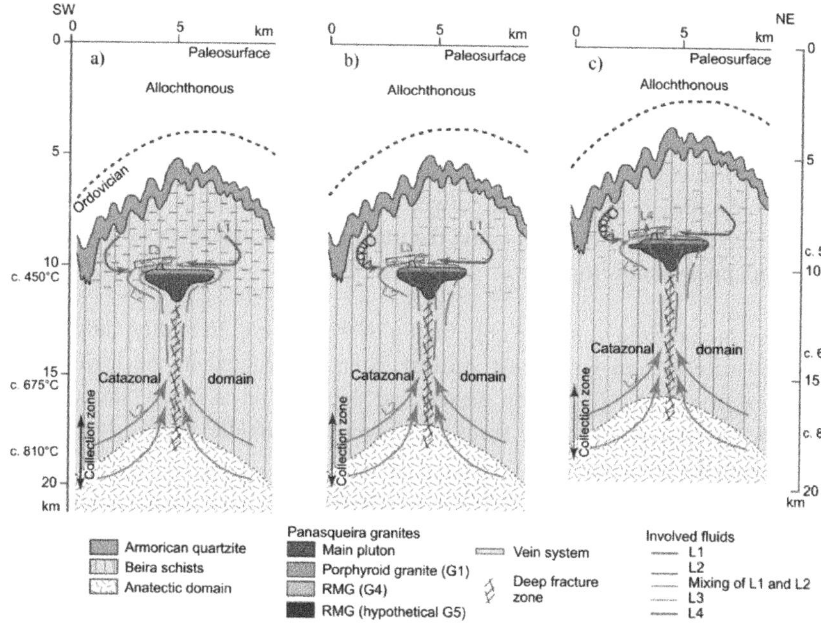

Figure 3.31. *Model for the early stages of the Panasqueira hydrothermal and magmatic system (adapted from Cathelineau et al. (2020) and Carocci et al. (2021)). The model is based on designs by Connelly (1989)*

COMMENT ON FIGURE 3.31.– *Throughout the system's evolution, two fundamental fluids participated in its evolution: L1 was an "inherited" metamorphic fluid trapped by the impermeable Armorican quartzite cover and set in motion by convection; L2 was a deeper fluid, issuing from an underlying anatectic domain, channeled by an area of crustal weakness, which also enabled the ascension of the Panasqueira granite; their mixing, in proportions that varied over time, nourished the creation and evolution of the vein system. (a) At stage I-A, L2 did not penetrate the main pluton, and only the G4 RMG G4 cupola was partially percolated by external fluids to supply the system on occasion with a fluid (L3) rich in Nb. (b) At stage I-B (and at stage II), the granitic massif was percolated and its alteration contributed to the tungsten budget of the fluids in the vein system; at the end of the alteration of the G4 cupola, a niobiferous fluid (L3) still contributed. (c) At stage III-A, a new RMG intrusion (the hypothetical G5) came to contribute significantly to the fluid budget with a magmatic-hydrothermal fluid (L4) that provided tin in particular. Additional explanations are in the text.*

The more or less continuous advection of heat determined by the magmatic and hydrothermal flows guided by the damaged zone explains the permanence of the high temperatures in the mineralized system, at the origin of the start of the forced convection of fluids trapped in the structural trap formed by the Ordovician quartzite vault. The recycling caused by such convection would reduce the total volume of fluids involved by the same amount, and the quantities calculated above would only represent an upper bound from this.

Thus, the mineralized veins are only the tip of the iceberg. In reality, they belong to a crustal-scale system with great vertical extension, rooted in the lower crust, prolonged over about 10 million years and located on a deep structure, or alternatively, fluids and magmas from the lower and middle crust combine and become involved.

In such a system, the origin of the fluids and metals is marked by a strong metamorphic imprint, with the fluids resulting from the dehydration of the crust playing a major role, whether they carry metals directly or whether they interact with the magmatic intrusions or with the fluids issuing directly from these intrusions to become mineralizers.

3.9. References

Antunes, I.M.H.R., Neiva, A.M.R., Silva, M.M.V.G. (2010). Isotopic geochronology of granitic rocks from the central Iberian zone: Comparison of methodologies. *Estud. Geol.*, 66, 45–50.

Audétat, A. (2019). The metal content of magmatic-hydrothermal fluids and its relationship to mineralization potential. *Econ. Geol.*, 114, 1033–1056.

Barbero, L., Glasmacher, U., Villaseca, A.C., García, J.A.L., Martín-Romera, C. (2005). Long-term thermo-tectonic evolution of the Montes de Toledo area (Central Hercynian Belt, Spain): Constraints from apatite fission-track analysis. *Int. J. Earth Sci.*, 94, 193–203.

Barton, P.B., Bethke, P.B., Philip, M. (1987). Chalcopyrite disease in sphalerite; pathology and epidemiology. *Amer. Mineral.*, 72, 451–467.

Bebout, G.E. and Fogel, M.L. (1992). Nitrogen-isotope compositions of metasedimentary rocks in the Catalina Schist, California: Implications for metamorphic devolatilization history. *Geochim. Cosmochim. Acta*, 56, 2839–2849.

Bloot, C. and de Wolf, L.C.M. (1953). Geological features of the Panasqueira tin-tungsten ore occurrence (Portugal). *Soc. Geol. Portugal Bol.*, 11, 1–58.

Boiron, M.C., Cathelineau, M., Banks, D.A., Yardley, B.W.D., Noronha, F., Miller, M.F. (1996). P-T-X conditions of late Hercynian fluid penetration and the origin of granite-hosted gold quartz veins in northwestern Iberia: A multidisciplinary study of fluid inclusions and their chemistry. *Geochim. Cosmochim. Acta*, 60, 43–57.

Burnard, P.A. and Polya, D.A. (2004). Importance of mantle derived fluids during granite associated hydrothermal circulation: He and Ar isotopes of ore minerals from Panasqueira. *Geochim. Cosmochim. Acta*, 68, 1607–1615.

Bussink, R.W. (1984). Geochemistry of the Panasqueira tungsten-tin deposit, Portugal. *Geologica Ultraiectina*, 33, 1–159.

Bussink, R.W., Kreulen, R., Arie, F.M., de Jong, A.F.M. (1984). Gas analyses, fluid inclusions and stable isotopes of the Panasqueira W-Sn deposits, Portugal. *Bull. Minéral.*, 107, 703–713.

Campbell, A.R. and Robinson-Cook, S. (1987). Infrared fluid inclusion microthermometry on coexisting wolframite and quartz. *Econ. Geol.*, 82, 1640–1645.

Carocci, E. (2019). Transport et dépôt du tungstène dans les environnements magmatiques-hydrothermaux : exemple de Panasqueira (Portugal). PhD Thesis, Université de Lorraine.

Carocci, E., Marignac, C., Cathelineau, M., Truche, L., Lecomte, A., Pinto, F. (2018). Rutile from Panasqueira (Central Portugal): An excellent pathfinder for wolframite deposition. *Minerals*, 9(9). doi: 10.3390/min9010009.

Carocci, E., Marignac, C., Cathelineau, M., Truche, L., Poujol, M., Boiron, M.-C., Pinto, F. (2021). Incipient wolframite deposition at Panasqueira (Portugal): W rutile and tourmaline compositions as proxies for the early fluid composition. *Econ. Geol.*, 115. doi: 10.5382/econgeo.4783.

Cathelineau, M., Boiron, M.C., Fourcade, S., Ruffet, G., Clauer, N., Belcourt, O., Coulibaly, Y., Banks, D.A., Guillocheau, F. (2012). A major Late Jurassic fluid event at the basin/basement unconformity in western France: 40Ar/39Ar and K–Ar dating, fluid chemistry, and related geodynamic context. *Chem. Geol.*, 322–323, 99–120.

Cathelineau, M., Marignac, C., Boiron, M.C., Carocci, E., Rolland, J.-M., Dejean, M., Dourk, M., Pinto, F. (2018). Nouvelle approche du chemin P-T-t des fluides dans le gisement à W–Sn–Cu de Panasqueira (Portugal) : persistence d'une anomalie thermique d'origine profonde en cours d'exhumation. In *Proceedings of the 26th Earth Science Meeting (RST)*, 694. Lille.

Cathelineau, M., Boiron, M.-C., Marignac, C., Dour, M., Dejean, M., Carocci, E., Truche, L., Filipe Pinto, F. (2020). High pressure and temperatures during the early stages of tungsten deposition at Panasqueira revealed by fluid inclusions in topaz. *Ore Geol. Rev.*, 126, 103741.

Cave, B.J., Pitcairn, I.K., Craw, D., Large, R.R., Thompson, J.M., Johnson, S.C. (2016). A metamorphic mineral source for tungsten in the turbidite-hosted orogenic gold deposits of the Otago Schist, New Zealand. *Mineral. Dep.*, 52(4), 515–537.

Černý, P., Blevin, P.L., Cuney, M., London, D. (2005). Granite-related ore deposits. *Econ. Geol.*, 100, 337–370.

Clark, A.H. (1964). Preliminary study of the temperatures and confining pressures of granite emplacement and mineralization, Panasqueira, Portugal. *Trans. Inst. Mining Metall.*, 73, 813–824.

Clark, A.H. (1965). Rhomboedral molybdenite from the Minas da Panasqueira, Beira Baixa, Portugal. *Min. Mag.*, 34, 69–71.

Clark, A.H. (1970). Potassium-argon age and regional relationships of the Panasqueira tin-tungsten mineralization. *Com. Serv. Geol. Portugal*, 54, 243–261.

Cloetingh, S., Ziegler, P.A., Beekman, F., Andriessen, P.A.M., Matenco, L., Bada, G., Garcia-Castellanos, D., Hardebol, N., Dèzes, P., Sokoutisa, D. (2005). Lithospheric memory, state of stress and rheology: Neotectonic controls on Europe's intraplate continental topography. *Quartern. Sci. Rev.*, 24, 241–304.

Codeço, M.S., Weis, P., Trumbull, R.B., Pinto, F., Lecumberri-Sanchez, P., Wilke, F.D.H. (2017). Chemical and boron isotopic composition of hydrothermal tourmaline from the Panasqueira W-Sn-Cu deposit, Portugal. *Chem. Geol.*, 468, 1–16.

Codeço, M.S., Weis, P., Trumbull, R.B., Glodny, J., Wiedenbeck, M., Rolf, L. Romer, R.L. (2019). Geothermometry indicates fluid cooling during magmatic-hydrothermal W-Sn ore formation boron isotope muscovite-tourmaline. *Econ. Geol.*, 114, 153–163.

Codeço, M.S., Weis, P., Trumbull, R.B., Van Hinsberg, V., Pinto, F., Lecumberri-Sanchez, P., Schleicher, A.M. (2020). The imprint of hydrothermal fluids on trace-element contents in white mica and tourmaline from the Panasqueira W–Sn–Cu deposit, Portugal. *Mineral. Dep.*, 55 (9). doi: 10.1007/s00126-020-00984-8.

Connolly, J.A.D. (1989). Mid-crustal focused fluid movement: Thermal consequences and silica transport. In *Fluid Flow and Transport in Rocks, Mechanisms and Effects*, Jamtveit, B. and Yardley, B.W.D. (eds). Chapman & Hall, London.

D'Orey, F.C. (1967). Tungsten-tin mineralization and paragenesis in the Panasqueira and Vale da Ermida mining districts, Portugal. *Com. Serv. Geol. Portugal*, 52, 117–167.

De Ascenção Guedes, R. (2002). Le Coteau minier de Panasqueira, Beira Baixa (Portugal). *Le Règne Minéral*, 43, 6–32.

Debon, F. and Le Fort, P. (1988) A cationic classification of common plutonic rocks and their magmatic associations: Principles, method, applications. *Bull. Minéral.*, 111, 493–510.

Derré, C., Lecolle, R., Roger, G., Tavares de Freitas Carvalho, J. (1986). Tectonics, magmatism, hydrothermalism and sets of flat joints locally filled by Sn-W aplite-pegmatite and quartz veins; southeastern border of the Serra de Estrela granitic massif (Beira Baixa, Portugal). *Ore Geol. Rev.*, 1, 43–56.

Dias, G., Leterrier, J., Mendes, A., Simões, P., Bertrand, J.M. (1998). U–Pb zircon and monazite geochronology of syn- to post-tectonic Hercynian granitoids from the central Iberian zone (northern Portugal). *Lithos*, 45, 349–369.

Díez Fernández, R. and Pereira, M.F. (2017). Strike-slip shear zones of the Iberian Massif: Are they coeval? *Lithosphere*, 9, 726–744.

Díez Fernández, R., Arenas, R., Pereira, M.F., Sánchez-Martínez, S., Albert, R., Martin Parra, L.M., Rubio Pascual, F.J., Matas, J. (2016). Tectonic evolution of Variscan Iberia: Gondwana–Laurussia collision revisited. *Earth-Sci. Rev.*, 162, 269–292.

Fernández-Suárez, J., Gutiérrez-Alonso, G., Johnston, S.T., Jeffries, T.E., Pastor-Galan, D., Jenner, G.A., Murphy, J.B. (2011). Iberian late-Variscan granitoids: Some considerations on crustal sources and the significance of "mantle extraction ages". *Lithos*, 123, 121–132.

Foxford, K.A., Nicholson, R., Polya, D.A. (1991). Textural evolution of W-Cu-Sn bearing hydro-thermal quartz veins at Minas da Panasqueira, Portugal. *Min. Mag.*, 55, 435–445.

Foxford, K.A., Nicholson, R., Polya, D.A., Hebblethwaite, R.P.B. (2000). Extensional failure and hydraulic valving at Minas da Panasqueira, Portugal: Evidence from vein spatial distributions, displacements and geometries. *J. Struct. Geol.*, 22, 1065–1086.

Gonçalves, M.A., Mateus, A., Pinto, F., Vieira, R. (2018). Using multifractal modelling, singularity mapping, and geochemical indexes for targeting buried mineralization: Application to the W-Sn Panasqueira ore-system, Portugal. *J. Geochem. Expl.*, 189, 42–53.

Harlaux, M., Mercadier, J., Wilédio, M.E.B., Kremer, V., Marignac, C., Cuney, M. (2017). Geochemical signature of magmatic-hydrothermal fluids exsolved from the Beauvoir rare-metal granite (Massif Central, France): Insights from LA-ICPMS analysis of primary fluid inclusions. *Geofluids*, 1925817.

Hebblethwaite, R.B. and Antão, A.M. (1982). A report on the study of dilation patterns within the Panasqueira orebody. Report, Beralt Tin and Wolfram S.A., Portugal.

Huizenga, J.M. (2001). Thermodynamic modelling of C-O–H fluids. *Lithos*, 55, 101–114.

Isaacs, A.M. and Peacor, D.R. (1981). Panasqueiraite, a new mineral: OH-equivalent of isokite. *Can. Mineral.*, 19, 389–392.

Isaacs, A.M., Peacor, D.R., Kelly, W.C. (1979). Thadeuite, Mg(Ca,Mn) $(Mg,Fe,Mn)_2$ $(PO4)_2(OH, F)_2$, a new mineral from Panasqueira, Portugal. *Am. Mineral.*, 64, 359–361.

Jaques, L. and Pascal, C. (2017). Full paleostress tensor reconstruction using quartz veins of Panasqueira Mine, central Portugal; Part I: Paleopressure determination. *J. Struct. Geol.*, 102, 58–74.

Jacques, D., Vieira, R., Muchez, P., Sintubin, M. (2018). Transpressional folding and associated cross-fold jointing controlling the geometry of post-orogenic vein-type W-Sn mineralization: Examples from Minas da Panasqueira, Portugal. *Mineral. Dep.*, 53, 171–194.

Juez-Larré, J. and Ter Voorde, M. (2009). Thermal impact of the break-up of Pangea on the Iberian Peninsula, assessed by thermochronological dating and numerical modeling. *Tectonophysics*, 474, 200–213.

Kelly, W.C. and Rye, R.O. (1979). Geologic, fluid inclusion, and stable isotope studies of the tin-tungsten deposits of Panasqueira, Portugal. *Econ. Geol.*, 74, 1721–1822.

Knutson, C., Peacor, D.R., Kelly, W.C. (1985). Luminescence, color and fission track zoning in apatite crystals of the Panasqueira tin-tungsten deposit, Beira-Baixa, Portugal. *Amer. Mineral.*, 70, 829–837.

Kretschmar, U. and Scott, S.D. (1976). Phase relations involving arsenopyrite in the system Fe-As-S and their application. *Canad. Mineral.*, 14, 364–386.

Kuhn, T. (2018). Al-in-Hornblende barometry of Southern New England intrusions and comparison with metamorphic bathograds. PhD Thesis, Yale University, New Heaven.

Launay, G. (2018) Hydrodynamique des systèmes minéralisés péri-granitiques : étude du gisement à W-Sn-(Cu) de Panasqueira (Portugal). PhD Thesis, Université d'Orléans.

Launay, G., Sizaret, S., Guillou-Frottier, L., Gloaguen, E., Pinto, F. (2018). Deciphering fluid flow at the magmatic-hydrothermal transition: A case study from the world-class Panasqueira W–Sn–(Cu) ore deposit (Portugal). *Earth Planet. Sci. Lett.*, 499, 1–12.

Launay, G., Sizaret, S., Guillou-Frottier, L., Fauguerolles, C., Champallier, R., Gloaguen, E. (2019). Dynamic permeability related to greisenization reactions in Sn-W ore deposits: Quantitative petrophysical and experimental evidence. *Geofluids*, 5976545.

Lecumberri-Sanchez, P., Vieira, R., Heinrich, C.A., Pinto, F., Wälle, M. (2017). Fluid-rock interaction is decisive for the formation of tungsten deposits. *Geology*, 45, 579–582.

Leuthold, J., Müntener, O., Baumgartner, L.P., Putlitz, B., Ovtcharova, M., Schaltegger, U. (2012). Time resolved construction of a bimodal laccolith (Torres del Paine, Patagonia). *Earth Planet. Sci. Lett.*, 325–326, 85–92.

Lima, L., Guedes, A., Noronha, F. (2019). Tungsten mineralization associated with the Argemela microgranite (Central Portugal). *J. Iber. Geol.*, 45, 625–640.

Linnen, R.L. and Cuney, M. (2005). Granite-related rare-element deposits and experimental constraints on Ta-Nb-W-Sn-Zr-Hf mineralization. In *Rare-element Geochemistry and Mineral Deposits*, Linnen, R.L. and Samson, I.M. (eds). Geological Association of Canada.

Lourenço, A.O. (2002). Paleofluidos e mineralizações associadas às fases tardias da Orogenia Hercínica. PhD Thesis, University of Porto.

Lourenço, A.O. (2006). Granito da Panasqueira (Cúpula da Barroca Grande). Estudo de um sistema granítico associado a mineralizações de estanho. The Panasqueira granite (Barroca Grande cupola) (Study of a Granite system associated with tin lodes). In *Proceedings of the VIIth Congresso Nac. Geol.* University of Évora.

Lüders, V. (1996). Contribution of infrared microscopy to fluid inclusion studies in some opaque minerals (wolframite, stibnite, bournonite): Metallogenic implications. *Econ. Geol.*, 91, 1462–1468.

Marignac, C. (1973). Analyse structurale de l'environnement du gisement à tungstène-étain de Panasqueira (Beira-Baixa, Portugal) ; implications génétiques. *C. R. Acad. Sci.*, 277(50), 269–272.

Marignac, C. and Cathelineau, M. (2009). The nature of ore-forming fluids in peri-batholitic Sn-W deposits and a classification. In *Proceedings of the Tenth Biennial SGA Meeting: Smart Science for Exploration and Mining*, Williams, P.J. (ed.). James Cook University, Townsville.

Marignac, C., Cuney, M., Cathelineau, M., Lecomte, A., Carocci, E., Pinto, F. (2020). The Panasqueira rare metal granite suites and their involvement in the genesis of the world-class Panasqueira W–Sn–Cu vein deposit: A petrographic, mineralogical, and geochemical study. *Minerals*, 10(6), 562.

Martins, I. (2017). Microanálise elementar de sulfuretos, sulfossais e carbonatos constituintes das associações polifásicas que preenchem os sistemas filonianos de W(-Sn) suporte da mina da Panasqueira (Portugal); reconhecimento de marcadores metalogenéticos. PhD Thesis, University of Lisbon.

Martins, I., Mateus, A., Figueiras, J., Rodrigues, P., Vieira, R., Filipe Pinto, F. (2017). Paragenetic meaning of the Bi- and Ag-rich phases from the Panasqueira lode system (Portugal). Conference, Goldschmidt.

Martins, I., Mateus, A., Figueiras, J., Rodrigues, P., Pinto, F. (2020). Thermal evolution of the W-Sn(-Cu) Panasqueira ore system (Portugal): Insights from pyrite-pyrrhotite and arsenopyrite geothermometers (Evolução térmica do sistema mineralizado W-Sn(-Cu) da Panasqueira (Portugal): Contributos dos geotermómetros pirite-pirrotite e arsenopirite). *Com. Geol.*, 107(2), 69–74.

Mateus, A. and Noronha, F. (2010). Sistemas mineralizantes epigenéticos na Zona Centro-Ibérica; ex-pressão da estruturação orogénica meso- a tardi-varisca (Epigenetic ore-forming systems in Central-Iberian Zone; products of an evolving orogenic framework in meso- to late-Variscan times). *Ciências Geológicas, Ensino, Investigação e sua História*, 2, 47–61.

Mateus, A., Figueiras, J., Martins, I., Rodrigues, P.C., Pinto, F. (2020). Relative abundance and compositional variation of silicates, oxides and phosphates in the W-Sn-rich lodes of the Panasqueira mine (Portugal): Implications for the ore-forming process. *Minerals*, 10(6), 551.

Michaud, J.A.S., Gumiaux, C., Pichavant, M., Gloaguen, E., Marcoux, E. (2020). From magmatic to hydrothermal Sn-Li-(Nb-Ta-W) mineralization: The Argemela area (central Portugal). *Ore Geol. Rev.*, 116, 103215.

Michel, J., Baumgartner, L., Putlitz, B., Schaltegger, U., Ovtcharova, M. (2008). Incremental growth of the Patagonian Torres del Paine laccolith over 90 ky, Patagonia. *Geology*, 36, 459–462.

Neiva, A.M.R. (2002). Portuguese granites associated with Sn-W and Au mineralizations. *Bul. Geol. Soc. Finl.*, 74, 79–101.

Neiva, A.M.R., Neiva, J.M.C., Parry, S.J. (1987). Geochemistry of the granitic rocks and their minerals from Serra da Estrela, Central Portugal. *Geochim. Cosmochim. Acta*, 51, 439–454.

Noronha, F. (2017). Fluids and Variscan metallogenesis in granite related systems in Portugal. Water-Rock Interaction International Symposium. *Procedia Earth Planet. Sci.*, 1, 1–4.

Noronha, F., Doria, A., Dubessy, J., Charoy, B. (1992). Characterization and timing of the different types of fluids present in the barren and ore-veins of the W-Sn deposit of Panasqueira. *Central Portugal. Mineral. Dep.*, 27, 72–79.

Oosterom, M.G., Bussink, R.W., Vriend, S.P. (1984). Lithogeochemical studies of aureoles around the Panasqueira tin-tungsten deposit, Portugal. *Mineral. Dep.*, 19, 283–288.

Orville, P.M. (1963). Alkali ion exchange between vapour and feldspar phases. *Amer. J. Sci.*, 261, 201–237.

Pattison, D.R.M. (2001). Instability of Al2O5 'triple point' assemblages in muscovite+biotite+quartz bearing metapelites, with implications. *Amer. Mineral.*, 86, 1414–1422.

Pereira, A.J.S.C., Carter, A., Hurford, A.J., Neves, L.P.J.F., Godinho, M.M. (1998). Evidence for the unroofing history of Hercynian granitoids in Central Portugal derived from Late Palaeozoic and Mesozoic sedimentary zircons. In *Advances in Fission-Track Geochronology*, Van den Haute, P. and de Corte, F. (eds). Springer, Dordrecht.

Pereira, M.F., Díez Fernández, R., Gama, C., Hofmann, M., Gärtner, A., Linnemann, U. (2018). S-type granite generation and emplacement during a regional switch from extensional to contractional deformation (central Iberian zone, Iberian autochthonous domain, Variscan Orogeny). *Int. J. Earth Sci.*, 107, 251–267.

Pinto, F. (2014). Estudo da distribuição do Estanho na Mina da Panasqueira. PhD Thesis, University of Porto.

Pinto, F., Vieira, R., Ferraz, P., Noronha, F. (2014). Estudo da distribuição espacial do Sn no Jazigo de W-Sn(Cu) da Panasqueira (Study of Sn spatial distribution on W-Sn(Cu) Panasqueira ore deposit). *Com. Geol.*, 101(2), 829–832.

Pinto, F., Vieira, R., Noronha, R. (2015). Different cassiterite generations at the Panasqueira deposit (Portugal): Implications for the metal zonation model. In *Mineral Resources in a Sustainable World, Proceedings of the 13th SGA Biennial Meeting*. Nancy.

Pirard, E. (1987). L'argent dans les sulfures des filons à Sn-W de Panasqueira (Portugal). *Ann. Soc. Géol. Belgique*, 110, 371–377.

Polya, D.A. (1988a). Compositional variation in wolframites from the Barroca Grande mine, Portugal: Evidence for fault-controlled ore formation. *Min. Mag.*, 52, 497–503.

Polya, D.A. (1988b). Efficiency of hydrothermal ore formation and the Panasqueira W-Cu(Ag)-Sn-vein deposit. *Nature*, 333, 838–841.

Polya, D.A. (1989) Chemistry of the main-stage ore-forming fluids of the Panasqueira W-Cu(Ag)-Sn deposit, Portugal: Implications for models of ore genesis. *Economic Geology*, 84, 1134–1152.

Polya, D.A., Foxford, K.A., Stuart, F., Boyce, A., Fallick, A.E. (2000). Evolution and paragenetic context of low δD hydrothermal fluids from the Panasqueira W-Sn deposit, Portugal: New evidence from microthermometric, stable isotope, noble gas and halogen analyses of primary fluid inclusions. *Geochim. Cosmochim. Acta*, 64, 3357–3371.

Ribeiro, R.F. (2017). Gravimetric modelling and geological interpretation of Argemela-Panas-queira area. PhD Thesis, University of Porto.

Ribeiro, A. and Pereira, E. (1982). Controles paleogeograficos, patrologicos e estructurais na genese dos jazigos portugueses de estanho e volframio. *Geonovas*, 1(3), 23–31.

Roda-Robles, E., Villaseca, C., Pesquera, A., Gil-Crespoa, P.P., Vieira R., Lima, A., Garate-Olave, I. (2018). Petrogenetic relationships between Variscan granitoids and Li-(F-P)-rich aplite-peg-matites in the central Iberian zone: Geological and geochemical constraints and implications for other regions from the European Variscides. *Ore Geol. Rev.*, 95, 408–430.

Sant'Ovaia, H., Olivier, P., Ferreira, N., Noronha, F., LeBlanc, D. (2010). Magmatic structures and kinematics emplacement of the Variscan granites from Central Portugal (Serra da Estrela and Castro Daire areas). *Geology*, 10, 1450–1465.

Schermerhorn, L.J.G. (1981). Framework and evolution of Hercynian mineralization in the Iberian Meseta. *Leidse Geol. Mededel.*, 52, 23–56.

Schoell, M. (1980). The hydrogen and carbon isotopic composition of methane from natural gases of various origins. *Geochim. Cosmochim. Acta*, 44, 649–661.

Sheppard, S.M.F. (1994). Stable isotope and fluid inclusion evidence for the origin and evolution of Hercynian mineralizing fluids. Metallogeny of collisional orogens. In *Proc. IAGOD Erzgebirge Meeting, Geyer, 4–6 June, 1993, Czech Geological Survey*, Seltmann, R., Kampf, H., Moller, P. (eds). Prague.

Sheppard, S.M.F. and Charef, A. (1986). Eau organique : caractérisation isotopique et évidence de son rôle dans le gisement Pb-Zn de Fedj-el-Adoum, Tunisie. *C.R. Acad. Sci. Paris*, 302(2), 1189–1192.

Simon, A.C., Pettke, T., Candela, P.A., Piccoli, P.M., Heinrich, C.A. (2004). Magnetite solubility and iron transport in magmatic-hydrothermal environments. *Geochim. Cosmochim. Acta*, 68, 4905–4914.

Snee, L.W., Sutter, J.F., Kelly, W.C. (1988). Thermochronology of economic mineral deposits: Dating the stages of mineralisation at Panasqueira, Portugal, by high-precision 40Ar-39Ar age spectrum techniques on muscovite. *Econ. Geol.*, 83, 335–354.

Suzuoki, T. and Epstein, S. (1976). Hydrogen isotope fractionation between OH-bearing minerals and water. *Geochim. Cosmochim. Acta*, 40, 1229–1240.

Thadeu, D. (1951). Geologia do couto mineiro da Panasqueira. *Com. Serv. Geol. Portugal*, 32, 1–64.

Tornos, F., Delgado, A., Casquet, C., Galindo, C. (2000). 300 million years of episodic hydro-thermal activity: Stable isotope evidence from hydrothermal rocks of the eastern Iberian Central System. *Miner. Dep.*, 35, 551–569.

Toulmin III, P.T. and Barton Jr., P.B. (1964). A thermodynamic study of pyrite and pyrrhotite. *Geochimica et Cosmochimica Acta*, 28, 641–671.

Van den Kerkhof, A., Kronz, A., Simon, K. (2014). Deciphering fluid inclusions in high-grade rocks. *Geosci. Front.*, 5(5), 683–695.

Vázquez-Vílchez, M., Jabaloy-Sánchez, A., Azor, A., Stuart, F., Persano, C., Alonso-Chaves, F.M., Martín-Parra, L.M., Matas, J., García-Navarro, E. (2015). Mesozoic and Cenozoic exhumation history of the SW Iberian Variscides inferred from low-temperature thermochronology. *Tectonophysics*, 663, 110–121.

Vityk, M.O. and Bodnar, R.J. (1995). Textural evolution of synthetic fluid inclusions in quartz during re-equilibration, with applications to tectonic reconstruction. *Contrib. Mineral. Petrol.*, 121(3), 309–323.

Vityk, M.O. and Bodnar, R.J. (1998). Statistical microthermometry of synthetic fluid inclusions in quartz during decompression reequilibration. *Contrib. Mineral. Petrol.*, 132, 149–162.

Walther, J.V. and Orville, P.M. (1983). Volatile production and transport in regional metamorphism. *Contrib. Mineral. Petrol.*, 79, 252–257.

Wang, X.S., Timofeev, A., Williams-Jones, A.E., Shang, L.B., Bi X.W. (2019). An experimental study of the solubility and speciation of tungsten in NaCl-bearing aqueous solutions at 250, 300, and 350°C. *Geochim. Cosmochim. Acta*, 265, 313–329.

Weisbrod, A. (1984). Utilisation des inclusions fluides en géothermométrie. In *Thermométrie et barométrie géologiques*, Lagache, M. (ed.). Soc. Fr. Min. Cristal., Paris.

Wimmers, D. (1985). Silver minerals of Panasqueira, Portugal: A new occurrence of Te-bearing canfieldite. *Min. Mag.*, 49, 745–748.

Wood, S.A. (1992). Experimental determination of the solubility of $WO_3(s)$ and the thermodynamic properties of $H_2WO_4(aq)$ in the range 300–600°C at 1 kbar: Calculation of scheelite solubility. *Geochim. Cosmochim. Acta*, 56, 1827–1836.

Wood, S.A. and Samson, I.M. (2000). The hydrothermal geochemistry of tungsten in granitoid environments I: Relative solubilities of ferberite and scheelite as a function of T, P, pH, and mNaCl. *Econ. Geol.*, 95, 143–182.

Zhang, R., Ramos, V., Leal, S., Fernando Noronha, F., Pinto, F. (2019). U-Pb geochronology of cassiterites from primary Sn mineralizations in Sn-W Variscan Metallogenic Province, Portugal. In *Life with Ore Deposits on Earth, Proc. 15th SGA Biennial Meeting*, Glasgow.

Appendix 1

The Rare Earth Resources of Europe and Greenland: Mining Potential and Challenges

Nicolas CHARLES[1], Johann TUDURI[1,2], Gaétan LEFEBVRE[1], Olivier POURRET[3], Fabrice GAILLARD[2] and Kathryn GOODENOUGH[4]

[1] BRGM, Orléans, France

[2] ISTO, CNRS, University of Orléans, France

[3] AGHYLE, UniLaSalle, Beauvais, France

[4] British Geological Survey, Edinburgh, United Kingdom

A1.1. Table summarizing rare earth deposits listed in the paper written for this book

The table below illustrates the rare earth minerals found in Europe and Greenland. The data presented here include the type of mineral deposit, its age, the main commodities presented and used there, potentially interesting additional commodities and known or inferred reserves/resources as well as bibliographical references.

Name of the deposit/locality	Country	Type of deposit	Age	Main commodities	Other commodities	Known/inferred reserves/resources	References
Afrikanda	Russia	Ultrabasic-carbonatite complex	Upper Devonian	Fe, Ti, REE	–	34 Mt @ 0.67% REO	(Kalashnikov et al. 2016)
Aghios Ioannis	Greece	Laterite–Bauxite	Upper Jurassic	Al	REE	–	(Maksimović and Pantó 1996)
Alnö	Sweden	Alkaline complex	Neoproterozoic/ Ediacaran	REE, Nb, Zr	–	–	(Andersen 1996)
Aksu Diamas	Turkey	Paleoplacer	Quaternary/ Pleistocene	REE, Zr, Ri	–	454 Mt @ 0.07 Mt REO	(EURARE 2017)
Arran	United Kingdom (Scotland)	Felsic plutonism	Palaeogene/ Palaeocene	REE	–	–	(Hyslop et al. 1999)
Araxa	Brazil	Carbonatite laterite	Upper Cretaceous	P, Nb	REE	28.3 Mt @ 4.21 Mt REO	(Itafos)
Ardennes	Belgium	Authigenic/placers	Ordovician, Silurian?	REE	–	–	(Burnotte et al. 1989)
Atalandi (Atalanti)	Greece	Laterite–Bauxite	Upper Jurassic	Al	REE	–	(Bárdossy et al. 1976)
Attu	Greenland (Denmark)	Pegmatite	Palaeoproterozoic/ Statherian	REE	–	–	(GEUS 2014)
Bailleul	France	Paleoplacer	Lower Ordovician	Ti, Zr, REE	Th	3.5Mt @ 0.45% REO	(Lemarchand 1970)
Bastnäs	Sweden	Skarn	Palaeoproterozoic/ Orosirian	Fe, Cu, REE	–	–	(Holtstam 2004)
Bayan Obo	China	Carbonatite	Mesoproterozoic/ Ectasian-Calymnian	REE, Nb, Fe	–	118 Mt @ 6% REO	(Verplanck et al. 2016)
Ben Loyal	United Kingdom (Scotland)	Alkaline granite	Silurian/Wenlockian	–	–	–	(Walters et al. 2012)

Name of the deposit/locality	Country	Type of deposit	Age	Main commodities	Other commodities	Known/inferred reserves/resources	References
Biggejav'ri	Norway	Hydrothermal-metasomatic	Lower Palaeoproterozoic	Sc	U, REE	0.05 Mt @ 0.2% REO	(Olerud 1988)
Browns Range	Australia	Hydrothermal vein	Palaeoproterozoic/ Statherian	REE	–	9.26 Mt @ 0.67% REO	(Nazari-Dehkordi et al. 2018)
Brtná	Czech Republic	Paleoplacer	Ordovician	Ti, Zr, Th, REE	–	–	(Goliáš 2001)
Buckton	Canada	Authigenic/ shales	–	Ni-U-Zn-Cu-Co	REE	4544 Mt @ 0.02% REO	(Weng et al. 2015)
Central Wales	United Kingdom (Wales)	Authigenic/ placers	Silurian/Pridoli	REE	–	–	(Evans and Zalasiewicz 1996)
Cnoc nan Cuilean	Scotland	Alkaline granite	Silurian/Wenlockian	–	–	–	(Walters et al. 2012)
Cuxhaven	Germany	Beach placer	Quaternary	REE, Ti, Zr	–	–	(Hedrick 1985)
Dalucao	China	Carbonatite	Neogene/Miocene	REE	–	0.76 Mt @ 5% REO	(Weng et al. 2015)
Delitzsch (Storkwitz)	Germany	Carbonatite	Upper Cretaceous	REE	–	4.5 Mt @ 0.45% REO	(Krüger et al. 2013)
Evje-Iveland	Norway	NYF-type pegmatite	Neoproterozoic/ Tonian	Ta, Nb, REE	–	–	(Müller et al. 2018)
Fen-Søve	Norway	Alkaline-carbonatite complex	Neoproterozoic/ Ediacaran	REE, Nb, Fe	–	83.7 Mt @ 1.08% REO	(Dietzel et al. 2019)
Froland–Gloserheia	Norway	NYF-type pegmatite	Neoproterozoic/ Tonian	Ta, Nb, REE	–	4 Mt @ 0.02% REO	(EURARE 2017)

Name of the deposit/locality	Country	Type of deposit	Age	Main commodities	Other commodities	Known/inferred reserves/resources	References
Galiñeiro	Spain	Alkaline complex	Lower Ordovician	REE, Th	–	–	(Montero et al. 2009)
Gardiner	Greenland (Denmark)	Ultramafic to differentiated alkaline complex	Palaeogene/ Eocene	REE	–	–	(Campbell et al. 1997)
Glamsland	Norway	NYF-type pegmatite	Neoproterozoic/ Tonian	Ta, Nb, REE	–	–	(Müller et al. 2018)
Grebnik	Kosovo	Laterite–Bauxite	Upper Cretaceous	Al	REE	–	(Bárdossy et al. 1976)
Grønnedal-Ika	Greenland (Denmark)	Alkaline complex	Mesoproterozoic/ Ectasian	REE	–	–	(Sørensen and Kalvig 2011)
Hanstholm	Denmark	Beach placer	Quaternary	REE, Zr, Ti	–	–	(Orris and Grauch 2002)
Honkilehto	Finland	Hydrothermal-metasomatic	–	REE, Co, U	–	–	(Sarapää et al. 2013)
Iivaara	Finland	Alkaline complex	Upper Devonian	P	REE	–	(Sarapää et al. 2013)
Isle of Skye	United Kingdom (Scotland)	Granite	Palaeogene/ Palaeocene–Eocene	REE	–	–	(Exley 1980)
Kaiserstuhl	Germany	Alkaline-carbonatite complex	Neogene/Miocene	Nb, Ta, REE	Th, U	–	(Walter et al. 2018)
Kap Parral	Greenland (Denmark)	Alkaline complex	Palaeogene	Nb, REE	–	–	(Nielsen 2002)
Kap Simpson (Bjørnedal-Traill)	Greenland (Denmark)	Alkaline complex	Palaeogene	Nb, REE	Pb, Zn, Mo	–	(Stembera 2012)

Name of the deposit/locality	Country	Type of deposit	Age	Main commodities	Other commodities	Known/inferred reserves/resources	References
Katajakangas	Finland	Meta-alkaline granite	Palaeoproterozoic/ Orosirien	Nb, Zr, Ti	REE, Th, U	0.46 Mt @ 2.71% REO	(Hugg 1985)
Kerala	India	Beach placer	Quaternary	Ti, Zr, REE	Grenats	118 Mt @ 0.09% REO	(Orris and Grauch 2002)
Khibiny (total)	Russia	Alkaline complex	Upper Devonian	P	Sr, REE	3200 Mt @ 0.36% REO	(Kalashnikov et al. 2016)
Kiruna	Sweden	IOA	Palaeoproterozoic/ Orosirian	Fe, P	REE	2000 Mt @ 0.5% REO	(EURARE 2017)
Kiruna-Malmberget	Sweden	Mining residue	Quaternary/ Anthropocene	P, REE	–	5.6 Mt apatite @ 0.5% REO	(Bru et al. 2015)
Koashva	Russia	Alkaline complex	Upper Devonian	P	Sr, REE	634 Mt @ 0.41% REO	(Kalashnikov et al. 2016)
Kodal	Norway	Ultramafic alkaline complex	Permian/Cisuralian	P, Ti, Fe	REE	49 Mt @ 0.12% REO	(Ihlen et al. 2001)
Kontioaho	Finland	Meta-alkaline granite	Palaeoproterozoic/ Orosirian	Nb, Zr, Ti	REE, Th, U	4 Mt @ 0.7% REO	(Kärenlampi et al. 2020)
Korsnäs	Finland	Carbonatite	Palaeoproterozoic/ Orosirian	Pb	REE	0.87 Mt @ 0.96% REO	(Tasman Metals Ltd. 2010)
Kortejärvi	Finland	Carbonatite	Lower Palaeoproterozoic	REE	–	–	(Al-Ani et al. 2011)
Kovdor	Russia	Ultrabasic-carbonatite alkaline complex	Upper Devonian	Fe, Zr, P	REE, Sc	500 Mt @ 0.001 to 0.09% REO	(Korsakova et al. 2012)
Kringlerne	Greenland (Denmark)	Alkaline complex	Mesoproterozoic/ Stenian	Ta, Nb, REE, Zr	–	4300 Mt @ 0.65% REO	(Sørensen and Kalvig 2011)

Name of the deposit/locality	Country	Type of deposit	Age	Main commodities	Other commodities	Known/inferred reserves/resources	References
Kukisvumchorr	Russia	Alkaline complex	Upper Devonian	P	Sr, REE	415 Mt @ 0.25% REO	(Kalashnikov et al. 2016)
Kvanefjeld	Greenland (Denmark)	Alkaline complex	Mesoproterozoic/ Stenian	U, REE, Zn	–	673 Mt @ 1.09 Mt REO	(GME 2014)
La Monnerie	France	Alluvial placer	Quaternary	REE	–	0.08 Mt @ 0.08% REO	(Cornet and Sapinart 1967)
Laivajoki	Finland	Alkaline-carbonatite complex	Lower Palaeoproterozoic	REE	–	–	(Nykänen et al. 1997)
Langesundfjord-Larvik	Norway	NYF-type pegmatite	Permian/ Cisuralian	REE	Nb, Th	–	(Larsen 2010)
Lesnoe	Russia	Alkaline complex	Neoarchaean	Zr	REE, Nb	1.51 Mt @ 0.22% REO	(Korsakova et al. 2012)
Loch Ailish	United Kingdom (Scotland)	Ultramafic alkaline complex	Silurian/ Llandoveryian	–	–	–	(Walters et al. 2012)
Loch Borralan	United Kingdom (Scotland)	Ultramafic alkaline complex	Silurian/ Wenlockian	–	–	–	(Walters et al. 2012)
Lofdal	Namibia	Alkaline-carbonatite complex	Neoproterozoic/ Tonian	REE	–	6.16 Mt @ 0.29% REO	(Zhou et al. 2017)
Longnan	China	Ionic clays	Quaternary	REE	–	48 Mt @ 0.1% REO	(Zhou et al. 2017)
Lovozero (total)	Russia	Alkaline complex	Upper Devonian	Nb	REE	1020 Mt @ 0.75% REO	(Kalashnikov et al. 2016)
Lužnice	Czech Republic	Alluvial placer	Quaternary	Ti, Zr, Th, REE, Nb	–	–	(Maòour et al. 1992)
Mäkärä-Vaulo	Finland	Ionic clays	–	Kaolin	REE	–	(Sarapää et al. 2013)

Name of the deposit/locality	Country	Type of deposit	Age	Main commodities	Other commodities	Known/inferred reserves/resources	References
Maoniuping	China	Alkaline-carbonatite complex	Palaeogene/Oligocene	REE, Pb, Ba, F	–	3.17 Mt @ 2.95% REO	(Verplanck et al. 2016)
Marmara	Greece	Laterite–Bauxite	Lower Cretaceous	Al	REE	–	(Maksimović and Pantó 1996)
Marmeiko	Greece	Laterite–Bauxite	Upper Jurassic	Al	REE	–	(Maksimović and Pantó 1996)
Maw	Canada	Hydrothermal-breccia	Mesoproterozoic/Calymnian	REE	–	0.46 Mt @ 0.21% Y_2O_3	(Rabiei et al. 2017)
Milne Land	Greenland (Denmark)	Palaeoplacer	Middle Jurassic	REE, Ti, Zr	–	5 Mt @ 0.5–1.9% REO	(Harpøth et al. 1986)
Misværdalen	Norway	Ultramafic alkaline complex	Silurian/Llandoveryian	P	REE	30 Mt @ 0.07% REO	(EURARE 2017)
Mont Parnasos	Greece	Laterite–Bauxite	Upper Jurassic	Al	REE	–	(Maksimović and Pantó 1996)
Monts Mourne	United Kingdom (Scotland)	Granite	Palaeogene/Eocene	REE	–	–	(Green et al. 2005)
Motzfeldt Sø	Greenland (Denmark)	Alkaline complex	Mesoproterozoic/Stenian	Th-U-Nb-Ta-Zr-REE	–	340 Mt @ 0.26% REO	(Ram Resources 2012)
Mount Weld	Australia	Carbonatite laterite	Cretaceous or Palaeogene	REE	–	23.9 @ 7.86 Mt REO	(Zhou et al. 2017)
Mountain Pass	USA	Carbonatite	Mesoproterozoic/Stenian–Ectasian	REE	–	16.7 Mt @ 7.98% REO	(Verplanck et al. 2016)
Nafpactos	Greece	Laterite–Bauxite	Upper Cretaceous	Al	REE	–	(Bárdossy et al. 1976)

Name of the deposit/locality	Country	Type of deposit	Age	Main commodities	Other commodities	Known/inferred reserves/resources	References
Nagyharsány (Villány)	Hungary	Laterite–Bauxite	Lower Cretaceous	Al	REE	–	(Bárdossy et al. 1976)
Nassuttutata tasia	Greenland (Denmark)	Pegmatite	Palaeoproterozoic/ Statherian	REE	–	–	(GEUS 2014)
Nazda-Vlasenica	Bosnia-Herzegovina	Laterite–Bauxite	Lower Cretaceous	Al	REE	–	(Bárdossy et al. 1976)
Nea Peramos	Greece	Beach placer	Quaternary	REE	–	–	(Melfos and Voudouris 2012)
Niaqornakassak–NIAQ	Greenland (Denmark)	Metasomatic, carbonatite flow	Palaeoproterozoic/ Orosirian	REE	Th	26.3 Mt @ 1.36% REO	(Mott et al. 2013)
Nyorkpakhk	Russia	Alkaline complex	Upper Devonian	P	Sr, REE	58 Mt @ 0.37% REO	(Kalashnikov et al. 2016)
Nisi (Nissi)	Greece	Laterite–Bauxite	Upper Jurassic	Al	REE	–	(Maksimović and Pantó 1996)
Nordre Isortoq	Greenland (Denmark)	Pegmatite	Palaeoproterozoic/ Statherian	REE	–	–	(GEUS 2014)
Norra Kärr	Sweden	Alkaline complex	Mesoproterozoic/ Calymnian	REE, Zr	–	58.1 Mt @ 0.59% REO	(EURARE 2017)
Oleny Ruchey	Russia	Alkaline complex	Upper Devonian	P	Sr, REE	325 Mt @ 0.38% REO	(Kalashnikov et al. 2016)
Olserum	Sweden	Hydrothermal vein and metasomatic	Palaeoproterozoic/ Statherian	REE	–	7.8 Mt @ 0.62% REO	(Tasman Metals Ltd. 2013)
Olympic Dam	Australia	IOCG	Mesoproterozoic/ Calymnian	Cu, U, Au, Ag	REE	9576 Mt @ 0.55% REO	(Weng et al. 2015)

Name of the deposit/locality	Country	Type of deposit	Age	Main commodities	Other commodities	Known/inferred reserves/resources	References
Påläng	Sweden	Phosphorite	Paleoproterozoic/ Rhyacian	P, U	REE	6 Mt	(Gustafsson 1979)
Palkiskuru-Palovaara	Finland	Hydrothermal-metasomatic or carbonatite	Unknown	REE, Th, U	–	–	(Sarapää et al. 2013)
Panormos (Skopelos)	Greece	Laterite–Bauxite	Lower Jurassic	Al	REE	–	(Bárdossy et al. 1976)
Partomchorr	Russia	Alkaline complex	Upper Devonian	P	Sr, REE	748 Mt @ 0.2% REO	(Kalashnikov et al. 2016)
Pennines	United Kingdom (England)	Hydrothermal vein	Permian/Cisuralian	Pb, Zn, F, Ba	REE	–	(Kraemer et al. 2019)
Pointe de l'Espiguette	France	Beach placer	Quaternary	REE, Zr	–	–	(Vazquez-Lopez and Jézéquel 2002)
Pretul	Austria	Hydrothermal vein	Upper Cretaceous	–	–	–	(Bernhard et al. 1998)
Pylos	Greece	Laterite–Bauxite	Palaeogene/Eocene	Al	REE	–	(Bárdossy et al. 1976)
Qaqarssuk (Qeqertaasaq)	Greenland (Denmark)	Carbonatite	Middle Jurassic	Nb, REE, P	–	–	(Stendal 2014)
Qassiarsuk	Greenland (Denmark)	Alkaline-carbonatite complex	Mesoproterozoic/ Stenian–Ectasian	REE	–	–	(Sørensen and Kalvig 2011)
Rasvumchorr	Russia	Alkaline complex	Upper Devonian	P	Sr, REE	340 Mt @ 0.35% REO	(Kalashnikov et al. 2016)
Ría Vigo	Spain	Beach-alluvial placer	Quaternary	REE	–	–	(Prego et al. 2012)

Name of the deposit/locality	Country	Type of deposit	Age	Main commodities	Other commodities	Known/inferred reserves/resources	References
Richards Bay	South Africa	Beach placer	Quaternary/Holocene	Fe, Zr, ti	REE	4980 Mt @ 0.01% REO	(Zhou et al. 2017)
Sæteråsen	Norway	Alkaline volcanism–differentiated	Permian/Cisuralian	Zr, Nb, REE	–	8 Mt @ 0.53% REO	(EURARE 2017)
Sakharjok	Russia	Alkaline complex	Neoarchaean	Zr	REE, Nb	35.8 Mt @ 0.16% REO	(Korsakova et al. 2012)
St Kilda	United Kingdom (Scotland)	Granite	Palaeogene/Eocene	REE	–	–	(Harding et al. 1982)
San Giovani Rotondo	Italy	Laterite–Bauxite	Upper Cretaceous	Al	REE	–	(Bárdossy and Pantó 1973)
Sarfartoq	Greenland (Denmark)	Carbonatite	Neoproterozoic/ Ediacarian	P, Nb, Ta, REE	–	8.3Mt @ 1.72% REO	(Druecker and Simpson 2012)
Siilinjärvi	Finland	Alkaline-carbonatite complex	Neoarchaean	P	REE	200 Mt @ 0.06% REO	(Decrée et al. 2020)
Singertât	Greenland (Denmark)	Alkaline-carbonatite complex	Neoarchaean	Sr, REE	–	–	(Blichert-Toft et al. 1995)
Smerna	Greece	Laterite–Bauxite	Paleogene/Eocene	Al	REE	–	(Bárdossy et al. 1976)
Sokli	Finland	Alkaline-carbonatite complex	Upper Devonian	P	REE, Nb, Ta, Zr, U	–	(Sarapää et al. 2013)
Sørensen	Greenland (Denmark)	Alkaline complex	Mesoproterozoic/ Stenian	U, REE, Zn	–	242 Mt @ 1.1 Mt REO	(GME 2014)
Steenkampskraal	South Africa	Hydrothermal vein	Mesoproterozoic/ Stenian	REE	Cu	0.6 Mt @ 14.4% REO	(Weng et al. 2015)

Name of the deposit/locality	Country	Type of deposit	Age	Main commodities	Other commodities	Known/inferred reserves/resources	References
Stenis Arahovis	Greece	Laterite–Bauxite	Upper Jurassic	Al	REE	–	(Bárdossy et al. 1976)
Štitovo	Montenegro	Laterite–Bauxite	Upper Cretaceous	Al	REE	–	(Bárdossy et al. 1976)
Strange Lake	Canada	Alkaline complex	Mesoproterozoic/ Ectasian	REE, Zr, Nb	–	492 Mt @ 0.89% REO	(Weng et al. 2015)
Tåsjö	Sweden	Phosphorite	Cambrian–Ordovician	P, U	REE	75/150 Mt @ 0.13/0.28% REO	(Armands 1970)
Tikiusaaq	Greenland (Denmark)	Carbonatite	Upper Jurassic	P, REE	–	0.2 Mt @ 1-5% REO	(Stendal 2014)
Timmiarmiut	Greenland (Denmark)	Alkaline-carbonatite complex	Archaean	Sr, REE	–	–	(Steensgaard et al. 2010)
Třebíč	Czech Republic	NYF-type pegmatite	Carboniferous/ Mississippian	REE	–	–	(Škoda and Novák 2007)
Trimouns-Luzenac	France	Hydrothermal-metasomatic	Lower Cretaceous	Talc	REE	–	(Moëlo et al. 1974)
Tupertalik	Greenland (Denmark)	Carbonatite	Mesoarchaean	Sr, Zr, REE	–	–	(Larsen et al. 1983)
Umiammakku Nunnaa–UMIA	Greenland (Denmark)	Hydrothermal-metasomatic	Palaeoproterozoic/ Orosirian	REE	Th	–	(Mott et al. 2013)
Uuniniemi	Finland	Hydrothermal-metasomatic		P, REE, Nb	–	–	(Sarapää et al. 2013)
Virtasalmi	Finland	Ionic clays	Mesoproterozoic/ Stenian	Kaolin	REE	–	(Sarapää et al. 2013)
Weishan	China	Alkaline-carbonatite complex	Lower Cretaceous	REE, F	–	0.19 @ 4.61% REO	(Jia and Liu 2020)

Name of the deposit/locality	Country	Type of deposit	Age	Main commodities	Other commodities	Known/inferred reserves/resources	References
Ytterby	Sweden	NYF-type pegmatite	–	–	–	–	(Romer and Smeds 1994)
Yuksporr	Russia	Alkaline complex	Upper Devonian	P	Sr, REE	540 Mt @ 0.39% REO	(Kalashnikov et al. 2016)
Zagrad	Montenegro	Laterite–Bauxite	Upper Cretaceous	Al	REE	–	(Bárdossy et al. 1976)
Zone 3	Greenland (Denmark)	Alkaline complex	Mesoproterozoic/Stenian	U, REE, Zn	–	95 Mt @ 1.16 Mt REO	(GME 2014)

Appendix 2

The Cornubian Batholith: Post-Collisional Variscan Granites and Resources

Robin SHAIL[1] **and Beth SIMONS**[2]

[1] *Camborne School of Mines, University of Exeter, United Kingdom*

[2] *St Agnes, Cornwall, United Kingdom*

A2.1. Ore and mineral deposits/occurrences associated with the Cornubian Batholith and mentioned in the chapter

Metallic Resources 1,

Name of the deposit/locality	Type of deposits	Age	Main commodities	Other commodities (Potential) byproducts	Known/inferred reserves/resources	Reference
Beam	Quartz-tourmaline vein	Permian	Sn	–	Unknown, historical extraction	(Dines 1956)
Bosworgey	Massive greisen	Permian	W, Sn	Bi, Nb, Ta	Unknown	(Ball and Basham 1984)
Botallack	Replacement; Tourmaline lodes; Polymetallic sulphide	Lode: Rb-Sr feldspar 267 ± 4.0 Ma	Sn, Cu, As	In, Fe	14.5 kt Sn, 20 kt Cu and 1.5 kt As historical extraction	(Dines 1956; Halliday 1980)
Castle-an-Dinas	Greisen-bordered sheeted vein complex	Tungsten-quartz lode: Ar-Ar muscovite 272.7 ± 0.8 Ma	W, Sn	–	Historical extraction	(Clark et al. 1993; Dines 1956)
Cligga (Head)	Greisen-bordered sheeted vein complex	U-Pb cassiterite c. 285 Ma	W, Sn	Cu, Zn, In	300t W and 200t Sn historical extraction	(Moore and Jackson 1997; Tapster and Bright 2000)
Ding Dong	Alluvial Sn; Tourmaline-dominated sheeted veins and stockworks	Permian	Sn	Cu	3.5 kt Sn historical extraction 1814–78	(Dines 1956)
Dolcoath	Tourmaline-dominated lodes	Permian	Sn, Cu, Zn	As, In	Substantial historical extraction	(Dines 1956)
Geevor	Tourmaline lodes, polymetallic sulfides	Sn lode: Ar-Ar muscovite 270.5 ± 0.8	Sn	Cu, Zn	50 kt Sn historical extraction	(Chesley et al. 1993; Dines 1956)
Goonbarrow	Greisen-bordered sheeted vein complex; China Clay	Mineralization: Ar-Ar muscovite 276.8 ± 1.2 Ma	W, Sn, China Clay	–	Historical extraction	(Bray and Spooner 1983; Chesley et al. 1993)
Great Flat Lode	Tourmaline lodes, polymetallic sulfide	Permian	Sn, Cu, Zn	In, As	Substantial historical extraction	(Dines 1956)

Name of the deposit/locality	Type of deposits	Age	Main commodities	Other commodities (Potential) byproducts	Known/inferred reserves/resources	Reference
Gryll's Bunny	Replacement deposit	Permian	Sn	W	Historical extraction	(Jackson 1974)
Haytor Iron Mine	Replacement deposit	Ar-Ar hornblende 280.3 ± 1.0 Ma	Fe	Sn, Mn	15t @ 57% Fe extracted	(Chesley et al. 1993)
Hemerdon	Greisen-bordered sheeted vein complex	Cassiterite: 291.8 ± 3.4 Ma	W, Sn	As, Cu	Inferred Resource 218 Mt @ 0.02% Sn, 0.18% WO3	(Moscati and Neymark 2019; Wolf Minerals 2010)
Kit Hill I	Quartz-tourmaline; Greisen bordered veins	Granite: 284.4 ± 1.0 Ma	Sn, W	Cu, As	Historical extraction, grades <1%	(Clark et al. 1994; Dines 1956)
Magdalen	Replacement	Permian	Fe	Sn	Historical extraction	(Dines 1956)
Red-a-Ven	Replacement	Permian	Cu	Sn, Zn, In	Historical extraction	(Dines 1956)
Redmoor	Greisen-bordered sheeted vein complex	Permian	Sn, W, Cu	Zn, In	Inferred Resource, 11.7 Mt @ 0.56% WO_3, 0.16% Sn, and 0.50% Cu, at a cut-off grade of 0.45% (SnEq)	(Cornwall Resources 2021)
South Crofty	Tourmaline-quartz veins; Polymetallic sulfide vein system	Mineralisation: Ar-Ar muscovite: 269.0 ± 0.8	Lower mine: Sn Upper mine: Cu, Zn	Upper mine: In	Inferred Resource, Lower mine: 738 kt @ 1.91% Sn Upper mine: 464 Kt @ 0.67% Sn, 0.62% Cu and 0.63% Zn	(Cornish Metals 2021; Chen et al. 1993)
South Terras	Uranium mineralization	U-Pb uraninite 225 ± 5 Ma	U, Fe	Pb, As	Historical extraction	(Pockley 1964)

Name of the deposit/locality	Type of deposits	Age	Main commodities	Other commodities (Potential) byproducts	Known/inferred reserves/resources	Reference
St Austell District	Kaolinisation; Li-rich micas	Permian granite	China Clay	Li, Nb, Ta, Sn, W	Clay: In excess of 60 Mt. Over 165 Mt extracted since the late 1800s	(Cornwall Council 2018)
St Michael's Mount	Greisen-bordered sheeted vein complex	Greisen margin: Ar-Ar muscovite 281.9 ± 1.4 Ma	W, Sn	–	Unknown	(Chen et al. 1993)
Treliver	Alluvial Sn, Tourmaline veins, replacement	Permian	Sn	In	Resource: 4–7 Mt @ 0.3–0.4% Sn	(Camm and Dominy 1997)
United Downs	Brine	Quaternary	Geothermal, Li	–	Unknown	(Cornish Lithium 2021)
Wheal Bray	Five-element association	U-Pb uraninite 165 ± 4 Ma	Cu, U	Ni	Historical extraction	(Damely et al. 1965)
Wheal Fortune	Greisen-bordered sheeted vein complex	Permian	Sn	Cu, W	Historical extraction 2992t Sn; 322t Cu	(Dines 1956)
Wheal Jane	Greisen, tourmaline breccia, polymetallic sulfide	Muscovite margin: Ar-Ar isochron <283.5 ± 0.3 Ma	Sn, Zn, Cu	Ag, As, Fe	Substantial historical production of Sn and Cu	(Chesley et al. 1993)
Wheal Remfry	Tourmaline breccia	Ar-Ar muscovite 270.3 ± 08 Ma	China Clay, Sn	W	Unknown	(Chesley et al. 1993)

Appendix 3

The Panasqueira W Deposit (Portugal): A Critical Bibliographic Review

Christian MARIGNAC and Michel CATHELINEAU
GeoResources, CREGU, University of Lorraine, Nancy, France

A3.1. Estimate of the tungsten endowment in Variscan Western Europe

The initial tungsten endowment in Western Europe (past production + reserves + resources) was estimated at 673 kt W by Cassard et al. (2015). As the Table A3.1 in Appendix 1 shows, this tonnage seems to have been underestimated, since the real value is probably closer to a million tons of tungsten. This value is doubtlessly a slight underestimate, since past production is not always fully known. In the case of the deposit at Panasqueira, for example, which has been mined since the end of the 19th century, the information available on past production does not date back beyond the end of the 1930s and would amount to 120 kt (Vigne et al. 2018), to which must be added 27 kt of reserves and resources (Wheeler 2016) to reach the 147 kt as in the table in Appendix 1. It is sensible to note that this value is close to the 150 kt W estimated directly by Polya (1989), so this estimate is probably close to the real value.

If only reserves and resources are considered, it nevertheless seems that the endowment of Variscan Western Europe is reduced to ~740 kt.

For a color version of all figures in this chapter, see www.iste.co.uk/decree/metallic1.zip.

Metallic Resources 1,
coordinated by Sophie DECRÉE. © ISTE Ltd 2023.

	Deposits	Tonnages (kt W)	Nature
Austria	Mittersill	82 (Raith 2010; Lauri et al. 2018)	Metamorphosed W-porphyry, scheelite
Cornwall	Past production	5 (Porter GeoConsultant 2016)	–
	Drakelands/Hemerdon	173 (Lauri et al. 2018)	Greisen stockwork (wolframite)
Total Cornwall	Redmoor	52 (Cornwall Resources 2017)	Perigranitic veins (wolframite)
	–	230	–
Erzgebirge	Past production	214 (Porter GeoConsultant 2016)	Greisens and veins (wolframite)
	Cinoviec/Zinnwald	102 (European Metals 2017)	Greisen (wolframite)
	Krasno	35 (Werner et al. 2014)	Greisen (wolframite)
	Antonsthal	23 (Lehmann 2010)	Skarn (scheelite)
Total Erzgebirge	Pöhla-Globenstein	34 (Lehmann 2010)	Skarn (scheelite)
	Other	19 (Lehmann 2010)	Greisens: Altenberg, Gottesberg, Sadisdorf (wolframite)
	–	426	–
France	Salau	23 (Fonteilles et al. 1989; Marcoux 2018)	Skarn (scheelite)
	Fumade	21 (Marcoux 2018)	Skarn (scheelite)
France total	Montredon	17 (Cassard et al. 2015)	Perigranitic vein field (wolframite)
	Other: scheelite	20 (Charles 2015; Marcoux 2018)	Skarn: Cost-an-Noz; calc silicate gneiss: La Favière; intragranitic stockwork: Auxelles-Haut

	Deposits	Tonnages (kt W)	Nature
France total	Other: wolframite	31 (Charles 2015; Marcoux 2018)	Perigranitic veins: Echassières, Enguialès, Leucamp, Montbelleux; breccia pipe: Puy-les-Vignes; perigranitic (wolframite)
	–	111	–
CIZ: Portugal	Panasqueira	147 (Polya 1989; Vigne et al. 2018; W Resources 2020a)	Perigranitic veins (wolframite)
	Lagoaça	26 (Filipe 1989)	Perigranitic stockwork (scheelite)
	Covas	12 (Blackheath Resources 2015)	Skarn (scheelite)
	Borralha	13 (Gonçalves et al. 2017)	Veins and intragranitic breccias (wolframite)
	Other	36 (Martins 2012; Filipe 2017; W Resources 2020a, 2020b)	Skarn: Riba d'Alva, Serra de Arga, Tabuaço (scheelite); perigranitic veins (Bejanca, Vale de Gatas (wolframite); Régua)
CIZ: Spain	Barrecuopardo	58 (Ormonde Mining 2018)	Intragranitic stockwork (scheelite)
	La Parrilla	34 (W Resources 2020b)	Intragranitic stockwork (scheelite)
	Los Santos	15 (Lauri et al. 2018)	Skarn (scheelite)
	Valtreixal	15 (Wheeler 2016)	Skarn (scheelite)
	Santa Comba	15 (Rafaella Resources 2020)	Aplo-pegmatite (wolframite)
Total CIZ	–	351	–
Variscan total	–	1,117	–

Table A3.1. *Ab initio estimate of the tungsten endowment of Variscan Western Europe*

	Deposits	Tonnages (kt W)	Nature
Austria	Mittersill	26 (Lauri 2018)	Metamorphosed W-porphyry, scheelite
Cornwall	Drakelands/Hemerdon	173 (Lauri 2018)	Greisen stockwork (wolframite)
	Redmoor	52 (Cornwall Resources 2017)	Perigranitic veins (wolframite)
Total Cornwall	–	225	–
Erzgebirge	Cinoviec/Zinnwald	102 (European Metals 2017)	Greisen (wolframite)
	Krasno	35 (Werner 2014)	Greisen (wolframite)
	Antonsthal	23 (Lehmann 2010)	Skarn (scheelite)
	Pöhla-Globenstein	34 (Lehmann 2010)	Skarn (scheelite)
	Other	19 (Lehmann 2010)	Greisens: Altenberg, Gottesberg, Sadisdorf (wolframite)
Erzgebirge total	–	212	–
France	Salau	12 (Marcoux 2018)	Skarn (scheelite)
	Fumade	21 (Marcoux 2018)	Skarn (scheelite)
	Montredon	17 (Cassard et al. 2015)	Perigranitic vein field (wolframite)
	Other: scheelite	20 (Charles 2015; Marcoux 2018)	Skarns (scheelite): Cost-an-Noz; calc silicate gneiss: La Favière; intragranitic stockwork: Auxelles-Haut

	Deposits	Tonnages (kt W)	Nature
France total	Other: wolframite	20 (Charles 2015; Marcoux 2018)	Perigranitic veins: Echassières, Enguialès, Leucamp, Montbelleux; perigranitic breccia pipe: Puy-les-Vignes
	–	62	–
CIZ: Portugal	Panasqueira	27 (Wheeler 2016)	Perigranitic veins (wolframite)
	Lagoaça	26 (Filipe 2017)	Perigranitic stockwork (scheelite)
	Covas	10 (Blackheath Resources 2015)	Skarn (scheelite)
	Other	35 (Martins 2012; Wheeler 2016; Filipe 2017)	Skarn: Riba d'Alva, Serra de Arga, Tabuaço (scheelite); Perigranitic veins (Bejanca, Vale de Gatas (wolframite); Régua)
CIZ: Spain	Barrecuopardo	56 (Ormonde Mining 2018)	Intragranitic stockwork (scheelite)
	La Parrilla	34 W (Resources 2020)	Intragranitic stockwork(scheelite)
	Los Santos	15 (Lauri 2018)	Skarn (scheelite)
	Valtreixal	15 (Resources 2020)	Skarn (scheelite)
	Santa Comba	15 (Rafaella Resources 2020)	Aplo-pegmatite (wolframite)
Total CIZ	–	214	–
Variscan total	–	739	–

Table A3.2. *Estimate of tungsten reserves and resources in Variscan Western Europe*

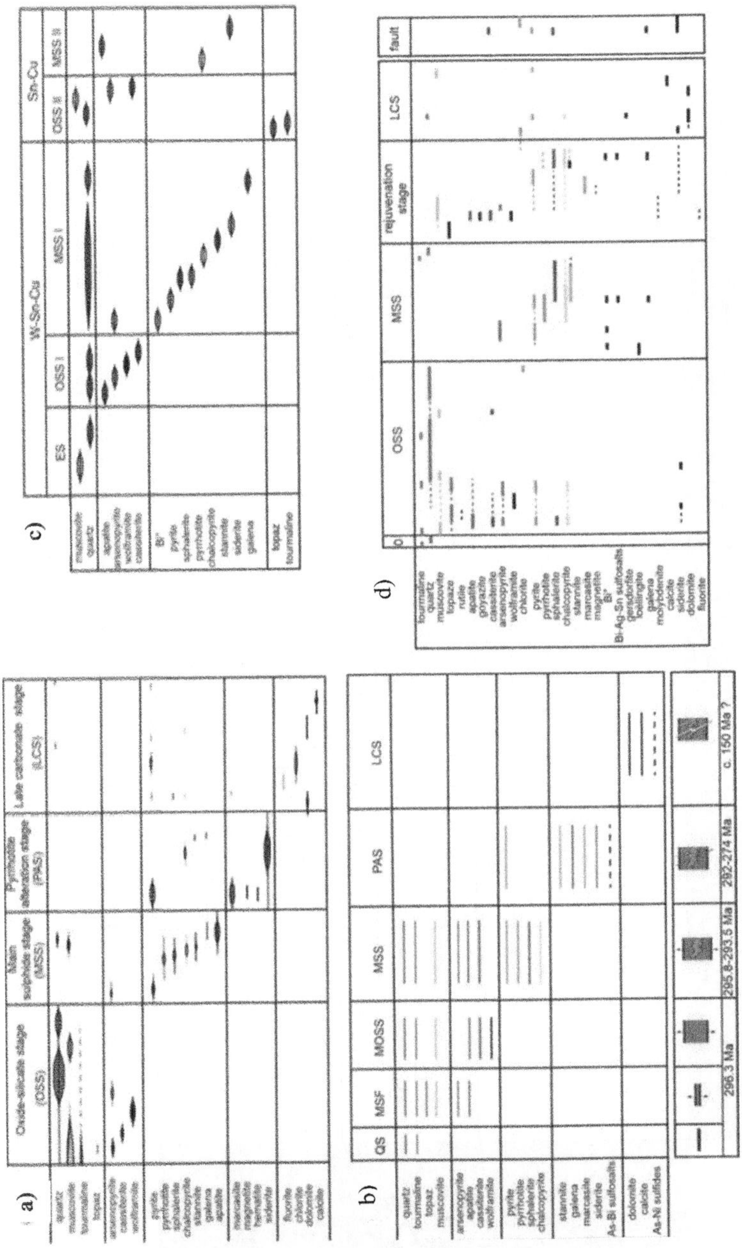

Figure A3.1. *History of paragenetic successions for the Panasqueira deposit*

COMMENT ON FIGURE A3.1.– *(a) The founding paragenesis: Kelly and Rye (1979). (b) Polya's paragenesis et al. (2000): first study of the succession of opening stages (diagrams under the chart). (c) The initial identification of two stages of cassiterite deposit (OSS1 and OSS2) (Lourenço 2002). (d) Martins' complex paragenesis (2017). Explanations are given in the text.*

These figures make the Western European Variscan province one of the world's great tungsten provinces. However, they still place it far behind the Yanshan super-province of south-eastern China, whose reserves alone amount to 1,900 kt W.

A3.2. Additional figures

The following figures summarize the evolution of knowledge/models on the mineral paragenesis described at Panasqueira (Figure A3.1), but also provide many specifics on the facies and textures encountered in this deposit (Figures A3.2–A3.10).

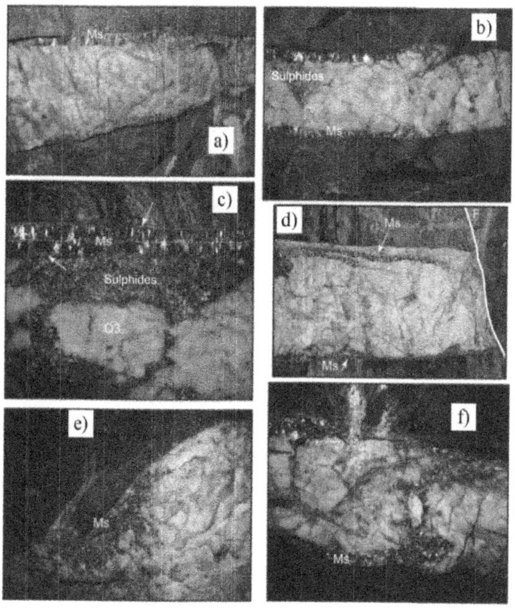

Figure A3.2. *Variability of the muscovite selvages; the sulfurs are essentially of pyrrhotite (more or less pyritized)*

COMMENT ON FIGURE A3.2.– *(a) Example of dissymmetry between the two wall rocks; here, the edge of the roof is more developed than that of the wall, but this is not systematic (L1-P6-D5-3). (b) Example of the lack of lateral continuity; note the symmetry between the wall rocks. (c) A case of internal symmetry in a selvage: on the roof of the band of muscovite and on its wall (in contact with the sulfurs), we observe an edging of fine-grained muscovite (arrows); the fan-shaped muscovite displays a double polarity, suggestive of a syntax fill (in dotted lines: the median line) (L2-D1-P4-I1S). (d) Example of multiplicity and absence of continuity of the selvages of palmar muscovite (L3-D9-AW36). (e) Irregular development of an aggregate of muscovite in the wall rock (L3-D9-R1-AW36). (f) Presence of a muscovite "connector" between the two wall rocks; note the sinuous character: it is not (not only?) a clogged fissure (L2-D1-P4-I15).*

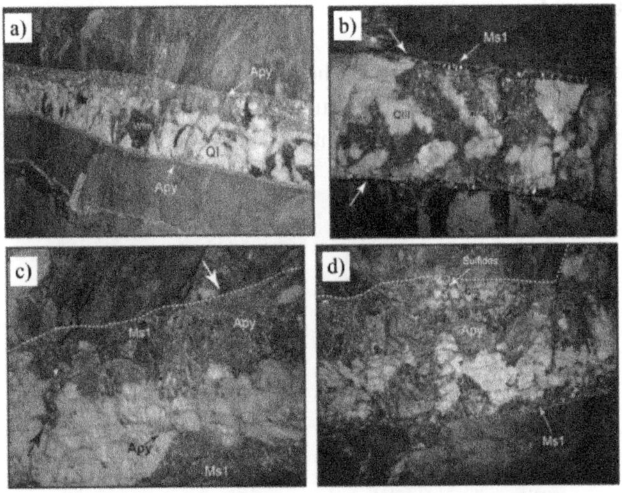

Figure A3.3. *Examples of arsenopyrite selvages*

COMMENT ON FIGURE A3.3.– *Wfm wolframite. (a) Asymmetrical development on the roof and wall of a vein of stage I quartz-wolframite I; note the block-shaped QI quartz auto-morphes ("alpine" style); note the flat join parallel with the vein's orientation (arrow) (L1-D29-R7-AW14). (b) Asymmetrical development on the roof (above all) and wall of a stage III vein; note the discontinuous selvages (arrows) of muscovite (L2-D1-P4-I1N). (c) Arsenopyrite in sheaf-shaped automorphic prisms, at the top of a vein of quartz with muscovite (Ms1) selvages on the roof and wall; note the gap of Ms1 at the roof, where the arsenopyrite is in direct contact with the surround (white arrow); note in the quartz the small cleft of arsenopyrite, and the "appendage" of muscovite (black arrow) (L3-D9-R1-AW36). (d) The arsenopyrite, in sheaf-shaped automorphic prisms, seems to obliterate a vein of stage III quartz-sulfurs (L3-D9-R1-AW36).*

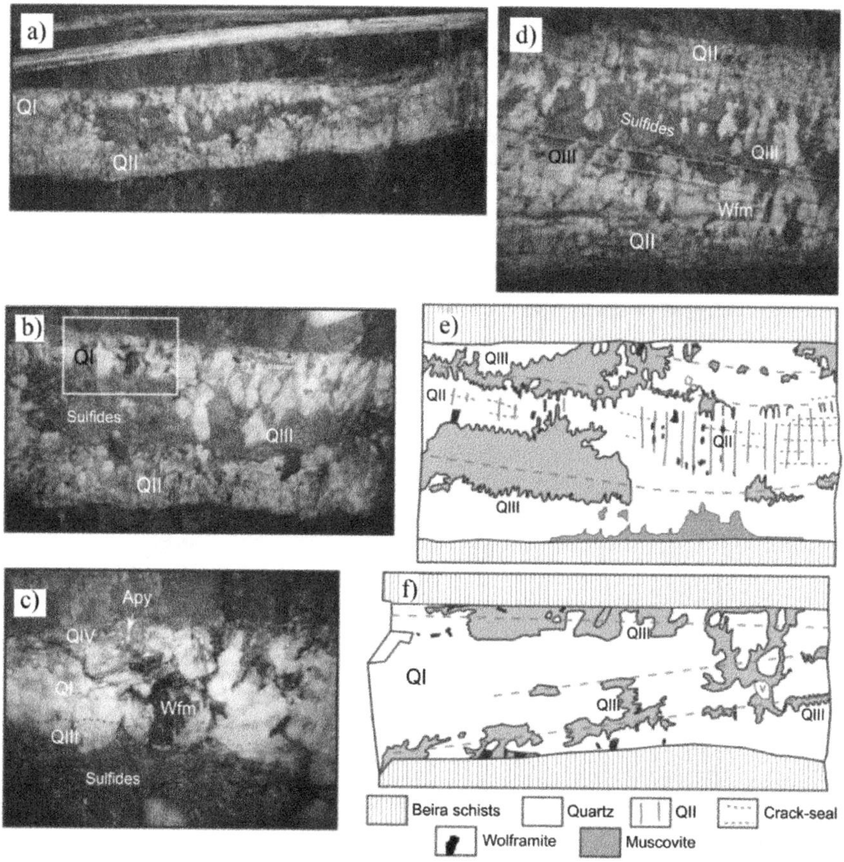

Figure A3.4. *Characterization of stage III geodes, clogged by sulfurs (III-B)*

COMMENT ON FIGURE A3.4.– *(a and c) Gradual zooming-in on a central geode (L1D29R13AW14E). (a) An overview: the geode is apparently implanted at the junction between a QI vein and a QII vein. (b) Enlargement, showing the deployment of neoformed quartz QIII at the bounds of the geode. (c) Detail of B, showing the increase in the quartz QIII (zoned, clear) supplied by a QI wall; the wolframite (Wfm) forms part of QI; note that the arsenopyrite in selvage position is also jointed by a clear quartz (QIV). (d) Large geode obliterating a stage II vein; the QIII quartzes are clearly visible, some cut parallel to the axis, the others (of which there are fewer) perpendicular to it: the geode is closed in all directions. (e) Multiple openings at stage III, more or less at the wall rocks of the stage II vein (orange dashed lines: axes of III openings) (drawing taken from the bottom [CM, 1970]; L0D9R17AW13).*

(f) Multiple openings, in directions clearly oblique to that of the re-opened veins (orange dashed lines) (drawing taken from the bottom [CM, 1970]; L0D9R17AW13).

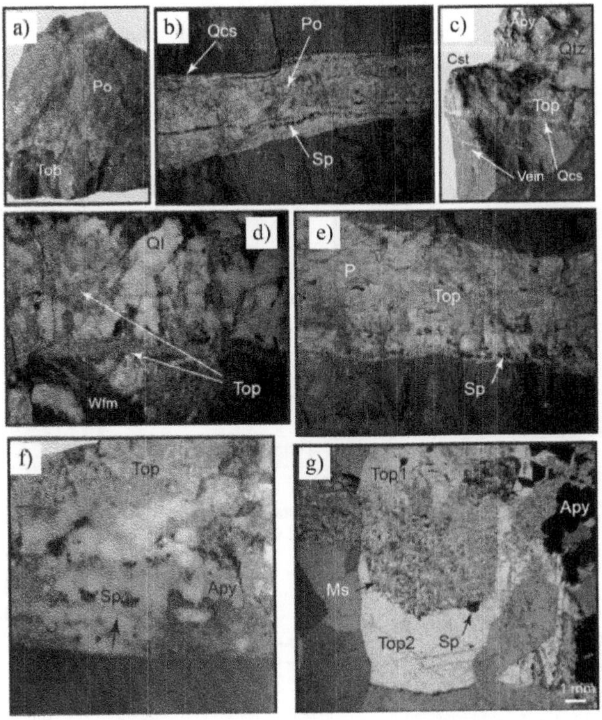

Figure A3.5. *The beginning of stage III: aspects of the development of topaz*

COMMENT ON FIGURE A3.5.– *(a) Sample of a topaz-pyrrhotite vein (Panasqueira museum). (b) Topaz vein, showing a banded structure ('crack-seal'); note the filling of "central" cavities, either by sphalerite, or by pyrrhotite; and the Qcs vein at the roof (L1D29R13AW14-E). (c) Topaz-cassiterite selvage; note the small Qcs veins (E8). d) Association of a topaz vein intersecting the wolframite within a stage I vein and bands of metasomatic topaz in the QI quartz (L1D29R13AW14-E). (e) Invasion of a vein of quartz (probable QI) by metasomatic topaz. (f) Metasomatic topaz and micro-geodes of sphalerite; note the fissure in-fills (arrow) (E9). (g) The two generations of topaz; here, the second generation (Top2) is in addition to the first (Top1); note the sphalerite at the interface; the topaz is changed to muscovite: note that if both generations are affected, the topaz 1 is almost entirely pseudo-morphed (LMP, polarized light E10).*

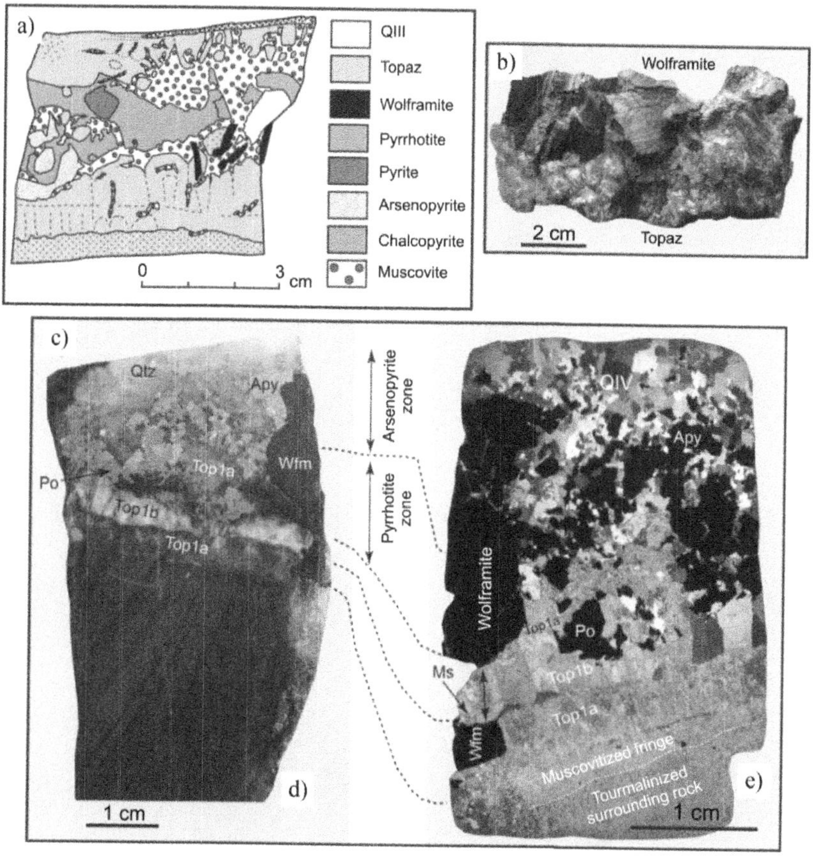

Figure A3.6. *Topaz-wolframite association (Wfm3-III-A)*

COMMENT ON FIGURE A3.6.– *(a) The wolframite prisms are encased in automorphic topaz; note the pyrrhotite in-filling (with an associated quartz III) of the syntactic topaz vein, as well as the late arsenopyrite and muscovite (cross-section, light-room drawing; E11). (b) Large wolframite associated with topaz (L0D 15R15AW10 "Bank"). (c) Vein of topaz-wolframite-pyrrhotite, with superimposition of a arsenopyrite assembly mosaic (Apy)-quartz (QIV); two successive generations of topaz (Top1d and Top1e) conveying a "crack-seal" process that also affects the wolframite (double arrow); Top1d is entirely muscovitized, whereas Top1e is less affected (Ms); the pyrrhotite (Po) invades the topaz; note the fringe of muscovitization at the wall of the vein. (d) Cross-section. (e) LMP, polarized light; E12.*

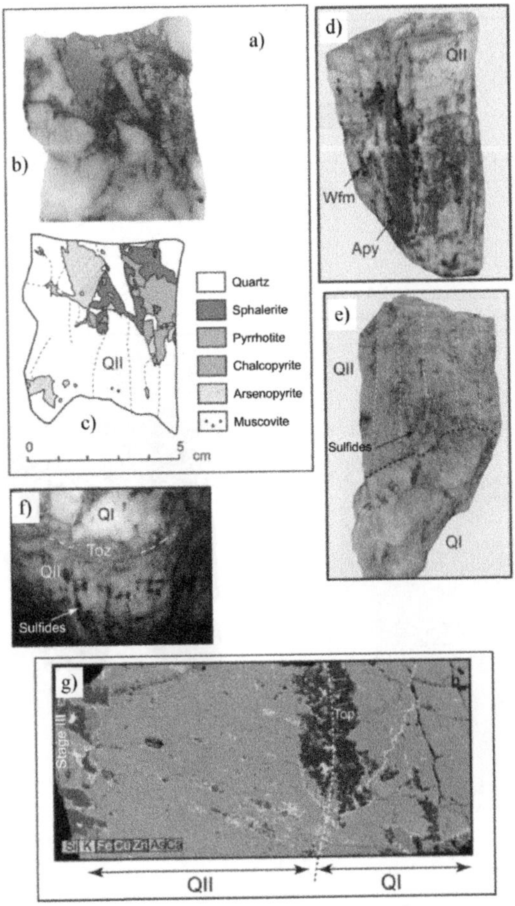

Figure A3.7. *Locating the implantation of later stages through earlier structures*

COMMENT ON FIGURE A3.7.– *(a) Sulfurs partly use the grain boundaries of quartz QII prisms. (b) Cross-section. (c) Light-room drawing of the obverse (E15). (d) Arsenopyrite (Apy) works its way into the grain boundaries and partly intersects the stage II wolframite prisms (Wfm2-II) (Panasqueira museum). (e) QI-QII contact and sulfurs superimposed onto QII prisms (L0D15R15AW10 "Bank"). (f) Obliteration by topaz of the QI-QII contact; note that the opening of stage II occurs at the wall of the stage I vein; and that the sulfurs occupy the interstices between the QII prisms (L1D29R13AW14-E). (g) The topaz, then the sulfurs (notably, arsenopyrite) and finally muscovite, are inserted along the line of contact between QI and QII; the sulfurs and the muscovite also use the grain boundaries between QII prisms, before vanishing in a stage III opening that truncates the QII quartz (image µ-XRF in artificial colors) for unknown reasons, image Si (Ka) reveals the quartz structures.*

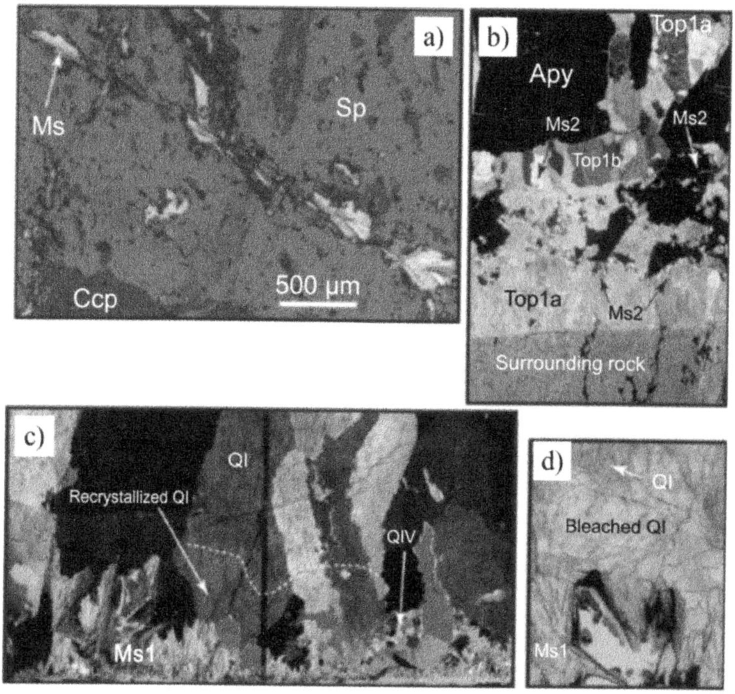

Figure A3.8. *Muscovite 1 and muscovite 2*

COMMENT ON FIGURE A3.8.– *(a) Healed muscovite cleft in stage III sulfurs; note the sub-automorphic character of the mica; dissolution–replacement is added to the fissuring (LMP, reflected light; E1). (b) Three stages of development: polyphase opening of a topaz selvage (Top1a, Top1b); superimposition of the outline by arsenopyrite; "crack-seal" opening divided by small muscovite clefts (Ms2), supplied by small veins; the topaz is changed by the muscovite, and even Top1a is pseudo-morphed at the wall (LMP, polarized light E25). (c and d) Arguments for a late deposition of the muscovite selvages (E26): (c) the recrystallization of the QI quartz on contact with the palmar muscovites of one selvage is revealed by the variations in the polarization color (LEp, polarized light); (d) the recrystallized zones correspond to the 'bleached' zones, hence the early IF of the QI quartz have been erased (LEP, natural light).*

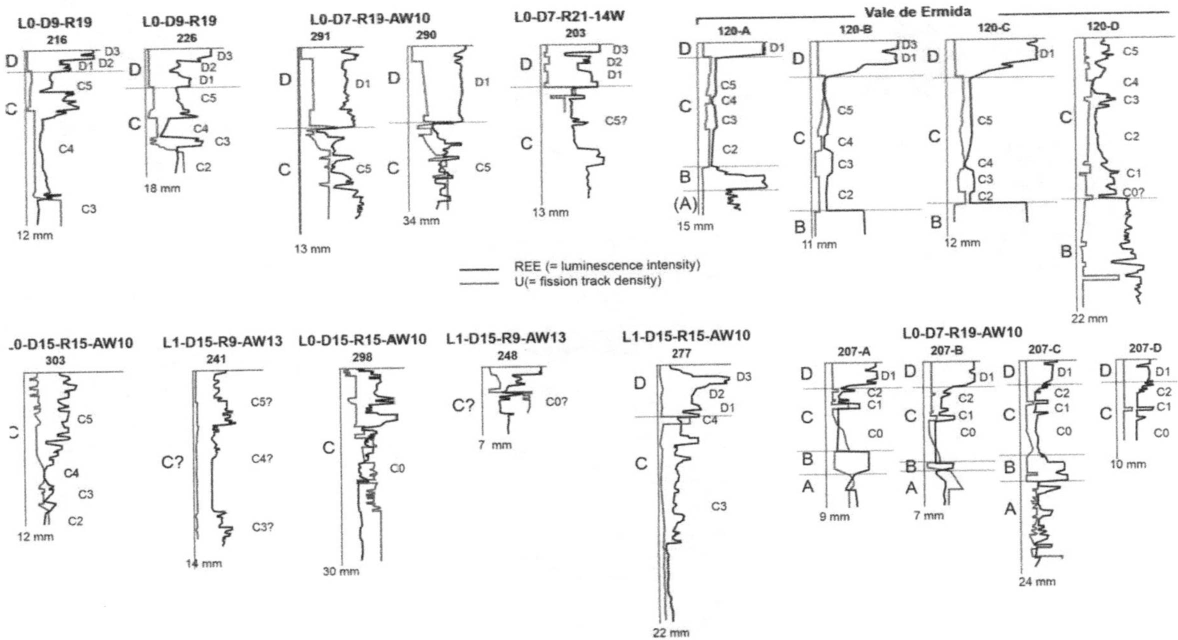

Figure A3.9. *Demonstration of a micro-stratigraphy in stage IV geodic aptites: re-interpretation of data from Knutson et al. (1987)*

COMMENT ON FIGURE A3.9.– *The figure shows that the profiles made by these authors make it possible to construct a zone system with four zones (A, B, C, D) from the core to the periphery, defining a micro-stratigraphy and permitting correlations at any scale, as soon as it is accepted that crystallization did not necessarily start or finish everywhere at the same instant (in the same zone) at the scale of the deposit, although this is the case with the elementary geode. In addition, some zones are susceptible to division into sub-zones (C0 to C5, D1 to D3), which are also correlable, although not necessarily everywhere, and the existence of lacuna in some geodes should be taken into account. These facts bear witness to the interconnection of geodes in space and time.*

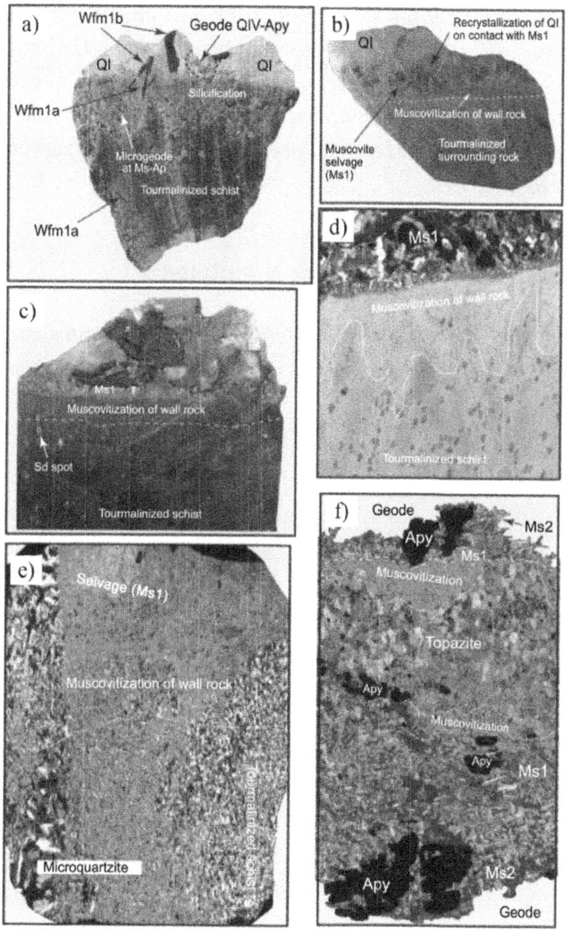

Figure A3.10. *Some aspects of changes in the wall rock at Panasqueira*

COMMENT ON FIGURE A3.10.– *(a) Silicification and tourmalinization of the wall rock; wolframite 1a, in the wall rock, which contain tourmaline inclusions (Tur1a), is truncated at the same time as the latter by the opening of the QI-Wfm1b vein (which can press on Wfm1a), synchronous with silicification; note the network of late micro-geodes of muscovite-apatite (sample cross-section E41). (b) Fringe of muscovitization at the wall rock of a muscovite selvage (Ms1); QI quartz recrystallizes on contact with Ms1 (E40). (c) Muscovitization (superimposed on the tourmalinization) at the wall of a muscovite selvage (Ms1); note the "spots" of carbonate (siderite) in the schist (L0D15R15AW10 "Banque"). (d) Muscovitization of the wall rock, linked to the muscovite selvage (Ms1); muscovitization affects the micro-quartzite in penetrative fashion (dissolution of the quartz) (LMP; polarized light, E34). (e) Muscovitization of the wall rock and muscovite selvage (Ms1); the carbonated "spots" disappear in the muscovitized fringe (LMP, polarized light; E42). (f) Screen of altered surrounding rock within a stage IV geode; the change is polyphase: the topazite was muscovitized at the time the muscovite selvages (Ms1) formed; arsenopyrite follows Ms1 and precedes Ms2 which surrounds the geodes (LMP, polarized light; E19).*

A3.3. Inventory of samples used for the plates

The samples are stored in the *Géoresources* laboratories' "Panasqueira" collection. "Bank" was a particularly rich site (vein thickness, content) in 1970.

No.	Samples	Site	No.	Sample	Site
E1	PAN-II-5	L0D15WR15AW10 “Bank”	E23	PAN-IV-11	L0D15WR15AW10 “Bank”
E2	Unnumbered	L0D15WR15AW10 “Bank”	E24	PAN-XVI-6	Unknown
E3	PAN-III-3	Unknown	E25	PAN-XIX-6	L1D23R13AW14
E4	PAN-XVIII-11	L1D13R10AW17	E26	PAN-XIV-181	L1D5R7AW15
E5	PAN-III-2	L0D9R15I2	E27	PAN-4	Route de Barroca Grande
E6	PAN-IV-3	L0D15WR15AW10 “Bank”	E28	PAN-VIII-3	L0D15WR15I2
E7	PAN-V-16	L0D15R15I2	E29	PAN-XIX-16	L1D29R13AW14E
E8	PAN-I-11	L0D3WR17I2W	E30	PAN-IIa	L0D15WR15AW10 “Bank”
E9	PAN-XIII-7	L1D29R7AW14	E31	PAN-XV-1	L1D15R17AW15
E10	PAN-V-8	L0D15R17	E32	PAN-VI-11	L0D3 advance
E11	PAN-IV-8	L0D15WR15AW10 “Bank”	E33	PAN-III-8	L0D15R15I2
E12	PAN-IV-13	L0D15WR15AW10 “Bank”	E34	PAN-XX-3	L3D9R1AW33
E13	PAN-V-4	L0D15R15I2	E35	PAN-XI-196	L1D15R7AW15
E14	PAN-VI-6	L0D15WR15AW10 “Bank”	E36	PAN-VI-8	L0D15WR15AW10 “Bank”
E15	PAN-V-6	L0D15WR15I2	E37	PAN-X-10	L0D15R15AW10I2
E16	PNQ-16-3	Greisenized cupola L2	E38	PAN-XIII-8	L1D29R17AW14
E17	PAN-IV-1	L0D15WR15AW10 “Bank”	E39	PAN-XIII-10	L1D29R17AW14
E18	PAN-VII-5	L0, joint D3-D19	E40	PAN-II-6	Unknown
E19	PAN-IV-5	L0D15WR15AW10 “Bank”	E41	PAN-VII-6	L0D15R15I2
E20	PAN-I-12	L0D15WR15AW10 “Bank”	E42	PAN-XX-14	L3D9R1AW32
E21	PAN-VI-14	L0D15WR15AW10 “Bank”	E43	PAN-XX-5	L3D9R1AW32
E22	PAN-VI-1	L0D15WR15AW10 “Bank”	E44	PAN-7	Route de Barroca Grande

Table A3.3. *Inventory of samples used for the plates*

A3.4. Mass balances and source of fluids

A3.4.1. *Magmatic source of fluids*

A3.4.1.1. *Silica mass balance*

Kelly and Rye (1979) calculated the volume of water needed to transport and deposit the quartz found in the veins (which, remember, represents the main mineral in the system) and compared it to the quantity of fluid and silica that the greisenized cupola could have provided. Without giving the details of their calculations, they conclude that "the known cupola could not have contributed more than 1/500 of the minimum amount of quartz proven by drilling" (Kelly and Rye 1979, p. 1740). The calculation can be repeated based on Polya's estimate (1989) of the tonnage of quartz contained in the veins, which is 10.5 Mt SiO_2. It can be considered that the majority of the quartz was deposited at the early stages (I–II), and therefore that this tonnage is a plausible basis for calculating the volumes of fluid involved in these stages. Considering a solubility of the quartz in an aqueous fluid at 10 wt% NaCl, 500°C and 200 MPa (Monecke et al. 2018), which is more coherent with the conditions reconstructed by Cathelineau et al. (2020) than those on which Kelly and Rye (1979) relied, we find a value of 85 mmoles of SiO_2 per kg of solution. Given that 1 mole of SiO_2 weighs 60 g, then 1 kg of fluid may contain $(85 \times 10^{-3}) \times (60 \times 10^{-3}) = 5.1 \times 10^{-3}$ kg of SiO_2. Consequently, the minimum fluid mass needed to transport and deposit the system's 10.5×10^9 kg of quartz is $(10.5 \times 10^9/5.1 \times 10^{-3}) = 2.1 \times 10^{12}$ kg of water. Using a density of 620 kg·m^{-3} for an aqueous fluid at 500°C and 10% NaCl (Sakuma and Ikichi 2016), a total fluid volume of 3.4×10^9 m^3, can be deduced from this, that is, 3.4 km^3 of water. This brings us back to Kelly and Rye's (1979) conclusion: the volume of the greisenized cupola (approximated by a cone, 150 m high and 250 m in diameter) being ~2.5×10^6 m^3, and the possibility that the water needed to form the quartz veins could have come from it is totally excluded. Indeed, according to Marignac et al. (2020), before greisenization the cupola was a rare metal granite (RMG), whose water content would consequently have been around 5% in weight. With a magmatic liquid density of 2.3 t·m^{-3} (Dingwell et al. 1993), a volume of potentially released water in the order of 4×10^5 m^3 is obtained for the cupola, which is clearly negligible compared to the volume calculated previously. Note moreover that even the sheet of altered granites which, according to Marignac et al. (2020), is found on the roof of the main granite mass (see Figure 3.4), would not be sufficient: with a power in the order of 100 m, and an extent of km^2, we can allocate it a volume of ~10^8 m^3, which could not therefore provide more than 2×10^7 m^3 of water.

A3.4.1.2. *Tungsten mass balance: consideration of the anomalous halo*

Polya (1989) for his part undertook to estimate the volume of water needed to transport the Tungsten and to precipitate the wolframite, an approach that might be considered more relevant if it involved evaluating the potential contribution of the Panasqueira granites. This calculation requires knowing the W content of the mineralizing fluid. Polya (1989) adopts a value of 0.2 ppm W, which leads to a colossal volume (750 km^3) of water needed to form a deposit of 150 kt of tungsten. If we include the 350 kt of tungsten that Polya estimates to have been stored in the deposit's anomalous halo, this is even more than the 2,300 km^3 of water that would need to be mobilized. This 0.2 ppm content is nevertheless much too low, in view of the most recent data on the solubility of tungsten in aqueous fluids from (Wood 1992; Wood and Samson 2000; Wang et al. 2019), who show that high contents are possible, from one ppm up to 1000 ppm, and even more so when the temperature of the fluid is high (since pressure does not play a significant role). To go further, the "true" tungsten content of the Panasqueira flows would have to be known. A value, a variable one moreover, comprised between 1 and 70 ppm W was indeed measured by Lecumberri-Sanchez et al. (2017), but it was in the fluid inclusions, which we now know are characteristic of the later stages, and not the precipitation of the bulk of the wolframite (Cathelineau et al. 2020). If we allow however the wolframite to be precipitated jointly with the quartz in the veins in which it is found, and so by the same volume of water, it can easily be calculated that the tungsten content is in the order of 70 ppm W (comparable to the upper bound of the interval found by Lecumberri-Sanchez et al. (2017)). With this content, a volume of 12 km^3 of water can be deduced, necessary to store 500 kt of tungsten contained as much in the veins as in the anomalous halo: this is much less than Polya (1989) calculated, but it is nevertheless much more than the cupola or the granites of the roof of the Panasqueira pluton can provide.

A3.4.1.3. *Boron mass balance*

For the remainder, it has long been known that mineralized veins intersect the greisen of the cupola (Kelly and Rye 1979), which is in itself enough to disqualify it as a source of mineralizing fluids for tungsten. It could still be envisaged however that the cupola may have been in place before the hydrothermal system began to operate and this could make it the source of the boron needed for the tourmalinization of the wall rocks that (slightly) preceded the tungsten mineralization, as Carocci et al. (2020) have shown. In fact, both Codeço et al. (2017) and Launay et al. (2018) adopted the view of a magma origin for the boron tourmalinization. However, Carocci et al. (2020) have shown from a mass balance of the tourmalinization that, in reality, this perspective is untenable. These authors calculated the boron tonnage needed and compared it to the tonnage that could have delivered the cupola. From an estimate of the total volume occupied by the tourmalinized wall rocks, modeling the elementary veins based on descriptions by Foxford et al. (2000), they found a quantity comprised between 410 and 810 kt of boron (clearly higher than Polya's estimates (1989)). In contrast, Carocci et al. (2020) demonstrate not only that no tourmaline is found in the cupola (or in the other granites in the Panasqueira suite), but that where contacts are observed with the Beira schists, there is no indication of tourmalinization (Marignac et al. 2020). It can be deduced from this that the magma was probably not saturated with boron. Nevertheless, if the saturation value of 0.5% B_2O_3 is used (London 2011 and the included references), which gives an upper bound, the cupola could not have contained more than 9 kt of boron: this is therefore far off the mark. The same is true if we consider the evolved granites of the pluton roof, which could not have contained more than 360 kt B. If, however, we consider no more than the average content in B of magmatic fluids, in the order of 1,000 ppm (Audétat 2019), which is ~500 ppm in magma, the available tonnages are much lower, which definitively disqualifies the cupola or the evolved pluton roof as possible sources of boron.

The approach can clearly be reversed, and a calculation sought of the volume of evolved granite magma that would have been needed to provide the volumes of water calculated. We thus find that no less than 18 km^3 of granite (~4 × 10^{13} kg of granitic liquid) would have been needed to give the 3.4 km^3 of water needed to precipitate all the quartz in the veins, whereas a volume in the order of 60 km^3 would have been needed to supply the vein system and its halo with tungsten (which requires 12 km^3 of water). Similarly, Carocci et al. (2020) focused on the quantity of water needed to achieve tourmalinization: the idea here is that, taking account of the low porosity of the schists to be transformed (10^{-1} to 10^{-2}; Bickle and McKenzie 1987), it is necessary repeatedly to renew the fluid that impregnated the rock in the course of tourmalinization. After calculation, the authors concluded that 2.2×10^9 to 4.3×10^9 m^3 (from 2.2 to 4.3 km^3) of water must have been involved in the process: the granitic volumes needed form part of the previous range.

These volumes are in the same order of size as the granitic pluton of Panasqueira as modeled by Ribeiro (2017), the volume of which is around 32 km^3 (Marignac et al. 2020). Unfortunately, we may legitimately believe that the installation of the pluton preceded the emergence of the Panasqueira hydrothermal system: it has long been known in fact (Bloot and de Wolf 1953) that the ferromagnesian minerals of the contact aureole are affected by the tourmalinization, which must therefore postdate them. In view of the importance of this halo, it seems that only a pluton with the dimensions estimated for the whole rock mass can be responsible for it: it therefore preceded the tourmalinization and consequently the tungsten mineralization.

A3.4.2. *Feasibility of a metamorphic fluid source*

Although the majority of the fluid quantities or boron necessary to the hydrothermal system cannot have been provided by Panasqueira granites, the whole problem of finding their source remains. Kelly and Rye (1979) and Polya (1989) considered meteoric water and its interaction with the Panasqueira granite mass. They relied in particular on the low depth at which the system would have formed according to them (see section 3.7.1.2.1), and, for Polya, on the enormity of the volume of water involved, which would allow recourse to meteoric fluids. Indeed, it appears that the stages of mineralization took place at 10–12 km in depth (Cathelineau et al. 2020), under the impermeable cover of more than 500 m ordovician quartzites (Carocci et al. 2020), which makes the participation of fluids coming from the surface very unlikely; indeed, no indications of their presence remain in the early IF (Cathelineau et al. 2020).

The participation of fluids of metamorphic origin, or balanced with metamorphic rocks, is indicated moreover by the nature of the volatile phases associated with the early hydrothermal fluid (Cathelineau et al. 2020) or by the chemistry of the tourmalines (Carocci et al. 2020) (see section 3.8.1). It is now necessary to evaluate quantitatively how far metamorphism has been able to produce the quantities of fluid required. Indeed, prograde metamorphism of pelitic sediments such as the protoliths of the Beira schists can release significant quantities of water: 2.6% in weight according to Walther and Orville (1983). To produce the volumes required for the working of the hydrothermal system (veins + halo), it would therefore be necessary to have $7.1 \times 10^{12}/2.6 \times 10^{-2} = 1.05 \times 10^{14}$ kg of schists, that is, (with a density of 2,600 $kg \cdot m^{-3}$) a volume of ~100 km^3 (this value reduces to 30 km^3 if only the quartz veins are considered). Although important, this volume remains realistic: a parallelepiped of 10 × 10 × 1 km subject to prograde metamorphism would be enough to supply the system, without the need to invoke the recycling of multiples of a smaller quantity of fluids, as proposed by Polya (1989).

There remains the problem of the boron. As we have seen, the isotope composition of the boron in the tourmalines is compatible both with a magmatic origin and with a metamorphic origin (Codeço et al. 2017). The magmatic origin seems excluded by consideration of the mass balances, it remains to evaluate the plausibility of a metamorphic source: which Carocci et al. (2020) have done. The Beira schists, whose composition is very constant, contain around 100 ppm of boron (Oosterom et al. 1984). With the same rate of devolatilization estimated previously and allowing for an extraction efficiency of 10%, the volumes of schist needed are within a range of 17–34 km^3, estimated by Carocci et al. (2020)· This is in the same order as the volumes calculated previously and confirms that metamorphism could effectively mobilize the boron needed for tourmalinization.

A3.5. References

Blackheath Resources (2015). Technical (Ni-43-101) report and resource estimate Covas tungsten deposit [Online]. Available at: http://www.blackkheathresources.com/i/pdf/reports/2015-05-07 Covas N143-101 [Accessed 15 October 2015].

Cassard, D., Bertrand, G., Billa, M., Serrano, J.J., Tourlière, B., Angel, J.M., Gaál, G. (2015). ProMine mineral databases: New tools to assess primary and secondary mineral resources in Europe. In *3D, 4D and Predictive Modelling of Major Mineral Belts in Europe*, Weihed, P. (ed.). Springer, Berlin.

Charles, N. (2015). Petits métaux dans les Deposits français [Online]. Available at: http://www.amicalebrgm.fr [Accessed 15 October 2020].

Cornwall Resources (2017). Redmoor project [Online]. Available at: http://www.cornwallresources.com [Accessed 15 October 2020].

European Metals (2017). Preliminary feasibility study confirms Cinovec as potentially low cost lithium carbonate producer [Online]. Available at: http://www. europeanmet.com [Accessed 15 October 2020].

Filipe, A. and de Oliveira, D.P.S. (2017). Tungsten: An historical perspective of this critical raw material in Portugal. *Geonovas*, 30, 51–66.

Fonteilles, M., Soler, P., Demange, M., Derré, C., Krier-Schellen, A.D., Verkaeren, J., Guy, B., Zahm, A. (1989). The scheelite skarn deposit of Salau (Ariege, French Pyrenees). *Econ. Geol.*, 84, 1172–1209.

Gonçalves, A., Ramos, V., Lima, L., Noronha, F. (2017). The Santa Helena breccia pipe deposit in the Iberian tin-tungsten metallogenic province (Borralha – North Portugal), a new type of W ore. *Comunicações Geológicas*, 104, 1.

Lauri, L.S., Eilu, P., Brown, T., Gunn, G., Kalvig, P., Sievers, H. (2018). Identification and quantification of primary CRM resources in Europe. Report, European Union's Horizon 2020 research and innovation programme (SCRREEN).

Lehmann, U. (2010). Reserves and resources of ores and fluorite/barite in Saxony. *World of Mining Surface & Underground*, 62(1), 38–47.

Marcoux, E. (2018). Le Tungsten. journées mines en france, Paris, 7 June [Online]. Available at: http://www.lasim.org [Accessed 15 October 2020].

Martins, L.P. (2012). Mineral resources of Portugal [Online]. Available at: http://www.dgeg.gov.pt/wwwbase/wwwinclude/ficheiro.aspx?access=1&id=11622 [Accessed 15 October 2020].

Ormonde Mining (2018). Barruecopardo tungsten mine, Spain. Corporate presentation [Online]. Available at: http://www.ormondemining.com [Accessed 15 October 2020].

Polya, D.A. (1989). Chemistry of the main-stage ore-forming fluids of the Panasqueira W-Cu(Ag)-Sn deposit, Portugal: Implications for models of ore genesis. *Econ. Geol.*, 84, 1134–1152.

Porter GeoConsultant (2016). Cornubian ore field, Cornwall – South Crofty, Geevor, Mt Wellington, Wheal Jane, Hemerdon (UK) [Online]. Available at: http://www.portergeo.com [Accessed 15 October 2020].

Rafaella Resources (2020). Santa Comba project [Online]. Available at: http://www.rafaellaresources.com.au [Accessed 15 October 2020].

Raith, J.G. and Schmidt, S. (2010). Tungsten deposit Felbertal, Salzburg, Austria. *Acta Mineralogica-Petrographica*, 3, 1–24.

Vigne, J-L., André, G., Kapal, F. (2018). Données industrielles, économiques, géographiques sur les principaux produits chimiques, métaux, matériaux [Online]. Available at: http://www.societechimiquedefrance.fr/extras/Donnees/acc.htm [Accessed 15 October 2020].

W Resources (2020a). Régua project [Online]. Available at: http://www.wresources.com [Accessed 15 October 2020].

W Resources (2020b). La Parrilla [Online]. Available at: http://www.wresources.com [Accessed 15 October 2020].

Werner, A.B.T., Sinclair, D., Amey, E.B. (2014). International strategic mineral issues summary report – Tungsten. Report, US Geological Survey Circular 930, 72.

Wheeler, A. (2016). Technical report on the mineral resources and reserves of the Panasqueira mine, Portugal. Report NI 43-101 prepared for Almonty Industries [Online]. Available at: http://almonty.com/_resources/Amended_Valtreixal_43-101_Tech_Rep_Oct15_V4.pdf [Accessed 15 October 2020].

List of Authors

Michel CATHELINEAU
GeoRessources
CREGU
University of Lorraine
Nancy
France

Nicolas CHARLES
BRGM
Orléans
France

Sophie DECRÉE
Royal Belgian Institute of Natural Sciences
Geological Survey of Belgium
Brussels
Belgium

Fabrice GAILLARD
ISTO
CNRS
University of Orléans
France

Kathryn GOODENOUGH
British Geological Survey
Edinburgh
United Kingdom

Gaétan LEFEBVRE
BRGM
Orléans
France

Christian MARIGNAC
GeoRessources
CREGU
University of Lorraine
Nancy
France

Olivier POURRET
AGHYLE
UniLaSalle
Beauvais
France

Robin SHAIL
Camborne School of Mines
University of Exeter
United Kingdom

Beth SIMONS
St Agnes
Cornwall
United Kingdom

Johann TUDURI
BRGM
and
ISTO
CNRS
University of Orléans
France

Index

A, B

C, D

E, F

G, H

I, J

K, L

M, N

O, P

Q, R

S

T

U, V, W

X, Y, Z

Summary of Volume 2

Chapter 4. North African Mississippi Valley-Type Deposit and Its Link with the Alpine Chain Evolution
Mohammed BOUABDELLAH